# Nuclear Physics in Astrophysics IV 2009

AA002642

Journal of Physics: Conference Series Volume 202

## Frascati, Italy
## 8-12 June 2009

ISBN: 978-1-61738-275-8

**Printed from e-media with permission by:**

Curran Associates, Inc.
57 Morehouse Lane
Red Hook, NY 12571

**Some format issues inherent in the e-media version may also appear in this print version.**



Printed by Curran Associates, Inc. (2010)



Institute of Physics
Dirac House, Temple Back
Bristol BS1 6BE UK

Phone: 44 1 17 929 7481
Fax: 44 1 17 920 0979

techtracking@iop.org

**Additional copies of this publication are available from:**

Curran Associates, Inc.
57 Morehouse Lane
Red Hook, NY 12571 USA
Phone: 845-758-0400
Fax: 845-758-2634
Email: curran@proceedings.com
Web: www.proceedings.com

# Nuclear Physics in Astrophysics IV 2009

Journal of Physics: Conference Series Volume 202

## Frascati, Italy
## 8-12 June 2009

ISBN: 978-1-61738-275-8

IOP Institute of Physics

# TABLE OF CONTENTS

## BIG--BANG NUCLEOSYNTHESIS AND FORMATION OF FIRST STARS



## EXPLOSIVE NUCLEOSYNTHESIS



## STELLAR NUCLEOSYNTHESIS

## NUCLEAR STRUCTURE FAR FROM STABILITY

## UNDERGROUND FACILITIES FOR NUCLEAR ASTROPHYSICS AND RARE-EVENT PHYSICS



## POSTER SESSION

# Preface

The NPA IV conference was the fourth in the NPA, Nuclear Physics in Astrophysics, biennial series. The first conference was held in the Institute of Nuclear Research of the Hungarian Academy of Sciences (ATOMKI), Debrecen, Hungary in 2002.
The NPA IV conference was scheduled to take place at the Laboratori Nazionali del Gran Sasso (LNGS) of INFN in Assergi (L'Aquila), Italy, from June 08 to 12, 2009. However, due to the earthquake in the L'Aquila area on April 6th, the NPA IV conference was moved to the Laboratori Nazionali di Frascati (LNF) of INFN in Frascati.
The purpose of the NPA conferences is to bring together Nuclear Physics and Astrophysics giving the scientific community the possibility to review and discuss the status and prospectives of the Nuclear Astrophysics field focussing those subjects of nuclear physics that impact astrophysics and constitute essential input for the understanding of various astrophysical processes and phenomena. In addition a number of talks of more general interest about double beta decay, dark matter, low background facilities has been included in the conference program to provide an overview of different aspects of underground physics. Thus the NPA IV conference represented a variety of interdisciplinary issues in nuclear, particle and astrophysics, including galactic chemical evolution, neutrino masses and mixing.

The program consisted of invited review talks on recent achievements in Nuclear Astrophysics and selected oral and poster contributions on important experimental and theoretical results in the following fields. The main topics discussed during the conference are:

- Big-Bang Nucleosynthesis and Formation of First Stars
- Stellar Nucleosynthesis
- Explosive Nucleosynthesis Nuclear Structure Far from Stability
- Underground Facilities for Nuclear Astrophysics and Rare-Event Physics

Improvements in laboratory measurements (i.e. reduced uncertainty of several astrophysical factors at the relevant energies) and in theory (i.e. the development of multidimensional model), which characterized last year and brought new and interesting interaction with astrophysics and cosmology, were presented at the conference.

The submitted posters have been of high standard and very relevant for the understanding of astrophysical processes. Among the 90 conference participants, one third were scientists under the age 35 years.

The vivid scientific atmosphere has stimulated the interchange of knowledge and techniques implied also in other research fields. Interesting and fruitful discussions have been carried out in a lively physicist environment.

## Committees

### IAB - International Advisory Board:

F. Avignone (USA) avignone@physics.sc.edu
P. Corvisiero (Italy) corvi@ge.infn.it
R. Diehl (Germany) rod@mpe.mpg.de
E. Grosse (Germany) e.grosse@fzd.de
K. Langanke (Germany) k.langanke@gsi.de
M. Leino (Finland) Matti.Leino@phys.jyu.fi
A. Mengoni (Austria) alberto.mengoni@cern.ch
A. Pedro De Jesus (Portugal) ajesus@fct.unl.pt
C. Spitaleri (Italy) spitaleri@lns.infn.it
C. Rolfs (Germany) rolfs@ep3.rub.de
V. Zamfir (Romania) zamfir@tandem.nipne.ro

### IPC - International Programme Committee:

C. Bucci (Italy) bucci@lngs.infn.it
P. Descuovemont (Belgium) pdesc@ulb.ac.be
I. Dominguez (Spain) inma@ugr.es
Z. Fulop (Hungary) fulop@atomki.hu
C. Gustavino (Italy) gustavino@lngs.infn.it
M. Hass (Israel) fnhass@wisemail.weizmann.ac.il
F. Strieder (Germany) strieder@ep3.rub.de
K. Sümmerer (Germany) K.Suemmerer@gsi.de

### LOC - Local Organizing Committee:

Prof. Dr. Eugenio Coccia, director of LNGS, chair
Dr. Alba Formicola (scientific secretary)
Dr. Carlo Gustavino
Dr. Matthias Junker
Conference secretary: Mr. Fausto Chiarizia

NPA IV

Measurements of α branching ratio at the ESR

Astrophysically Important States in
$^{30}$S via $^{32}$S(p,t)$^{30}$S
McMaster

All papers published in this volume of *Journal of Physics: Conference Series* have been peer reviewed through processes administered by the proceedings Editors. Reviews were conducted by expert referees to the professional and scientific standards expected of a proceedings journal published by IOP Publishing

# Big-Bang Nucleosynthesis with updated nuclear data

**Alain Coc**

Centre de Spectrométrie Nucléaire et de Spectrométrie de Masse (CSNSM), CNRS/IN2P3, Université Paris Sud 11, UMR 8609, Bâtiment 104, F–91405 Orsay Campus, France

E-mail: Alain.Coc@csnsm.in2p3.fr

**Elisabeth Vangioni**

Institut d'Astrophysique de Paris, UMR-7095 du CNRS, Université Pierre et Marie Curie, 98 bis bd Arago, 75014 Paris (France)

E-mail: vangioni@iap.fr

**Abstract.** Primordial nucleosynthesis is one of the three evidences for the Big-Bang model together with the expansion of the Universe and the Cosmic Microwave Background. There is a good global agreement over a range of nine orders of magnitude between abundances of $^4$He, D, $^3$He and $^7$Li deduced from observations and calculated primordial nucleosynthesis. This comparison was used to determine the baryonic density of the Universe. For this purpose, it is now superseded by the analysis of the Cosmic Microwave Background (CMB) radiation anisotropies. Big-Bang nucleosynthesis remains, nevertheless, a valuable tool to probe the physics of the early Universe. However, the yet unexplained, discrepancy between the calculated and observed lithium primordial abundances, has not been reduced, neither by recent nuclear physics experiments, nor by new observations.

## 1. Introduction

There are presently three evidences for the Big-Bang Model : the universal expansion, the Cosmic Microwave Background (CMB) radiation and Primordial or Big-Bang Nucleosynthesis (BBN). The third evidence for a hot Big-Bang comes from the primordial abundances of the "light elements": $^4$He, D, $^3$He and $^7$Li. They are produced during the first $\approx$20 minutes of the Universe when it was dense and hot enough for nuclear reactions to take place. These primordial abundances are compared to astronomical observations in primitive astrophysical sites. The number of free parameters in Standard BBN have decreased with time. The number of light neutrino families is known from the measurement of the $Z^0$ width by LEP experiments at CERN: $N_\nu$ = 2.9840±0.0082 [1]. The lifetime of the neutron (entering in weak reaction rate calculations) $\tau_n$ = 885.7±0.8 s [2] and the nuclear reaction rates have been measured in nuclear physics laboratories. The last parameter to have been independently determined is the baryonic density of the Universe, which is now deduced from the observations of the anisotropies of the CMB radiation. It is usual to introduce $\eta$, the number of photons per baryon which remains constant during the expansion, and is directly related to $\Omega_b$ by $\Omega_b{\cdot}h^2$=3.65$\times10^7\eta$ with $\Omega_b{\cdot}h^2$=0.02273±0.00062 and $\Omega_b$=0.0441±0.0030 ("WMAP only" [15]). The parameter $h$ represents the Hubble constant in units of 100 km/s/Mpc (i.e. $h$= 0.719).

## 2. Nuclear reactions

Unlike in other sectors of nuclear astrophysics, nuclear cross sections have usually been directly measured at BBN energies (~100 keV). There are 12 nuclear reactions responsible for the production of $^4$He, D, $^3$He and $^7$Li in Standard BBN. There are many other reactions connecting these isotopes, but their cross sections are too small and/or reactants too scarce to have any significant effect.

The weak reactions involved in n↔p equilibrium are an exception; their rates [13] come from the standard theory of the weak interaction, normalized to the experimental neutron lifetime[2]. The $^1$H(n,$\gamma$)$^2$H cross section is also obtained from theory [3] but in the framework of Effective Field Theory. For the ten remaining reactions, $^2$H(p,$\gamma$)$^3$He, $^2$H(d,n)$^3$He, $^2$H(d,p)$^3$H, $^3$H(d,n)$^4$He, $^3$H($\alpha,\gamma$)$^7$Li, $^3$He(d,p)$^4$He, $^3$He(n,p)$^3$H, $^3$He($\alpha,\gamma$)$^7$Be, $^7$Li(p,$\alpha$)$^4$He and $^7$Be(n,p)$^7$Li, the cross sections have been measured in the laboratory at the relevant energies. Formerly, we used the reaction rates from the the evaluation performed by Descouvemont *et al* . [12] . However, more recent experiments and analysis have lead to improved reaction rates for several important reactions.

To point out the most important reactions, we display in Table 1 the sensitivity of the calculated abundances ($Y_i$ with $i$ = $^4$He, D, $^3$He and $^7$Li) w.r.t. to a change in the 12 reaction rates by a constant factor. We define the sensitivity as $\partial \log Y/\partial \log < \sigma v >$, based on the assumption that the nuclear cross section uncertainties are now dominated by systematic uncertainties that affect their normalization rather than by statistics. These values were obtained, at the WMAP baryonic density, by a parabolic fit of $\Delta Y_i/Y_i$ for $\pm 15\%$ variations of the reaction rates. The last column represents the Gamow window at BBN typical temperatures.

**Table 1.** Abundance sensitivity: $\partial \log Y/\partial \log < \sigma v >$ at WMAP baryonic density.

| Reaction | $^4$He | D | $^3$He | $^7$Li | $E_0$( $\Delta E_0/2$) (MeV) |
|---|---|---|---|---|---|
| n↔p | -0.73 | 0.42 | 0.15 | 0.40 | |
| $^1$H(n,$\gamma$)$^2$H | 0 | -0.20 | 0.08 | 1.33 | |
| $^2$H(p,$\gamma$)$^3$He | 0 | -0.32 | 0.37 | 0.57 | 0.11(0.11) |
| $^2$H(d,n)$^3$He | 0 | -0.54 | 0.21 | 0.69 | 0.12(0.12) |
| $^2$H(d,p)$^3$H | 0 | -0.46 | -0.26 | 0.05 | 0.12(0.12) |
| $^3$H(d,n)$^4$He | 0 | 0 | -0.01 | -0.02 | 0.13(0.12) |
| $^3$H($\alpha,\gamma$)$^7$Li | 0 | 0 | 0 | 0.03 | 0.23(0.17) |
| $^3$He(n,p)$^3$H | 0 | 0.02 | -0.17 | -0.27 | |
| $^3$He(d,p)$^4$He | 0 | 0.01 | -0.75 | -0.75 | 0.21(0.15) |
| $^3$He($\alpha,\gamma$)$^7$Be | 0 | 0 | 0 | 0.97 | 0.37(0.21) |
| $^7$Li(p,$\alpha$)$^4$He | 0 | 0 | 0 | -0.05 | 0.24(0.17) |
| $^7$Be(n,p)$^7$Li | 0 | 0 | 0 | -0.71 | |

This table can be used as a guide for further experimental efforts. We see for instance that *at WMAP baryonic density*, the $^3$H($\alpha,\gamma$)$^7$Li and $^7$Li(p,$\alpha$)$^4$He reactions play a negligible role. The sensitivity to the weak rates is high but (within standard theory), the uncertainty is governed by the neutron lifetime which is now known with a sufficient precision. The influence of the $^1$H(n,$\gamma$)$^2$H rate is unexpected. The $^7$Li final abundance depends strongly on the rate of this reaction while other isotopes are little affected. This unexpected effect can be traced to the increased neutron abundance at $^7$Be formation time for a low $^1$H(n,$\gamma$)$^2$H rate making its destruction by neutron capture, $^7$Be(n,p)$^7$Li(p,$\alpha$)$^4$He, more efficient. However, the few experimental informations available for this cross section at BBN energies are in good agreement with the calculations estimated to be reliable to within 1% error[3]. The next most important reaction (Table 1) is $^3$He($\alpha,\gamma$)$^7$Be as it is the path for the formation of $^7$Li at high density. Hence, the $^7$Li abundance is directly proportional to this rate, which has long been a subject of debate. Systematic differences in the measured cross section were found according to the experimental technique : prompt or activation measurements. Thanks to the recent experimental efforts [6], in particular at LUNA at the Laboratori Nazionali del Gran Sasso, the two methods provide now results in agreement, within each others error bars. With this new experimental data, Cyburt & Davids [10] calculated the S–factor which is significantly higher than the Descouvemont *et al* one [12]. The $^2$H(d,p)$^3$H reaction, also influential on $^7$Li was re-measured [together with $^2$H(d,n)$^3$He] by Leonard *et al* [20] and the very precisely measured cross section is in perfect agreement with the R-matrix fit[12] used.

## 3. Abundances of the cosmological elements

During the evolution of the Galaxy, complex nucleosynthesis takes place mainly in massive stars which release matter enriched in heavy elements into the interstellar medium when they explode as supernovae. Accordingly, the abundance of heavy elements in the gas, at the origin of star formation, increases with time. The observed abundance of *metals* (chemical elements beyond helium) is hence an indication of its age: the oldest stars have the lowest *metallicity*. To derive the most primitive abundances one has first, to extract them from observations of astrophysical sites which are thought to be non evolved and second, extrapolate them to zero *metallicity*.

Primordial lithium abundance is deduced from observations of low metallicity stars in the halo of our Galaxy where the lithium abundance is almost independent of metallicity, displaying a plateau. This constant Li abundance is interpreted as corresponding to the BBN $^7$Li yield. This interpretation assumes that lithium has not been depleted at the surface of these stars so that the presently observed abundance is supposed to be equal to the initial one. The small scatter of values around the "Spite plateau" is an indication that depletion may not have been very effective. Astronomical observations of 24 metal poor halo stars [25] have led to a relative primordial abundance of Li/H = $(1.23^{+0.68}_{-0.32}) \times 10^{-10}$ (95% c.l.), obtained by extrapolation to zero metallicity. New observations of a subset of these stars, followed by an improved analysis, have reduced the uncertainty to Li/H = $(1.10 \pm 0.1) \times 10^{-10}$[18] . Note also that observationally challenging detections of $^6$Li has been reported[4] to a level of $\sim 10^{-2}$ below the Spite plateau value. The presence of such a $^6$Li plateau should, however, be taken with caution.

Contrary to $^7$Li which can be both produced (spallation, asymptotic giant branch (AGB) stars, novae) and destroyed (in the interior of stars), deuterium, a very fragile isotope, can only be destroyed after BBN. Its most primitive abundance is determined from the observation of clouds at high redshift, on the line of sight of distant quasars. Very few observations of these cosmological clouds [19] are available and the adopted primordial D abundance is given by the average value $(2.78^{+0.44}_{-0.38}) \times 10^{-5}$ of D/H. However, new observations [24] have confirmed the convergence of the D/H observed values towards an improved mean of $(2.82^{+0.20}_{-0.19}) \times 10^{-5}$.

After BBN, $^4$He is produced by stars. Its primitive abundance is deduced from observations in HII (ionized hydrogen) regions of compact blue galaxies. Galaxies are thought to be formed by the agglomeration of such dwarf galaxies which are hence considered as more primitive. The primordial $^4$He abundance $Y_{\rm p}$ ($^4$He mass fraction) is given by the extrapolation to zero metallicity but is affected by systematic uncertainties [22, 16] such as plasma temperature or stellar absorption. As the most recent determinations, based on almost the same set of observations, but different atomic physics, lead to a large scatter of values, here, we will use a safe interval, $0.232 < Y_{\rm p} < 0.258$ [22], to account for systematic uncertainties.

Contrary to $^4$He, $^3$He is both produced and destroyed in stars so that the evolution of its abundance as a function of time is not well known and has only been observed in our Galaxy, $^3$He/H = $(1.1 \pm 0.2) \times 10^{-5}$ [5]. Consequently, the baryometric status of $^3$He [27] is not firmly established.

**Table 2.** Yields at WMAP baryonic density.

| | Cyburt el al 2008[11] | This work | Observations | Factor |
|---|---|---|---|---|
| $^4$He | 0.2486±0.0002 | 0.2476±0.0004 | 0.232–0.258[22] | $\times 10^0$ |
| D/H | 2.49±0.17 | 2.68±0.15 | $2.82^{+0.20}_{-0.19}$ [24] | $\times 10^{-5}$ |
| $^3$He/H | 1.00±0.07 | 1.05±0.04 | (0.9–1.3)[5] | $\times 10^{-5}$ |
| $^7$Li/H | $5.24^{+0.71}_{-0.67}$ | 5.14±0.50 | 1.1±0.1[18] | $\times 10^{-10}$ |

## 4. BBN primordial abundances compared to observations

Table 2 displays the comparison between BBN abundances deduced from the WMAP results and the spectroscopic observations. Figure 1 shows the abundances of $^4$He (mass fraction), D, $^3$He and $^7$Li (in number of atoms relative to H) as a function of the baryonic density. The thickness of the curves reflects the nuclear uncertainties. They were obtained by a Monte-Carlo calculation using for the nuclear rate uncertainties those obtained by [12] with the notable exception of $^3$He$(\alpha,\gamma)^7$Be [10] and $^1$H$(n,\gamma)^2$H [3]. (The dashed curves [7] do not include these two new rates.) The horizontal lines represent the limits on the $^4$He, D and $^7$Li primordial abundances deduced from spectroscopic observations. The vertical stripe represents the baryonic density deduced from CMB observations by [15]. The concordance between BBN and observations is in perfect agreement for deuterium. Considering the large uncertainty associated with $^4$He observations, the agreement with CMB+BBN is fair. The calculated $^3$He value is close to its galactic value showing that its abundance has little changed during galactic chemical evolution. On the contrary, the $^7$Li, CMB+BBN calculated abundance is significantly higher than the spectroscopic observations : from a factor of ≈3[7] when using the Descouvemont et al. library[12] only and the Ryan et al. observations [25], to a factor of ≈5[11] when using the new rates and Li observations [18]. The origin of this discrepancy between CMB+BBN and spectroscopic observations remains an open question. For $^6$Li, now that an increased $^2$H$(\alpha,\gamma)^6$Li cross section is excluded[17], the BBN $^6$Li yield ($^6$Li/H≈ $10^{-14}$) at WMAP baryonic density is about two orders of magnitude below the reported observations in some halo stars that nevertheless have to be confirmed.

## 5. BBN as a probe of the early Universe

As the baryonic density of the Universe is now deduced from the observations of the anisotropies of the CMB radiation with a precision that cannot be matched by BBN, one may wonder whether primordial nucleosynthesis studies are still useful. In the Report by the ESA-ESO Working Group on Fundamental Cosmology [23] it is shown that two of the most important issues in cosmology are modified gravity and varying constants. Since, except for the baryonic density, BBN parameters have *all been determined by laboratory measurements*, BBN can be used to test these two important issues within the first minutes of the Universe.

Gravity could differ from its general relativistic description, for instance a scalar field, in addition to the tensor field of general relativity (GR), appears naturally in superstring theories. The effect of the scalar field on BBN is to modify the rate of expansion. Constraints from BBN [8] can then put limits on the implied parameters (initial scalar field value, attraction strength towards GR,....).

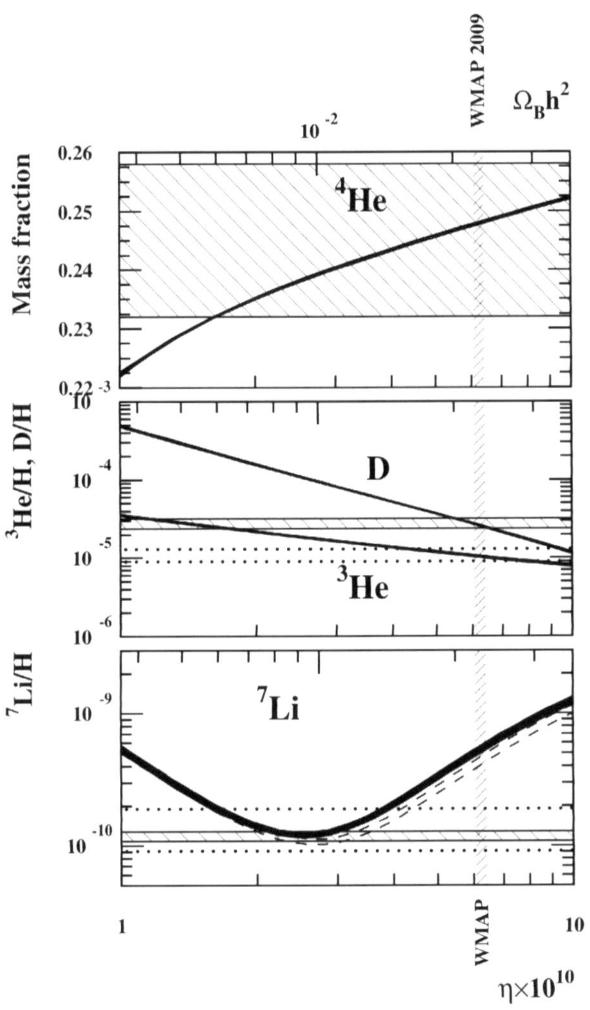


**Figure 1.** Abundances of $^4$He (mass fraction), D, $^3$He and $^7$Li (by number relative to H) as a function of the baryon over photon ratio $\eta$ (or $\Omega_b{\cdot}h^2$) showing the effect of nuclear uncertainties. The dashed curves (only discernable for Li) correspond to an earlier calculation [7] with older rates for $^3$He$(\alpha,\gamma)^7$Be and $^1$H$(n,\gamma)^2$H. The hatched bands represent the primordial abundances deduced from observations. (For lithium, the former observational limits are shown by dotted lines.) The vertical stripe is the WMAP baryonic density.

Coupled variation of the fundamental couplings is also motivated by superstring theories (see Ref. [26] for a review). Hence, this should induce variations of the masses of the particles, the energy scale of the strong interaction, $\Lambda_{QCD}$, the Fermi constant of the weak interactions.... The impact of these variations on the nuclear reaction rates is very difficult to estimate, as in general, nuclear physics uses phenomenological models whose parameters are not explicitly linked to fundamental constants. However, it is possible to estimate the variation of $Q_{np}$ (neutron-proton mass difference), $\tau_n$ (lifetime of the neutron), $B_D$ (binding energy of deuterium) and consequently the variation of the weak and $^1$H$(n,\gamma)^2$H rates. In particular, a relatively small ($\approx$4%) change of $B_D$ *at BBN epoch* would induce a larger variation [14] of the $^1$H$(n,\gamma)^2$H reaction rate sufficient to reconcile all primordial abundances deduced from observations with BBN calculations [9] including $^7$Li .

## 6. Conclusions

The baryonic density of the Universe as determined by the analysis of the CMB anisotropies is in very good agreement with Standard BBN compared to D primordial abundance deduced from cosmological cloud observations. However, it disagrees with lithium observations in halo stars by a factor that has increased with the availability of improved nuclear data and astronomical observations. Presently, the most favored explanation is lithium stellar depletion, but the larger needed depletion factor is hardly compatible whit the thin observed plateau. It is hence essential to determine precisely the *absolute* cross sections important for $^7$Li nucleosynthesis (Table 1). It is for instance surprising that for $^2$H(p,$\gamma$)$^3$He and $^3$He($\alpha,\gamma$)$^7$Be a larger number of experimental data is available at low energies than at BBN energies. (For $^3$He($\alpha,\gamma$)$^7$Be, see also Mohr 2009 [21].)

Nevertheless, primordial nucleosynthesis remains an invaluable tool for probing the physics of the early Universe. When we look back in time, it is the ultimate process for which we *a priori* know all the physics involved. Hence, departure from its predictions provide hints for new physics or astrophysics.

## 7. References

[1] The ALEPH Collaboration and The DELPHI Collaboration and the L3 Collaboration and The OPAL collaboration and The SLD Collaboration and the LEP Electroweak Working Group and the SLD Electroweak and Heavy Flavour Groups, *Phys. Rep.* 2006 **427** 257
[2] Amsler C *et al* (Particle Data Group) 2008 *Phys. Lett.* B **667** 1
[3] Ando S, Cyburt R H, Hong S W and Hyun C H 2006 *Phys. Rev.* C **74** 025809
[4] Asplund M, Lambert D L, Nissen Pl E, Primas F & Smith V V 2006 *Astrophys. J.* **644** 229
[5] Bania T, Rood R and Balser D 2002 Nature **415** 54
[6] Brown T A D *et al* 2007 *Phys. Rev.* C **76** 055801; Confortola F *et al* 2007 *Phys. Rev.* C **75** 065803; Gyürky G *et al* 2007 *Phys. Rev.* C **75** 035805; Nara Singh B S *et al* 2004 *Phys. Rev. Lett.* **93** 262503; Costantini H *et al* 2008 *Nucl. Phys.* A **814** 144
[7] Coc A, Vangioni-Flam E, Descouvemont P, Adahchour A and Angulo C 2004 *Astrophys. J.* **600** 544
[8] Coc A, Olive K A, Uzan J Ph and Vangioni E 2006 *Phys. Rev.* D **73** 083525
[9] Coc A, Nunes N, Olive K A, Uzan J Ph and Vangioni E 2007 *Phys. Rev.* D **76** 023511
[10] Cyburt R H and Davids B 2008 *Phys. Rev.* C **78** 064614
[11] Cyburt R H, Fields B D and Olive K A 2008 *JCAP* **11** 12
[12] Descouvemont P, Adahchour A, Angulo C, Coc A and Vangioni-Flam E 2004 *At. Data Nucl. Data Tables* **88** 203
[13] Dicus D, Kolb E, Gleeson A, Sudarshan E, Teplitz V and Turner M 1982 *Phys. Rev.* D **26** 2694
[14] Dmitriev V F, Flambaum V V and Webb J K 2004 *Phys. Rev.* D **69** 063506
[15] Dunkley J *et al* 2009 *Astrophys. J. S.* **180** 306
[16] Fukugita M and Kawasaki M 2006 *Astrophys. J.* **646** 691
[17] Heil M, *these conference proceedings*
[18] Hosford A, Ryan S G, Garcia-Perez A E, Norris J E and Olive K A 2009 *Astron. Astrophys.* **493** 601
[19] Kirkman D, Tytler D, Suzuki N, O'Meara J M and Lubin D 2003 *Astrophys. J. S.* **149** 1
[20] Leonard D S, Karwowski H J, Brune C R, Fisher B M and Ludwig E J 2006 *Phys. Rev.* C **73** 045801
[21] Mohr P 2009 *Phys. Rev.* C **79** 065804
[22] Olive K A and Skillman E 2004, *Astrophys. J.* **617** 290
[23] Peacock J, Schneider P, Efstathiou G, Ellis J, Leibundgut B, Lilly S and Mellier Y Report by the ESA-ESO Working Group on Fundamental Cosmology, ESA-ESO Working Groups Report No. 3, astro-ph/0610906
[24] Pettini M *et al* 2008 *Mon. Not. R. Astron. Soc.* **391** 1499
[25] Ryan S G, Beers T C, Olive K A, Fields B D and Norris J E 2000 *Astrophys. J.* **530** L57
[26] Uzan J Ph 2003 *Rev. Mod. Phys.* **75** 403
[27] Vangioni-Flam E *et al* 2003 *Astrophys. J.* **585** 611

# Presupernova Evolution and Explosion of Massive Stars

**Marco Limongi[1,3,4,5] and Alessandro Chieffi[2,3]**

1. INAF-Osservatorio Astronomico di Roma. Via Frascati 33, I-00040, Monteporzio Catone (Roma), Italy

Email: marco@oa-roma.inaf.it

2. INAF- Istituto di Astrofisica Spaziale e Fisica Cosmica. Via Fosso del Cavaliere, I-00133 Roma, Italy

Email: alessandro.chieffi@iasf-roma.inaf.it

3. School of Mathematical Sciences, P.O. Box 28M, Monash University, Victoria 3800, Australia

**Abstract**. We review our recent progresses on the presupernova evolution of massive stars in the range 11-120 $M_{\odot}$ of solar metallicity. Special attention will be devoted to the effect of the mass loss rate during the Wolf-Rayet stages in determining the structure and the physical properties of the star prior the supernova explosion. We also discuss the explosive yields and the initial mass-remnant mass relation in the framework of the kinetic bomb induced explosion and hence the contribution of these stars to the global chemical enrichment of the interstellar medium

## 1. Introduction

Massive stars, i.e. those stars exploding as core collapse supernovae, play a pivotal role in the chemical and dynamical evolution of the galaxies. In fact, among the other things, they provide most of the mechanical energy input into the interstellar medium via strong stellar winds and supernova explosions [1] and also synthesize most of the elements (with $4<Z<38$), especially those necessary to life. The interiors of massive stars, also, constitute invaluable laboratories with physical conditions not seen elsewhere in the universe. All these issues make the understanding of the evolution and the explosion of massive stars of paramount importance in astrophysics.

We review here our recent progresses on the presupernova evolution, the explosion and mostly the nucleosynthesis of massive stars. We mainly focus on the advanced evolutionary phases prior to the explosion, i.e. from the core He exhaustion up to the onset of the iron core collapse. However, since the advanced burning phases strongly depends on the evolutionary history of the star during the H- and He-burning stages, here we address only the aspects of these phases relevant for the further evolution of the star up to the explosion. In particular we discuss in some detail the effect of mass loss during H- and He-burning and its implications on the more advanced burning phases. The explosive nucleosynthesis as well as the contribution of massive stars to the global enrichment of the interstellar

[5] Invited Talk

medium and their role in the chemical evolution of the galaxies is, on the contrary, described in some detail.

## 2. Stellar Models

The models presented in this paper are the ones discussed in [2] and the computations were done with the latest release of the stellar evolutionary code FRANEC (5.050218). In particular, the interaction between convection and local nuclear burning has been taken into account by the coupling together and the simultaneous solution of the set of equations controlling the local nuclear burning and those describing the convective mixing. The convective mixing is treated by means of a diffusion equation in which the diffusion coefficient is computed by means of the mixing-length theory. The nuclear network is the same as the one adopted in [3] while the nuclear cross sections have been updated whenever possible (see Table 1 in [2]). A moderate amount of overshooting of 0.2 $H_P$ has been included in the calculation only for central H burning. Mass loss has been taken into account following the prescriptions of [4] for the blue supergiant phase ($T_{eff}$>12000 K) and of [5] for the red supergiant phase ($T_{eff}$<12000 K). The Wolf-Rayet (WR) phase has been followed in the Late Nitrogen rich phase (WNL) by adopting [6] and [7] in the following Early Nitrogen rich (WNE) and Carbon and Oxygen rich (WCO) Wolf-Rayet stages.

We asssume that the star enters the WR phase when Log($T_{eff}$)>4 and the surface H mass fraction is $H_{surf}$<0.4, and we adopt the following usual definitions for the various WR phases: WNL ($10^{-5}$<$H_{surf}$<0.4), WNE ($H_{surf}$<$10^{-5}$ and $(C/N)_{surf}$<0.1), Carbon and Nitron rich (WNC - 0.1<$(C/N)_{surf}$<10), and WCO ( $(C/N)_{surf}$>10).

## 3. Core H and He burning

The core H burning is the first nuclear burning stage and, in massive stars, it is powered by the CNO cycle. For these stars the central temperature is high enough (7.5<log $T_c$<7.8) for both the Ne-Na and the Mg-Al cycles to become efficient. The strong dependence of this cycle on the temperature implies the presence of a convective core, which reaches its maximum extension just at the beginning of the central H burning and then recedes in mass as the central H is burnt. Mass loss is rather efficient during this phase and increases substantially with the luminosity of the star (i.e. the initial mass). In stars more massive than 40 $M_\odot$ it leads to a significant reduction of the total mass. The H exhausted core (He core) that forms at the central H exhaustion scales (in mass) directly with the initial mass. It is worth noting here that the presently adopted mass loss rates in the blue supergiant phase (BSG) do not alter too much the initial mass-He core mass relation at the core H exhaustion, compared to the one obtained in absence of mass loss. On the contrary the size of the H convective core, which in turn may be affected by the overshooting and/or semiconvection, has a strong impact on the initial mass-He core mass relation and hence constitutes the greatest uncertainty in the computation of the core H burning phase.

At core H exhaustion all the models expand and move in the red portion of the HR diagram. At the same time the more internal zones contract until the core He burning is activated. The further fate of the models is largely driven by the competition between the efficiency of mass loss in reducing the H rich envelope and the core He burning timescale. In stars with initial mass M<30 $M_\odot$ mass loss is not so efficient hence these stars do not loose most of their H rich mantle. As a consequence they will likely explode as red supergiants (RSGs). On the contrary in stars with M>30 $M_\odot$ mass loss is efficient enough to induce the total ejection of the H rich envelope. These stars will eventually explode as WR stars. In particular, we find that (1) stars with initial mass $M \geq 30$ $M_\odot$ become (and explode as) WNL stars; (2) stars with initial mass $M \geq 35$ $M_\odot$ become (and explode as) WNE stars; (3) stars with initial mass $M > 40$ $M_\odot$ become (and explode as) WC stars.

Once the full H rich mantle is lost, the further evolution of the stellar model depends on the actual He core mass. Stars initially less massive than the minimum mass that completely loses its H-rich mantle develop a central He burning within the He core that never recedes in mass; it either advances or, at least, remains constant in mass. In this case, the He convective core never recedes and the reason is that the conversion of He to C and O increases the opacity in the entire convective core so that its outer border is continuously pushed outward rather than inward. Hence, a strong chemical discontinuity forms at the border of the He convective core where the He abundance changes from roughly zero to roughly one at central He exhaustion. Hence, a He convective shell will form outside the maximum extension of the convective core in a region whose chemical composition is still the one left by H burning.

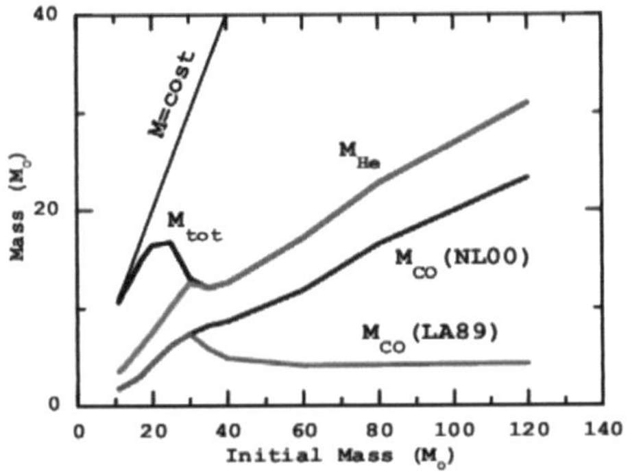


**Figure 1.** Total mass (black), He core mass (red) and CO core mass (blue for NL00 and grenn for LA89) at core He exhaustion as a function of the initial mass for two different prescription of the mass loss rate during the WNE/WCO stages (see text).

Vice versa, stars whose initial mass exceeds this threshold value not only lose their whole H-rich envelope but also a fraction of their He core mass. In this mass interval, therefore, central He burning occurs, partly or even completely, in the He core mass, which progressively decreases. Such an occurrence has a profound impact on the central He burning itself and, therefore, on all advanced burnings. As the He core mass decreases because of the mass loss, the central He burning readjusts on a physical structure and, hence, on a convective core size, which would compete for the current He core mass. Since the size of the convective core scales directly with that of the He core, the convective core recedes progressively in mass, leaving behind a He profile. In this case, a He convective shell will form in a region with a He profile whose steepness reflects the speed at which the He core mass has been eroded in He burning. In addition to that, the amount of carbon left in the He-exhausted core is affected by the mass loss because the central physical conditions readjust continuously to the ones proper for a star of progressively smaller He core mass. Since the C abundance left by He burning scales inversely with the He core mass, the smaller the final He core mass, the larger the abundance of C. The relevance of these effects, is very sensitive to the mass loss rate during the Wolf-Rayet stage.

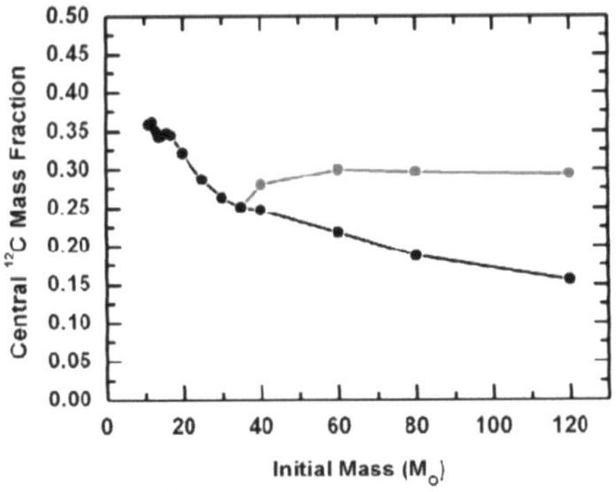


**Figure 2.** $^{12}C$ mass fraction at core He exhaustion as a function of the initial mass: the black line corresponds to NL00 models while the red line to LA89 models.

Figure 1 (left panel) shows the CO core mass at core He exhaustion as a function of the initial mass for two different prescriptions of the mass loss during the WNE/WCO WR stages, i.e., the one provided by [6] (NL00) and the one provided by [7] (LA89), the second one being higher by about 0.2-0.6 dex on average compared to the first one. Inspection of the figure shows that while in the case of the NL00 mass loss rate, the CO core mass preserves a clear trend with the initial mass, in the case of the LA89 mass loss rate, all the models develop a very similar structure that resembles that of a lower mass model. Moreover, Figure 2 shows that the $^{12}C$ left by over at core He exhaustion by the LA89 models is similar to that of the star with a similar CO core mass (in this case a 20 $M_{\odot}$). Since the evolution of a massive star after core He burning is mainly driven by both the CO core mass and its chemical composition (C and O constitute the basic fuel for all the further nuclear burning), it is clear that mass loss during the WR stage is fundamental in determining the evolutionary properties of these stars during the more advanced burning stages and also their final fate (see below).

## 4. Advanced burning stages: C, Ne, O and Si burning

The evolution of a massive star during the advanced burning stages is mainly driven by the size and by the composition of the CO core. The CO core mass determines the thermodynamical history of the centre, while $^{12}C$ and $^{16}O$ constitute the basic fuel for all the following nuclear burning.

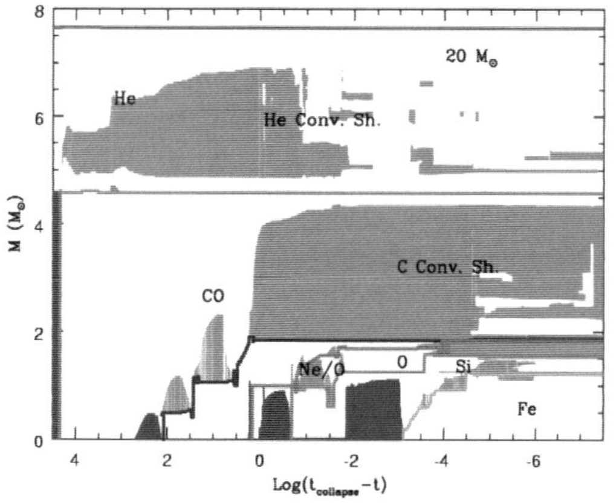


**Figure 3.** Convective history of a 20 M$_\odot$ model. The grey areas correspond to the convective regions. The solid line mark the nuclear burning shells.

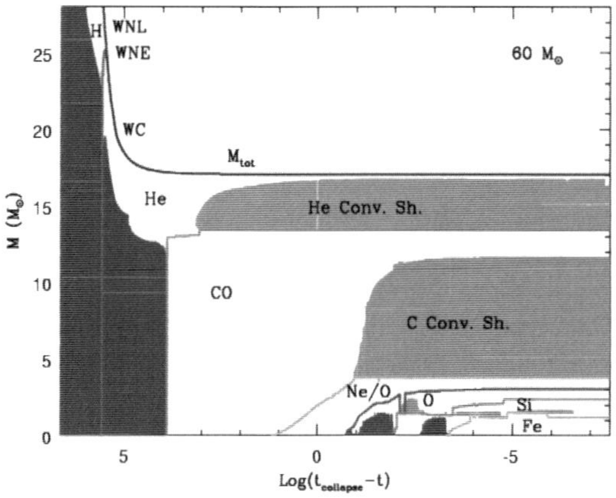


**Figure 4.** Convective history of a 60 M$_\odot$ model. The grey areas correspond to the convective regions. The solid line mark the nuclear burning shells.

In general each burning stage, from the C burning up to the Si burning, occurs first at the center and then, as the fuel is completely burnt, it shifts in a shell. The nuclear burning shell, then, may induce the formation of one of more successive convective zones, which can partially overlap. The general trend is that the number of convective shells scales inversely with the mass size of the CO core (Figure 3 and Figure 4).

The complex interplay among the shell nuclear burnings and the timing of the convective zones determines in a direct way the distribution of the chemical composition and the mass-radius (M-R) relation of the star at the presupernova stage (the relevance of the M-R relation for the explosive nucleosynthesis is discussed in the next section). Figure 5 shows that for the NL00 models the larger is the initial mass the more compact is the presupernova structure while, on the contrary, all the LA89 models have a very similar M-R relation at the presupernova stage. This is the consequence of the fact that the NL00 mass loss preserves a direct scaling between the initial mass and the CO core mass while, on the contrary, the LA89 mass loss is so strong that all the LA89 models converge toward a very similar structure

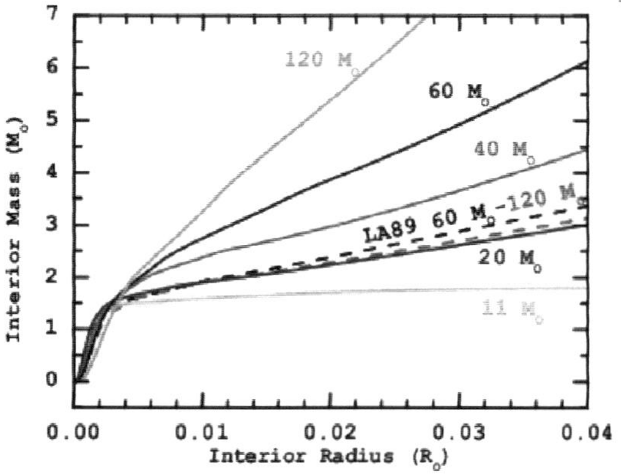


**Figure 5.** Mass-Radius relation for a subset of massive star models at the presupernova stage. The solid lines refer to the NL00 models while the dashed lines to the LA89 models.

(similar to that of a 20 $M_\odot$). The distribution of the most abundant chemical species at the presupernova stage is shown in Figure 6 for some selected NL00 models.

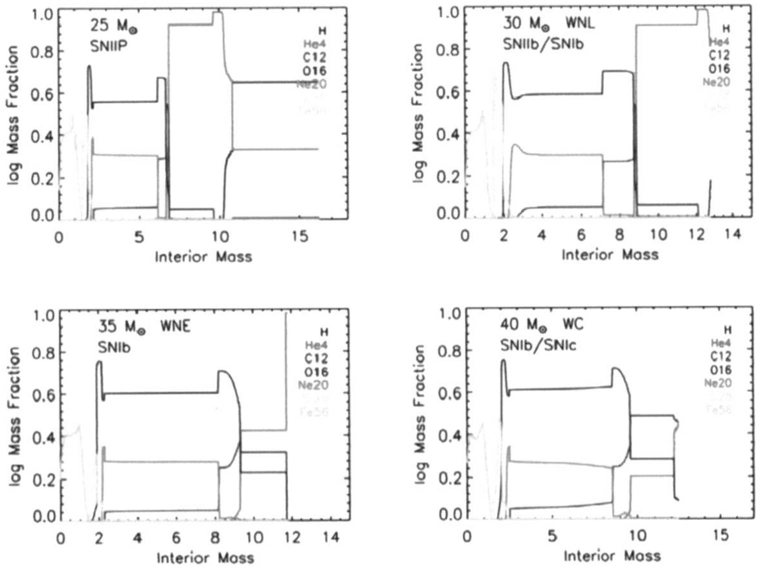


**Figure 6.** Presupernova distribution of the most abundant chemical species for a selected number of NL00 massive star models.

In the 25 $M_\odot$ the mass loss does not play a crucial role, most of the H rich envelope is retained and the star will eventually explode as SNII. On the contrary, the effect of mass loss is readily evident in the more massive stars that will explode as SNIIb, SNIb or SNIc depending on the efficiency of the mass loss in removing the H rich envelope as well as the He rich zones. In general the presupernova star consists of an iron core of mass in the range between 1.2 and 1.8 $M_\odot$, the higher is the mass of the star higher is the iron core mass, surrounded by active burning shells located at the base of zones loaded in the main products of silicon, oxygen, neon, carbon, helium and hydrogen burnings, i.e., the classical "onion structure". Thus, each zone keeps memory of the nucleosynthesis produced by the various central and/or shell burnings occurring either in a radiative environment or in a convective zone.

## 5. Explosive nucleosynthesis and initial mass-remnant mass relation

The chemical composition left by the hydrostatic evolution is partially modified by the explosion, especially that of the more internal zones. At present there is no self consistent hydrodynamical model for core collapse supernovae and consequently, at present, explosive nucleosynthesis calculations are based on simulated explosions. The basic idea is to deposit a given amount of energy at the base of the exploding envelope and to follow the propagation of the shock wave that forms by means of a hydro code. The initial amount of energy is fixed by requiring a given amount of kinetic energy at the infinity (typically of the order of $10^{51}$ erg=1 foe). Whichever is the technique adopted to deposit the energy into the presupernova model (piston, kinetic bomb or thermal bomb), in general the result is that some amount of material (the innermost one) will fall back onto the compact remnant while most of the envelope will be ejected with the desired final kinetic energy. The mass separation between remnant and ejecta is always referred to as the mass cut. This quantity strongly depends on the details of the explosive calculations, i.e., the mass location and the way in which the energy is deposited, the inner and the outer boundary conditions, and so on, hence it constitutes the most uncertain and free parameter in the explosive nucleosynthesis calculations for core collapse supernovae. The mass cut strongly affects not only the chemical yields of all those isotopes that are produced in the innermost zones of the exploding mantle, mainly $^{56}$Ni and also all the iron peak elements, but also the relation between the initial mass and the final remnant mass, i.e., which is the mass limit or the mass interval between stars forming neutron stars and stars forming black holes after the explosion.

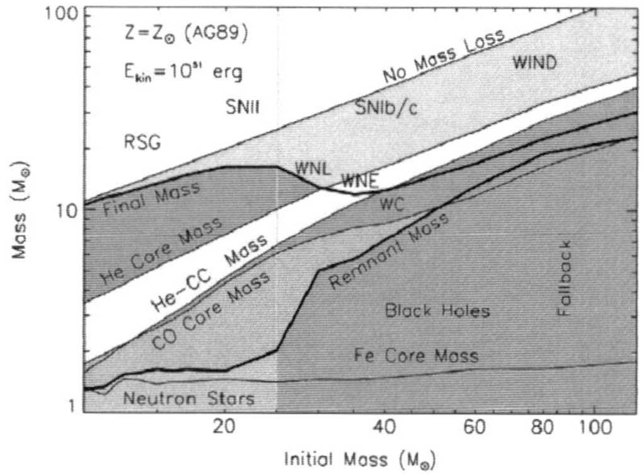


**Figure 7.** Characteristic masses as a function of the initial mass for the NL00 models. The final kinetic energy at infinity is set to Ekin=1 foe (see text).

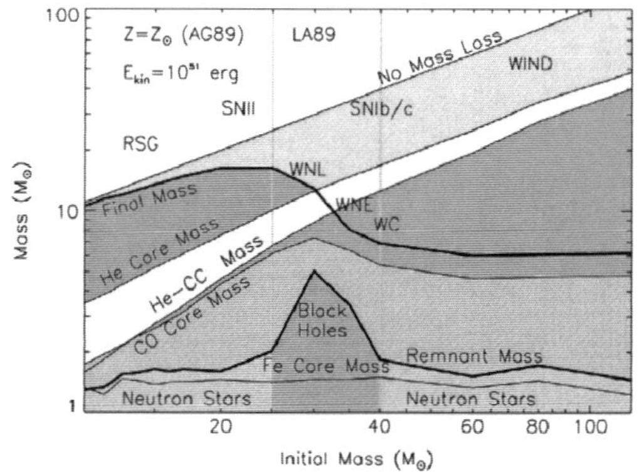


**Figure 8.** Same as Figure 7 but for the LA89 models.

Figure 7 shows the the initial-final mass relation for the NL00 models, in the assumption that the ejecta have 1 foe of kinetic energy at infinity (the explosions have been computed by means of the hydro code described in [2]). This choice implies that in stars with initial mass larger than 25-30 $M_{\odot}$ all the CO core, or a great fraction of it, fall back onto the compact remnant. This is due to the fact that the higher is the mass of the star the steeper is the mass-radius relation (i.e. the more compact is the structure) (Figure 5), the higher is the binding energy and hence the larger is, in general, the mass falling back onto the compact remnant. As a consequence these stars would not eject any product of the explosive burnings, as well as those of the C convective shell, and will leave, after the explosion, black holes with masses ranging between 3 and 11 $M_{\odot}$. In figure 7 are also shown the limiting masses that enter the various WR stages (WNL, WNE and WC) as well as the limiting mass (30- 35$M_{\odot}$) between stars exploding as Type II SNe and those exploding as Type Ib/c supernovae. The behavior of the LA89 models (Figure 8) is completely different because their binding energy is much smaller than their corresponding NL00 models due to the much smaller He core masses (Figure 1). In this case, the choice of a final kinetic energy of 1 foe allows, even in the more massive stars, the ejection of a substantial amount of the CO core, and hence heavy elements, leaving neutron stars as remnants.

## References

[1] Abbott, D.C. 1982, ApJ, 263, 723
[2] Limongi, M., Chieffi, A. 2006, ApJ, 647, 483
[3] Limongi, M., Chieffi, A. 2003, ApJ, 592, 404
[4] Vink, J.S., de Koter, A., Lamers, H. J. G. L. M. 2000, A&A, 362, 295
[5] de Jager, C., Nieuwenhuijzen, H., van der Hucht, K. A. 1988, A&AS 72, 259
[6] Nugis, T., Lamers, H. J. G. L. M. 2000, A&A, 360, 227
[7] Langer, N. 1989, AA, 220, 135

# Towards the Equation of State for Dense Stellar Matter

**N. Buyukcizmeci**[1], **A. S. Botvina**[2], **I. N. Mishustin**[3,4], **R. Ogul**[1]

[1] Department of Physics, University of Selçuk, 42079 Konya, Turkey
[2] Institute for Nuclear Research, Russian Academy of Sciences, 117312 Moscow, Russia
[3]Frankfurt Institute for Advanced Studies, J.W. Goethe University, D-60438 Frankfurt am Main, Germany
[4]Kurchatov Institute, Russian Research Center, 123182 Moscow, Russia

E-mail: nihal@selcuk.edu.tr

**Abstract.** The thermal nuclear multifragmentation reaction can be used to investigate the nuclear liquid-gas phase coexistence region and its relation to the astrophysical processes such as the collapse of massive stars, and the supernova type-II explosions. We have performed some calculations to estimate stellar equation of state (EOS) on the basis of statistical multifragmentation model.

## 1. Introduction

Recently, the nuclear multifragmentation has attracted a great interest due to its similarity to the collapse and explosion of massive stars [1]. By this reason, the thermal multifragmentation reaction is considered as a fundamental experimental tool to study astrophysical processes such as the collapse of massive stars, and the supernova type-II explosions, where stellar matter reaches the nuclear densities in the range of $\rho \approx 10^{-5}\rho_0 - 2\rho_0$, where $\rho_0$ is the nuclear saturation density, and the temperatures of T $\approx 0.5 - 10$ MeV. It was already demonstrated [2, 3] that the modified properties of hot nuclei (in particular, their symmetry energy) extracted from the multifragmentation data can essentially influence nuclear composition of stellar matter. The inputs for estimating the nuclear equation of state, that gives a relation between the pressure, the density(or the volume) and the temperature at extreme conditions, can be obtained by analysing the experimental data on multifragmentation obtained in intermediate and high energy heavy-ion collisions. The EOS of stellar matter in supernova has been under investigation since the years 1980's [4, 5], and modified by several recent studies [6, 7, 8, 9, 10, 11, 12, 13] with references therein. In this short communication, we have introduced a brief outline of some systematic calculations of stellar EOS on the basis of statistical multifragmentation model generalized for astrophysical conditions [2].

## 2. Calculations and Results

Statistical multifragmentation model [14] assumes a statistical equilibrium at a low-density freeze-out stage. All the channels composed of nucleons and excited fragments are taken into

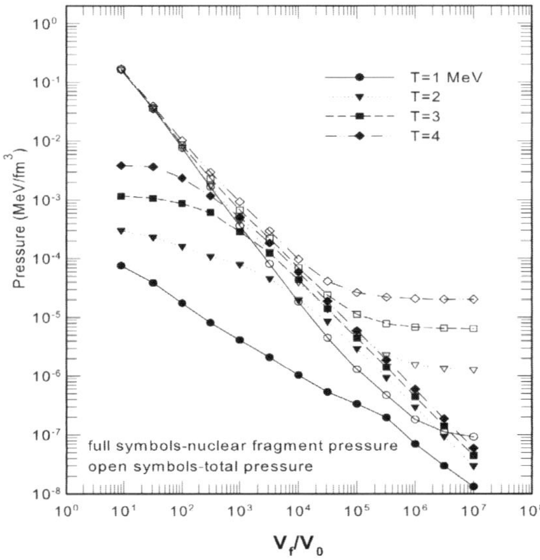


**Figure 1.** Variation of the nuclear fragment pressure and total pressure with relative freeze-out volume $V_f$ in the temperature range of 1-4 MeV.

account with the conservation of baryon number, electric charge and energy. Light nuclei with mass number $A \leq 4$ are treated as elementary particles of nuclear gas, with only translational degrees of freedom. Nuclei with $A > 4$ are treated as heated liquid drops. In this way one may study the nuclear liquid-gas coexistence in the freeze-out volume. The Coulomb interaction of all fragments is included by using the Wigner-Seitz approximation. Different channels are generated by Monte Carlo sampling according to their statistical weights. After the break-up the Coulomb repulsion and the secondary deexcitation of primary hot fragments are taken into account. When compared to the nuclear matter, the stellar matter should be electrically neutral, therefore, the electrons are included to balance the positive nuclear charge. We also consider the change of nuclear composition due to the photonuclear reactions, and irradiation of the matter by a strong neutrino wind from the protoneutron star [2].

In Fig. 1 we show the results for the nuclear fragment pressure and the total pressure as function of relative freeze-out volume $V_f$ in the temperature range of 1-4 MeV. One may see from this figure that the pressure is dominated by the relativistic electrons with high chemical potentials at high densities (low volume). In the intermediate densities, the nuclear contributions are important because of the fact that the composition of the free nucleons increases with the temperature. In this case, 50 per cent of the total pressure is produced by the nucleons. At the low density region, however, the total pressure is mainly dominated by the radiation pressure, which is proportional to $T^4$, and it is almost not depending on baryon density. In Fig. 2 we show the mass distributions of hot primary fragments at various subnuclear densities for T=1 and T=4 MeV for fixed lepton fractions $Y_L$ =0.4 and $Y_L$ =0.2 per nucleon. It is seen that larger and more neutron rich nuclei are produced for lower lepton fraction at T=1 MeV. We also show the results for the reduced symmetry energy parameter $\gamma$ =15 MeV, which gives the best description of experimental data [15, 16]. One may see that the modification of the symmetry energy term intensifies the increase in the mass of the heavy fragments at low temperatures(upper panels). It should be pointed out that when we increase temperature from T=1 MeV (upper panels) to T=4 MeV (lower panels) the large fragments disapear and it tends to U-shape and then to the

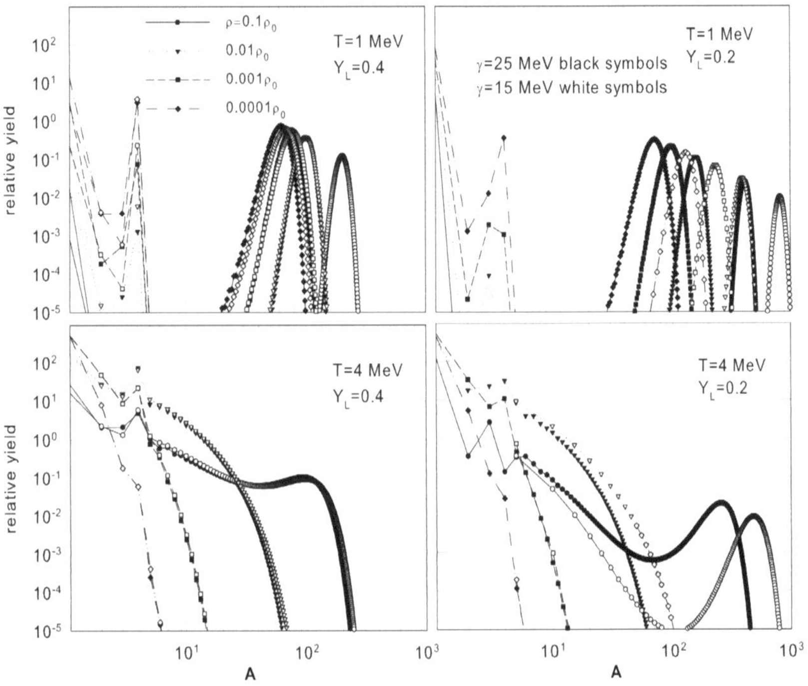


**Figure 2.** Mean mass distributions of hot primary fragments calculated with the SMM generalized for supernova conditions. Left panels show the results for the temperatures at T=1 and T=4 MeV and the fixed lepton fraction $Y_L$ =0.4 per nucleon, and right panels are for $Y_L$ =0.2 per nucleon. Different lines correspond to the baryon densities shown in the figure. Black symbols are for the symmetry energy parameter $\gamma$ =25 MeV and, white symbols are for the reduced value of $\gamma$ =15 MeV.

exponential fall-off distributions. This is in agreement with the mass distribution shapes for the nuclear multifragmentation reactions [17, 18, 19].

## 3. Conclusions

In order to describe properties of supernova matter the statistical multifragmentation model has been modified by including electrons and neutrinos in the statistical ensemble [2]. By this approach, it is confirmed that the properties of stellar matter at extreme conditions are very similar to those obtained for nuclear matter during the multifragmentation of heavy nuclei. First of all, the present results support the fact that decreasing the symmetry energy has a strong influence on the nuclear composition of stellar matter formed during the collapse of massive stars, and the supernova type-II explosions. For example, by using the reduced symmetry energy parameter $\gamma$ =15 MeV, which gives the best description of experimental data [15, 16], we get very large and very neutron-rich nuclei, as shown in Fig. 2. The mean mass distributions of hot primary fragments calculated by means of the same approach also exhibit similar trends as those obtained for nuclear multifragmentation. It has been demonstrated in Ref. [2] that a great variety of hot neutron-rich nuclei can be formed during the explosion. The present calculations confirmed the same trend for the temperatures T=1 and 4 MeV at different densities and for different assumptions on the lepton fraction. It was seen that the produced nuclei are quite large and rich in neutron. To sum up, the present results for EOS of stellar matter strongly support the applicability of statistical multifragmentation model for stellar matter as well.

**Acknowledgements:** N. Buyukcizmeci thanks BAP (Coord. of Scientific Research Projects of Selcuk University-09701267), and A. S. Botvina thanks TUBITAK-BIDEB for support that made his visit possible, to Selcuk University, Konya.

## References

[1] Botvina A S and Mishustin I N 2006 Eur. Phys. J. A 30 121
[2] Botvina A S and Mishustin I N 2004 Phys. Lett. B 584 233
[3] Botvina A S and Mishustin I N 2005 Phys. Rev. C 72 048801
[4] Bethe H A 1990 Rev. Mod. Phys. 62 801
[5] Cowan J J, Thielemanand F -K and Truran J W 1991 Phys. Rep. 208 267
[6] Janka H -T, Buras R, Kifonidis K, Rampp M and Plewa T 2001 Review in "Core Collapse of Massive Stars", Fryer, C.L. (ed.) , *astro-ph/0212314*; Janka H -T and Mueller E 1996 Astron. Astrophys. 306 167
[7] Liebendorfer M et al 2003 Nucl. Phys. A 719 144c
[8] Horowitz C J 1997 Phys. Rev. D 55 4577
[9] Sutaria F K, Ray A, Sheikh J A and Ring P 1999 Astron. Astrophys. 349 135
[10] Langanke K and Martinez-Pinedo G 2000 Nucl. Phys. A 673 481
[11] Hix W R et al 2003 Phys. Rev. Lett. 91 201102
[12] Langanke K et al 2003 Phys. Rev. Lett. 90 241102
[13] Frohlich C et al 2006 Phys. Rev. Lett. 96 142502
[14] Bondorf J P, Botvina A S, Iljinov A S, Mishustin I N and Sneppen K 1995 Phys. Rep. 257 133
[15] Le Fevre A et al 2005 Phys. Rev. Lett. 94 162701
[16] Shetty D V et al 2005 Phys. Rev. C 71 024602
[17] Buyukcizmeci N, Ogul R and Botvina A S 2005 Eur. Phys. J. A 25 57
[18] Eren N, Buyukcizmeci N and Ogul R 2007 Phys. Scripta 76 657
[19] Erdogan M, Buyukcizmeci N and Ogul R 2007 Tr. J. Chem. 31 403

# Alpha-induced reactions for the astrophysical p-process: the case of $^{151}$Eu

**Gy Gyürky[1], Z Elekes[1], J Farkas[1], Zs Fülöp[1], G G Kiss[1], E Somorjai[1], T Szücs[1], R T Güray[2], N Özkan[2], C Yalçın[2] and T Rauscher[3]**

[1] Institute of Nuclear Research (ATOMKI), H-4001 Debrecen, POB.51., Hungary
[2] Kocaeli University, Department of Physics, TR-41380 Umuttepe, Kocaeli, Turkey
[3] University of Basel, Department of Physics, CH-4056 Basel, Switzerland

E-mail: gyurky@atomki.hu

**Abstract.** The cross sections of $^{151}$Eu($\alpha,\gamma$)$^{155}$Tb and $^{151}$Eu($\alpha$,n)$^{154}$Tb reactions have been measured with the activation method. Some aspects of the measurement are presented here to illustrate the requirements of experimental techniques needed to obtain nuclear data for the astrophysical p-process nucleosynthesis. Preliminary cross section results are also presented and compared with the predictions of statistical model calculations.

## 1. Introduction

The chemical elements with atomic number higher than about 30 are very rare in nature. In the Solar System these trans-iron elements represent only roughly 1 atom in every 10 million. The synthesis of about 99 % of the nuclei in this mass region can be explained by neutron capture reactions in the astrophysical s- and r-processes under various astrophysical conditions [1, 2]. There are, however, more than 30 isotopes on the proton rich side of the nuclear chart which can be synthesized by none of these two processes. There are still many open questions regarding the synthesis of these so called p-nuclei. Their production mechanism is the astrophysical p-process which may involve several different sub-processes [3]. The most important mechanism is thought to be the $\gamma$-process which proceeds by $\gamma$-induced reactions starting from pre-existing s- or r-process seed nuclei. Consecutive ($\gamma$,n) reactions drive the material towards the neutron-deficient region, though charged-particle emitting ($\gamma,\alpha$) and ($\gamma$,p) reactions also contribute to the process. Figure 1 schematically shows the path of the $\gamma$-process in the case of the Tin isotopic chain.

Modelling of the p-process in the whole relevant mass region involves a reaction network of thousands of reactions. The aim of a p-process model calculation is to reproduce the p-isotope abundances observed in the Solar System, however, even the state-of-the-art models fail to give a satisfactory reproduction of the measured abundances (see e.g. [5]). The models use various input parameters. On one hand the astrophysical conditions where the process takes place must be known. This involves the knowledge of the stellar composition (seed abundances), temperature, time scale, etc. The presently favoured site for the $\gamma$-process is the O/Ne layers of massive stars in explosive or pre-explosive phase where temperatures of typically 3 GK are reached for a period of the order of one second [6]. On the other hand, the models require the knowledge of different nuclear physics parameters. The masses and decay half-lives of the

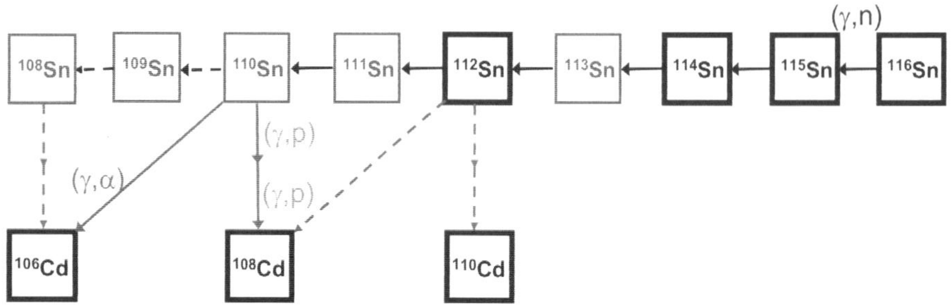


**Figure 1.** (Color online) Path of the $\gamma$-process around the Tin isotopes. Thick blue and thin red boxes indicate stable and unstable isotopes, respectively. The main reaction flow according to [4] is shown by solid arrows while dashed arrows show weaker branches of the path. Not all possible reactions are indicated, the figure is intended only to show the different reactions contributing to the synthesis of p-nuclei.

(mainly proton-rich) isotopes involved in the reaction network are relatively well known. The reaction rates of the dominant $\gamma$-induced reactions are, however, much more uncertain. Since the networks include thousands of reactions on mainly unstable nuclei, the calculations have to rely on theoretical reaction rate calculations obtained typically from Hauser-Feshbach statistical models. The calculated p-isotope abundances are very sensitive to the variation of the reaction rates (see e.g. [7]), therefore the experimental study of the reactions relevant for a $\gamma$-process network is important to test the theoretical reaction rates.

## 2. Cross section measurement for the $\gamma$-process

As discussed above, $\gamma$-induced reactions play a key role in the synthesis of p-nuclei. The measurement of $\gamma$-induced reaction cross sections is technically challenging, especially in the case of charged particle emitting $(\gamma,\alpha)$ and $(\gamma$,p) reactions. Recently, there has been a significant development in this field (see e.g. [8] and references therein). However, due to the huge difference between the reaction rate measured in the laboratory and the actual rate in a stellar plasma (stellar enhancement effect, for details see [9]), the experimental $\gamma$-induced reaction cross section data cannot be used directly in a nucleosynthesis model.

The experimental study of radiative capture reactions provides an alternative way to investigate the reactions relevant for the $\gamma$-process, since the rate of the $\gamma$-induced reactions can be inferred from the rate of the inverse capture reaction using the detailed balance theorem. Moreover, the stellar enhancement effect is much less pronounced in the case of capture reactions. While the neutron capture cross sections are rather well known experimentally, at least along the line of stability [10], there are only very few $(\alpha,\gamma)$ and – compared to the size of the reaction network – a limited number of (p,$\gamma$) cross sections measured in the relevant mass and energy range leaving the statistical model calculations for these reactions largely untested.

**Table 1.** List of A$\geq$70 isotopes for which experimental proton or alpha capture reactions cross sections are available

| (p,$\gamma$) | $(\alpha,\gamma)$ |
|---|---|
| $^{74,76,77}$Se, $^{84,86,87,88}$Sr, $^{89}$Y, $^{90,96}$Zr, $^{93}$Nb, $^{92,94,95,98}$Mo, $^{96,98,99,100,104}$Ru, $^{102,104,105,106}$Pd, $^{106,108}$Cd, $^{112,114,116,119}$Sn, $^{120}$Te | $^{70}$Ge, $^{96}$Ru, $^{106}$Cd, $^{112,117}$Sn, $^{113}$In, $^{144}$Sm, $^{197}$Au |

Table 1 lists those isotopes with mass number higher than 70 for which experimental proton or alpha capture reactions cross sections are available. It is apparent that in the case of $(\alpha,\gamma)$

reactions the available experimental database is very limited especially in the heavier mass region where the p-process models show the highest sensitivity to the $(\alpha,\gamma)$ reaction rates. Therefore it is highly needed to extend the experimental database of $(\alpha,\gamma)$ cross sections to the heavier mass region.

## 3. Experimental methods for charged particle capture cross section measurements

The conventional method for capture cross section measurements is the in-beam $\gamma$-spectroscopy. In the mass and energy range relevant for the $\gamma$-process, however, severe problems are encountered in the application of this technique. As it is usual in nuclear astrophysics, low cross sections from millibarn down to the nanobarn range must be measured and the beam-induced background produced on low Z target impurities may hinder the cross section measurement. Moreover, the compound nucleus created in a capture process is typically excited to an energy range where the nuclear energy level density is very high. This results in a complicated decay scheme involving many transitions. All these transitions must be measured with their complete angular distributions which requires huge experimental effort (see e.g. [11, 12]). With the application of a $4\pi$ summing crystal some of these problems can be avoided [13], but none of the in-beam $\gamma$-spectroscopy methods succeeded so far to reach the heavier (A > 110) mass region of the p-nuclei.

The activation method has been proved to be a very fruitful technique for the determination of cross sections relevant for the $\gamma$-process. Most of the isotopes listed in table 1 have been studied with activation. This method is applicable only if the reaction product of the capture leads to a radioactive isotope with convenient half-life and strong enough decay signature (like a high intensity $\gamma$-radiation). Most of the problems encountered in in-beam $\gamma$-spectroscopy can be avoided in an activation experiment. Since the off-line measurement of the induced activity leads to the determination of the total number of capture reactions, there is no need to consider individual $\gamma$-transitions, angular distributions, etc. The activation technique is also less sensitive to target impurities and in certain cases more isotopes of a given element can be studied simultaneously.

In the present work the cross section of the $^{151}$Eu$(\alpha,\gamma)^{155}$Tb and $^{151}$Eu$(\alpha$,n$)^{154}$Tb reactions have been measured with the activation method. (The study of $(\alpha$,n) reactions – although not directly relevant to the p-process – provides additional information to the investigation of different optical potentials, see below.) In the next section some details of the measurement are given emphasizing the special requirements of an activation experiment.

## 4. Cross section measurement of $^{151}$Eu$(\alpha,\gamma)^{155}$Tb and $^{151}$Eu$(\alpha$,n$)^{154}$Tb reactions

As can be seen in table 1, the only isotope for which $\alpha$-capture cross section measurement has been carried out near the A = 150 mass region is $^{144}$Sm [14]. For this isotope it was found, that the statistical model calculations using different $\alpha$-nucleus optical potentials give extremely different cross section results. One can choose the best optical model parametrization for a given reaction only if experimental data are available for comparison. Therefore, it is important to collect more experimental data in this mass region to have a larger experimental database to check the performance of different global optical potentials. Our next step towards this goal is the study of $^{151}$Eu.

Alpha capture on $^{151}$Eu (Q = -1.98 MeV) leads to $^{155}$Tb which decays to $^{155}$Gd by electron capture with a half-life of 5.32 d. The decay is followed by a few relatively strong $\gamma$-radiations which can be used for the cross section measurement. Moreover, the $(\alpha$,n) reaction on $^{151}$Eu (Q = -10.14 MeV) also produces a radioactive residual nucleus, $^{154}$Tb, so the $^{151}$Eu$(\alpha$,n) cross section can also be measured by activation. The decay scheme is, however, more complicated. Besides its ground state decaying to $^{154}$Gd with 21.5 h half-life, it has two long-lived isomeric states with half-lives of 9.4 h and 22.7 h decaying by $\beta^+$, electron capture and/or internal

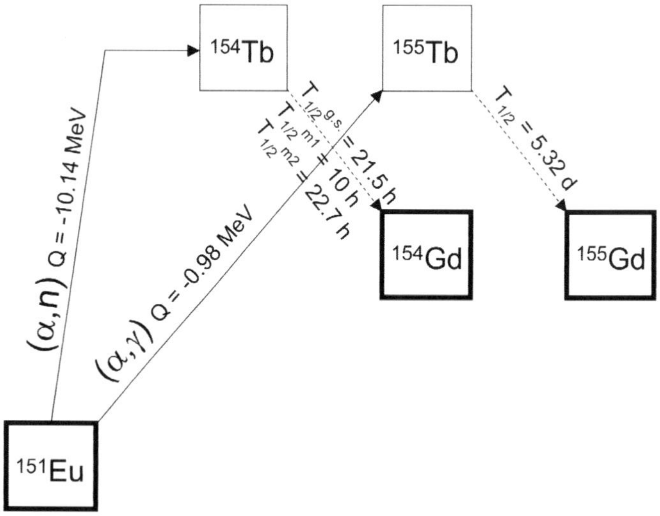


**Figure 2.** The studied reactions and the decay of the reaction products.

transition. Fortunately all these different decays are followed by many strong $\gamma$-radiations, a few of which can be attributed exclusively to the decay of a given state. Thus, partial cross sections to the ground and the two isomeric states can be determined separately and their sum gives the total ($\alpha$,n) cross section. Figure 2 shows the relevant part of the chart of nuclides with the studied reactions and the decay of the reaction product.

It should be noted that the precision of the decay parameters (half-life, $\gamma$-intensity) directly influences the precision of the half-life determination. In the case of the m1 isomer in $^{154}$Tb, the half-life has a relatively large uncertainty ($9.4 \pm 0.4$ h) and ambiguous values can be found in the literature. Therefore, we have performed a high precision experiment to determine this half-life value. We have obtained $T_{1/2} = 9.994 \pm 0.039$ h [15]. With this measurement we reduced the uncertainty of this half-life by one order of magnitude and avoided one possible source of systematic uncertainty in the $^{151}$Eu($\alpha$,n)$^{154}$Tb cross section determination.

### *4.1. Experimental technique*

Targets have been prepared by evaporating $Eu_2O_3$ enriched to 99.2 % in $^{151}$Eu onto thin Al foil. Although the natural abundance of $^{151}$Eu is relatively high (47.8 %), enriched target material was necessary to avoid the high background coming from the decay of $^{156}$Tb, the ($\alpha$,n) reaction product of the heavier Eu isotope. $^{156}$Tb has a half-life of 5.4 d, similar to that of $^{155}$Tb and the $^{153}$Eu($\alpha$,n) cross section is much higher than that of $^{151}$Eu($\alpha,\gamma$). Test experiments with natural isotopic composition targets showed that in the presence of the $^{156}$Tb decay, the weak $\gamma$-radiation from $^{155}$Tb is not measurable.

The number of target atoms enters directly to the calculation of the cross section. Using a thin Al foil as target backing, it was possible to obtain the number of target atoms by measuring the weight of the Al foil before and after the evaporation. It is also important that the target remains stable under $\alpha$-bombardment. The target stability can be monitored by detecting the backscattered $\alpha$-particles from the target continuously during the irradiation. For the detection of the backscattered particles, low Z backing material is preferred in order to have a clear separation of the scattering events from Eu and from the backing. Thus, Al was an ideal choice.

The irradiations have been carried out at the cyclotron accelerator of ATOMKI. The typical beam current was $1 - 1.5\,\mathrm{p}\mu\mathrm{A}$ and based on the backscattering spectra, no target deterioration was observed with this beam intensity. The energy range between 11.5 and 17.5 MeV was covered

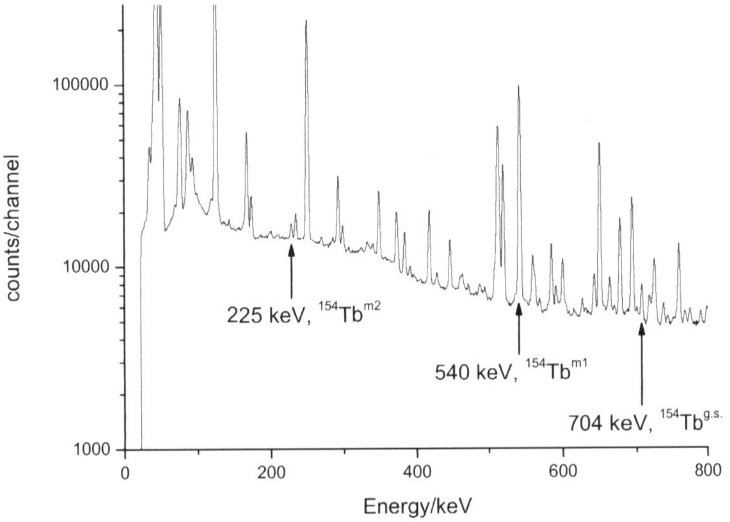


**Figure 3.** Gamma-spectrum measured for one hour on a target irradiated by 17 MeV alphas. The strongest peaks uniquely assigned to a given state in $^{154}$Tb are marked with arrows.

with about 0.5 MeV steps. Owing to the relatively short half-life of the reaction products, the targets could be used more than once at different energies. At some energies the irradiations were carried out with two different targets and consistent results were obtained. The length of the irradiations varied between 5 and 24 hours.

The induced $\gamma$-activity has been measured with a calibrated 40 % relative efficiency HPGe detector equipped with lead shielding. The countings were started typically one hour after the end of the irradiation and lasted for at least 12 hours. In this period the $\gamma$-spectra were dominated by the peaks from the $^{151}$Eu($\alpha$,n)$^{154}$Tb and the $\gamma$-lines from the $^{151}$Eu($\alpha,\gamma$)$^{155}$Tb reaction were buried under the Compton background. Therefore, a repeated $\gamma$-counting was carried out about one week after the irradiation. At this time the shorter-lived $^{154}$Tb isotope was essentially absent and $^{155}$Tb $\gamma$-peaks could be observed on a strongly reduced background. As an example, figure 3 shows the measured $\gamma$-spectrum directly after the irradiation where many peaks from the decay of the three states of $^{154}$Tb can be seen.

### *4.2. Preliminary results*

Cross sections of the $^{151}$Eu($\alpha,\gamma$)$^{155}$Tb and $^{151}$Eu($\alpha$,n)$^{154}$Tb reactions have been measured in the center of mass energy range between 11.2 and 17.1 MeV. At the lowest two energy points the yield from the ($\alpha,\gamma$) reaction product was too low and therefore only the ($\alpha$,n) cross section could be obtained. The Gamow window at T = 3 GK temperature for this reaction is roughly between 8 and 12 MeV, so the measurements were carried out just above the astrophysically relevant energy range. In the measured energy range the cross section values are between about 5 microbarns and 100 millibarns. The final analysis of the data is still in progress. Figure 4 shows the preliminary results of both measured reaction in the form of the astrophysical S-factor. Also shown are the statistical model calculations performed with the NON-SMOKER$^{\rm WEB}$ code version 5.8w [16] using two different global $\alpha$-nucleus optical potentials. One is the standard potential of the NON-SMOKER$^{\rm WEB}$ code from McFadden and Satchler [17] and the other from Fröhlich and Rauscher [18, 19]. The calculations reproduce well the energy dependence of the measured cross sections, while it seems that the absolute magnitude is better reproduced by the Fröhlich and Rauscher potential. The final analysis and the interpretation of the data are in progress.

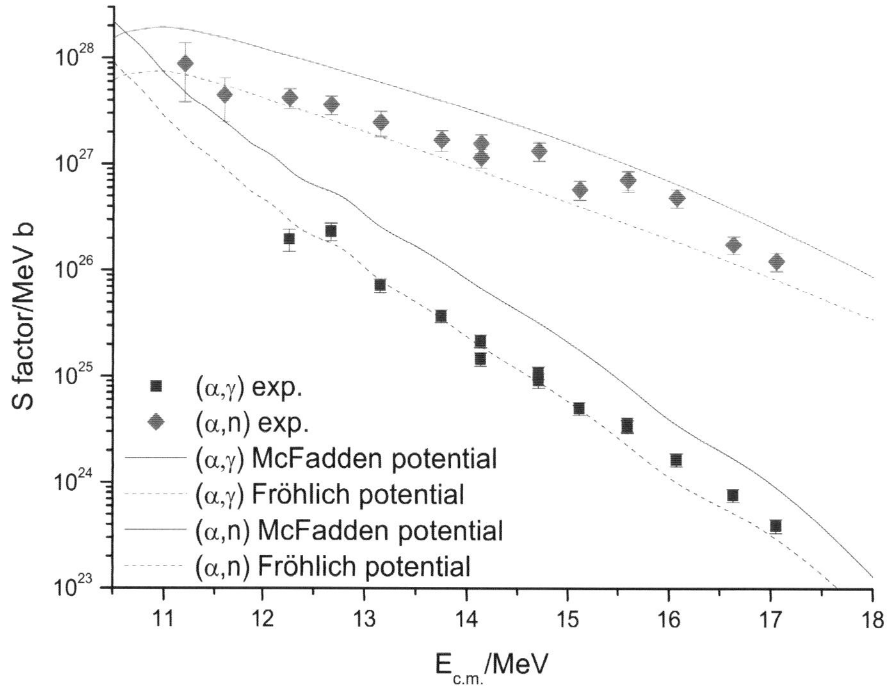


**Figure 4.** (Color online) Measured and calculated cross section of the $^{151}$Eu$(\alpha,\gamma)^{155}$Tb and $^{151}$Eu$(\alpha$,n$)^{154}$Tb reactions (preliminary results). The two theoretical curves obtained with the NON-SMOKER$^{\rm WEB}$ code version 5.8w correspond to two different global $\alpha$-nucleus optical potentials of refs. [17] and [18, 19].

**Acknowledgments**
This work was supported by ERC Grant 203175, GVOP-3.2.1.-2004-04-0402/3.0., OTKA (K68801, T49245), TUBITAK-Grant-108T508 and Kocaeli University BAP-Grant-2007/37.

## References

[1] Busso M, Gallino R and Wasserburg G 1999 *J Ann. Rev. Astron. Astrophys.* **37** 239
[2] Arnould M, Goriely S and Takahashi K 2007 *Phys. Rep.* **450** 97
[3] Arnould M and Goriely S 2003 *Phys. Rep.* **384** 1
[4] Gyürky Gy *et al* 2006 *Phys. Rev. C* **74** 025805
[5] Rapp W, Görres J, Wiescher M, Schatz H and Käppeler F 2006 *Astrophys. J.* **653** 474
[6] Rauscher T, Heger A, Hoffman R D and Woosley S E 2002 *Astrophys J.* **576** 323
[7] Rauscher T 2006 *Phys. Rev. C* **73** 015804
[8] Hasper J *et al* 2009 *Phys. Rev. C* **79** 055807
[9] Mohr P, Fülöp Zs and Utsunomiya H 2007 *Eur. Phys. J. A* **32** 357
[10] Bao Z Y, Beer H, Käppeler F, Voss F and Wisshak K 2000 *At. Data Nuc. Data Tables* **76** 70
[11] Galanopulos S *et al* 2003 *Phys. Rev. C* **67** 015801
[12] Hasper J *et al* this volume
[13] Spyrou A, Becker H-W, Lagoyannis A, Harissopulos S and Rolfs C 2007 *Phys. Rev. C* **76** 015802
[14] Somorjai E *et al* 1998 *Astron. Astrophys.* **333** 1112
[15] Gyürky Gy *et al* 2009 *Nucl. Phys. A* **828** 1
[16] NON-SMOKER$^{\rm WEB}$ version 5.8w http://www.nucastro.org
[17] McFadden L and Satchler G R 1966 *Nucl. Phys.* **84** 177
[18] Fröhlich C diploma thesis 2002 University of Basel, Switzerland
[19] Rauscher T 2003 *Nucl. Phys. A* **719** 73c; Erratum: Rauscher T 2003 *Nucl. Phys. A* **725** 295

# In-beam experiments on $(p, \gamma)$ and $(\alpha, \gamma)$ reactions for the astrophysical $p$ process

**Jens Hasper, Marc Büssing, Michael Elvers, Janis Endres, and Andreas Zilges**

Institut für Kernphysik, Universität zu Köln, Zülpicher Straße 77, 50937 Köln, Germany

E-mail: hasper@ikp.uni-koeln.de

**Abstract.** We have recently performed an experiment for the $(p, \gamma)$ and $(\alpha, \gamma)$ reactions on the nucleus $^{92}$Mo at energies of relevance for the astrophysical $p$ process. The reactions were investigated by the in-beam technique using the $\gamma$-ray detector array HORUS at the ion TANDEM accelerator of the University of Cologne. The experimental results are compared to data stemming from activation measurements and to statistical-model calculations.

## 1. Introduction

The $p$ process describes the nucleosynthesis of 35 proton-rich nuclei which are bypassed by the neutron-capture reactions of the $s$ and $r$ processes, the so-called *p nuclei*. This process is mainly driven by photodisintegration reactions in explosive astrophysical environments such as Type-II supernovae and involves a huge reaction network of about 2000 nuclei and 20000 reactions [1, 2]. To account for such a large number of reactions, the corresponding reaction rates are usually derived from statistical-model calculations performed within the framework of the Hauser-Feshbach theory. Experimental data for $p$-process relevant reactions are strongly required to test the reliability and robustness of these calculations. Whereas the Hauser-Feshbach theory itself is well established, significant uncertainties are introduced by uncertainties of the nuclear models which enter the calculations. Therefore, one of the major aim of experimental studies is to improve the descriptions for nuclear properties like photon-strength functions, nuclear level densities and optical-model potentials on the basis of reliable experimental data.

Increasing experimental efforts have been made in the last decade to study $p$-process relevant reactions. An overview of the various experimental approaches is given, e. g., in Ref. [3]. It has been shown that proton and $\alpha$-particle capture reactions are particularly well suited to determine stellar photodisintegration reaction rates via the Detailed Balance theorem [4, 5]. Most of the experiments on these type of reactions have so far applied the activation technique in which the reaction yield is measured offline under low-background conditions by tracing the $\beta$ decays of the radioactive reaction products, e. g., see Ref. [6]. However, since activation experiments are restricted to reactions that produce radionuclides with appropriate half-lives, complementary in-beam experiments have been performed in which the reaction is identified by the prompt $\gamma$ decays of the excited reaction products rather than by their subsequent $\beta$ decays. So far, in-beam experiments have either made use of highly-efficient large-volume $4\pi$ NaI detectors [7] which are available at the National Centre for Scientific Research Demokritos, Athens, and at

RUBION, University of Bochum, or have applied a setup of up to four high-resolution HPGe detectors installed at the former DYNAMITRON accelerator, University of Stuttgart [8].

In this article, we report on another experimental setup for in-beam measurements available at the ion TANDEM accelerator of the University of Cologne, which is dedicated to study $p$ process relevant reactions at energies far below the Coulomb threshold. This setup will be presented in the following in the context of a recent experiment.

## 2. Experiment

We have measured the $(p,\gamma)$ and $(\alpha,\gamma)$ reaction on $^{92}$Mo at several energies within the so-called *Gamow window*, i. e., the energy window of astrophysical relevance for charged-particle capture reactions under stellar conditions. In this experiment, metallic isotopically-enriched molybdenum targets of about 400 $\frac{\mu g}{cm^2}$ thickness each were bombarded for several hours with protons and $\alpha$ particles at beam currents of about 200 nA and 30 nA, respectively. The beam current was determined from the charge deposited in a Faraday cup behind the target. In addition, a silicon detector was placed under a backward angle of 135° to monitor the target thickness during the experiment by measuring the Rutherford scattering off the target.

The $(p,\gamma)$ and $(\alpha,\gamma)$ reactions were measured by detecting the prompt $\gamma$ decays of the reaction products with 13 HPGe detectors at the Cologne $\gamma$-ray detector array HORUS. Each detector features a relative efficiency between 40% and 80% compared to a 7.62×7.62 cm$^2$ NaI detector. The detectors were mounted as close as possible around the target chamber, so that an excellent absolute photopeak efficiency of almost 5% for photons of energy $E_\gamma$ = 1332.5 keV could be provided.

Figure 1 depicts the lower-energetic part of the summed photon spectrum observed for the $(p,\gamma)$ reaction at a proton energy of $E_p$ = 3500 keV. The spectrum clearly reveals most of the low-energetic $\gamma$ transitions from the reaction product $^{93}$Tc. In general, the total capture cross section

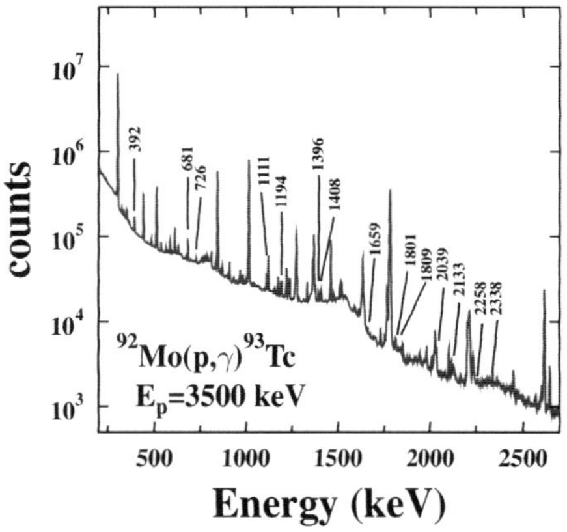


**Figure 1.** Measured photon spectrum of the proton capture reaction $^{92}$Mo$(p,\gamma)^{93}$Tc at an incident proton energy of $E_p$ = 3500 keV. The spectrum was obtained from the sum of 13 HPGe detectors at the $\gamma$-ray detector array HORUS. Observed transitions from the reaction product $^{93}$Tc are marked by black lines.

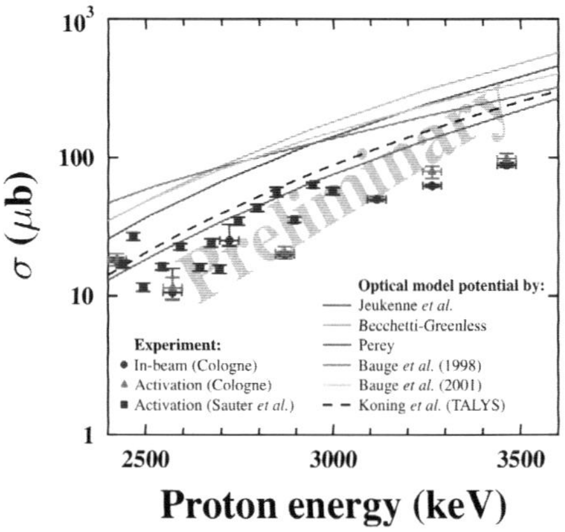


**Figure 2.** Experimental data for the ground-state cross section of the proton capture reaction $^{92}$Mo$(p,\gamma)^{93}$Tc$^{g.s.}$ measured in different experiments are compared to statistical-model calculations. The calculations were carried out by applying different optical-model potentials (see colored lines).

can be derived by adding up the reaction yields of all prompt $\gamma$ transitions going to the ground state of the reaction product. However, since $^{93}$Tc has an isomeric state with a rather long half-life of $t^{m}_{1/2} = 43.5$ min, this state does not show a prompt $\gamma$ decay to the ground state, and the cross section for capture reactions populating the ground state and isomeric state, respectively, had to be analyzed separately. Figure 2 shows preliminary results for the $(p,\gamma)$ ground-state cross section as derived from the in-beam technique. As a cross-check for the in-beam method, we also determined the cross-section from the $\beta$-decay activity of the radioactive reaction product $^{93}$Tc, which has an appropriate half-life of $t^{g.s.}_{1/2} = 2.7$ h. As depicted in the figure, the two different methods are in excellent agreement to each other. In addition, we compared our results to a previous activation experiment performed by Sauter and Käppeler [9]. Although the agreement between the two data sets looks quite reasonable, some discrepancies can be seen which require further investigation in the ongoing data analysis. Finally, we compared the experimental data to theoretical predictions from statistical-model calculations using different proton-nucleus optical-model potentials as an input. The calculations were performed with the TALYS [10] and NON-SMOKER$^{\mathrm{WEB}}$ code [11], respectively. The best agreement with the experimental data was found for the models by Jeukenne *et al.* and Koning *et al.*, respectively. Nevertheless, the cross section is still slightly overestimated for both descriptions.

Figure 3 shows a spectrum measured for the $(\alpha,\gamma)$ reaction on $^{92}$Mo for incident $\alpha$ particles of energy $E_\alpha = 9300$ keV. In comparison to the $(p,\gamma)$ experiment, $\alpha$-particle capture experiments are faced with a much larger background from competing reactions, in particular from $\alpha$-particle capture reactions on light elements like carbon and oxygen. Due to the dominating background, only the strongest transitions in the reaction product $^{96}$Ru could be observed, and a reliable

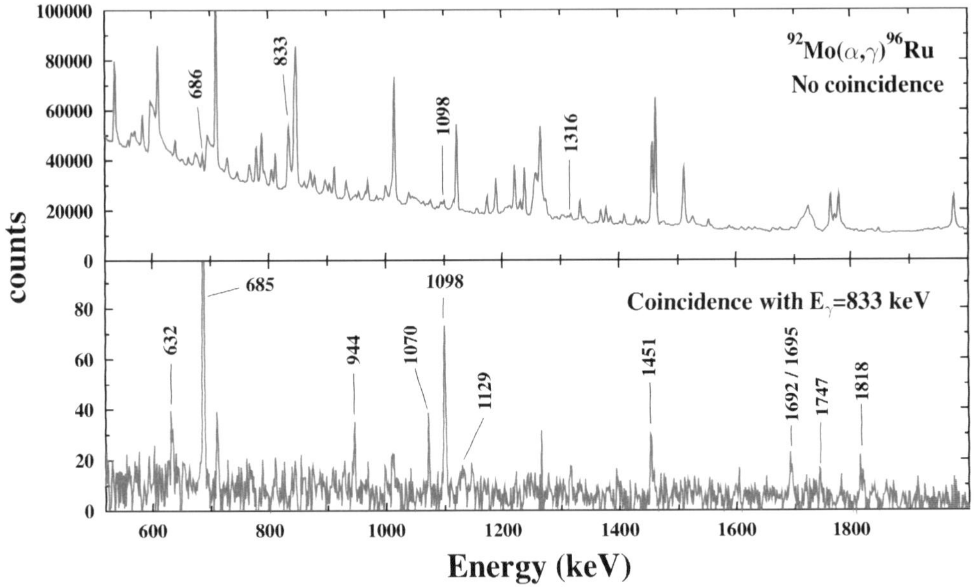


**Figure 3.** Measured photon spectrum of the $\alpha$-particle capture reaction $^{92}\mathrm{Mo}(\alpha,\gamma)^{96}\mathrm{Ru}$ at an incident $\alpha$-particle energy of $E_\alpha = 9300$ keV. *Upper panel:* Due to the large background from $\alpha$-particle capture reactions only a few low-energetic transitions in the reaction product $^{96}$Ru can be observed (marked by black lines). *Lower panel:* Background can be strongly suppressed by requiring a $\gamma$ coincidence with the $\gamma$ transition of energy $E_\gamma = 833$ keV. The coincidence spectrum reveals many more more $\gamma$ transitions of the reaction product $^{96}$Ru.

determination of the $(\alpha,\gamma)$ cross section from the raw $\gamma$ spectra was strongly hampered. A powerful tool, however, to eliminate background lines from the spectra is to make use of $\gamma$-coincidence techniques. A first analysis of the data applying this technique looks very promising and shows that an almost background-free spectrum can be obtained which reveals many more low-lying $\gamma$ transitions of the reaction product (see Fig. 3). We are confident that this technique will allow an accurate determination of the $\alpha$-particle capture cross section on $^{92}$Mo in the ongoing analysis.

## 3. Conclusion and Outlook

For the first time, the $\gamma$-ray detector array HORUS was used to measure absolute cross sections for proton and $\alpha$-particle capture reactions at astrophysically relevant energies. Although the data analysis for the $(p,\gamma)$ and $(\alpha,\gamma)$ experiment is still ongoing, first results are already in reasonable agreement to previous experimental data proving that the experimental setup and the in-beam method are working well. Compared to activation experiments, in-beam techniques have the major advantage that they can be used to study reactions on almost any stable nucleus. Thus, the number of experimentally accessible reactions can be widely extended. The advantage of using HPGe detectors instead of large-volume NaI detectors is that the measured high-resolution photon spectra provide detailed information about the decay pattern of the reaction product. This, e. g., allows to determine the partial cross sections of many single transitions, although this could not be discussed explicitly in this article. Information on partial cross sections are very valuable to further constrain the nuclear models adopted for statistical-model calculations. Thus, the $\gamma$-ray detector array HORUS provides an excellent tool to improve the experimental data base for $p$-process relevant reactions in the future.

## 4. Acknowledgments

This project has been supported by the Deutsche Forschungsgemeinschaft under Contract ZI 510/5-1 and by the BMBF under contract 06KY 9136 I.

[1] D. L. Lambert, Astron. Astroph. Rev. **3** (1992) 201.
[2] M. Arnould, S. Goriely, Phys. Rep. **384** (2003) 1.
[3] J. Hasper, M. Büssing, J. Endres, M. Elvers and A. Zilges, Proceedings of the International Conference on *Nuclear Structure and Related Topics, Dubna, Russia, June 30 - July 4 2009*, in print.
[4] W. A. Fowler, G. E. Caughlan, and B. A. Zimmermann, Annu. Rev. Astron. Astrophys. **5** (1967) 525.
[5] T. Rauscher, Proceedings of the International Conference on *Capture Gamma-Ray Spectroscopy and Related Topics, 13th International Symposium, Cologne, Germany, August 25 - August 29 2008*, AIP Conf. Proc. **1090** (2008) 48.
[6] C. Yalçin, R. T. Güray, N. Özkan, S. Kutlu, Gy. Gyürky, J. Farkas, G. G. Kiss and Zs. Fülöp, A. Simon, E. Somorjai and T. Rauscher, Phys. Rev. C **79** (2009) 065801.
[7] A. Spyrou, H.-W. Becker, A. Lagoyannis, S. Harissopulos, and C. Rolfs, Phys. Rev. C **76** (2007) 015802.
[8] S. Galanopoulos, P. Demetriou, M. Kokkoris, S. Harissopulos, R. Kunz, M. Fey, J. W. Hammer, Gy. Gyürky, Zs. Fülöp, E. Somorjai, and S. Goriely, Phys. Rev. C **67** (2003) 015801.
[9] T. Sauter and F. Käppeler, Phys. Rev. C **55** (1997) 3127.
[10] A. Koning, S. Hilaire, M. C. Duijvestijn, Proceedings of the International Conference on *Nuclear Data for Science and Technology - ND2004, Santa Fe, USA, September 26 - October 1 2004*, AIP Conf. Proc. **769** (2005) 1154.
[11] NON-SMOKER$^{\mathrm{WEB}}$, Version 5.8w, http://nucastro.org/nonsmoker.html.

# The r-, p-, and $\nu$p-Process

**F.-K. Thielemann[1,6], I. Dillmann[2], K. Farouqi[3], T. Fischer[1], C. Fröhlich[3], A. Kelic-Heil[6], I. Korneev[4], K.-L. Kratz[5], K. Langanke[6], M. Liebendörfer[1], I.V. Panov[4], G. Martinez-Pinedo[6], T. Rauscher[1]**

[1]Departement Physik, Universität Basel, CH-4056 Basel, Switzerland
[2]Physik Department, Technische Universität München, D-85748 Garching, Germany
[3] Departement of Astrophysics and Astronomy, University of Chicago, Chicago, IL 60637, USA
[4]ITEP Moscow, 117259 B. Cheremushkinskaja 25, Russia
[5]Max-Planck-Institut für Chemie (Otto-Hahn-Institut), D-55128 Mainz, Germany
[6]GSI Helmholtz Center, Planckstr. 1, 64291 Darmstadt, Germany

E-mail: f-k.thielemann@unibas.ch

**Abstract.** The processes discussed in this review are three of the four nucleosynthesis processes involved in producing heavy nuclei beyond Fe (not counting the rp-process in X-ray bursts). Opposite to the fourth process (the s-process), which operates in stellar evolution during He- and C-burning, they are all related to explosive burning phases, (presumably) linked to core collapse supernova events of massive stars. The (classical) p-process is identified with explosive Ne/O-burning in outer zones of the progenitor star. It is intitiated by the passage of the supernova shock wave and acts via photodisintegration reactions like a spallation process which produces neighboring (proton-rich) isotopes from pre-existing heavy nuclei. The reproduction of some of the so-called lighter p-isotopes with $A < 100$ faces problems in this environment. The only recently discovered $\nu$p-process is related to the innermost ejecta, the neutrino wind expelled from the hot proto-neutron star after core collapse in the supernova explosion. This neutrino wind is proton-rich in its early phase and reactions with neutrinos permit to overcome decay/reaction bottlenecks for the flow beyond the Fe-group, thus permitting the production of those p-isotopes, which face problems in the classical p-process scenario. The understanding of the r-process, being identified for a long time with rapid neutron captures - and passing through nuclei far from stability - is still experiencing major problems. These are on the one hand related to nuclear uncertainties far from stability (masses and half-lives), affecting the process speed and abundance peaks, on the other hand the site is still not definitely located, yet. Later neutron-rich, high entropy phases of the neutrino wind could permit its operation, other options include the ejection of very neutron-rich neutron star matter. Two different environments are required for a weak and a main/strong r-process, witnessed by observations of low metallicity stars.

## 1. Introduction

Two of the three processes discussed in this overview are classical in the sense that they have long been introduced by their abundance features. The p-process is easily defined by all isolated stable isotopes on the proton-rich side of stability with typically 1% of the total element abundance (except for lower mass numbers). This process is identified by solar abundances features and the question remains whether only one astrophysical site is responsible. The paradigm relates it to zones of explosive Ne/O burning in the outer part of core collapse supernovae, acting on pre-existing heavy nuclei and making it a secondary process in terms of galactic evolution. The observational indications for a lighter element primary process [91](LEPP) at low metallicities points towards a primary origin of the lighter p-isotopes, which would indicate a direct production closer to the supernova core.

The r-process has been identified by the double peak structure near closed neutron shells in solar heavy element abundances. When subtracting s-process abundances (quite well understood via neutron captures in stellar evolution and nuclear physics at and close to stability [41]), it emerges as a process with a path far on the neutron-rich side of stabilty, requiring explosive environments with large neutron to seed nuclei ratios. There exist major advances in the nuclear physics involved, while many open questions remain and will be related to future rare isotope beam facilities. Within the present nuclear physics uncertainties, the necessary astrophysical environment conditions have been identified. The main problem seems the apparent (non-)realization in astrophysical simulations/models. Observations of low metallicity stars indicate also here the probable splitting in two types of events: (a) a rare event, reproducing the heavy r-process abundances always in solar proportions, and (b) a more frequent event, responsible for the lighter r-abundances [35, 82, 15, 38, 37].

The $\nu$p-process has only been discovered in recent years [26, 70, 95] and resulted from progress in core collapse supernova efforts. While it was previously expected that the innermost ejected layers, close to the freshly formed neutron star, are neutron-rich and just automatically the site of the r-process, the latter expectation has actually been tempered. This seemed mainly due to the fact that sufficiently high entropies could not be attained [88, 93]. On the other hand, recent explosion calculations, with careful accounting of the interaction with neutrinos, led to slightly proton-rich conditions in the early phase of the neutrino wind [51, 8, 21]. This results in a proton and alpha-rich freeze-out producing nuclei up to $^{64}$Ge, where a long beta-decay half-life is encountered. Anti-neutrino capture on the remaining protons creates neutrons and the reaction $^{64}\mathrm{Ge}(n,p)^{64}\mathrm{Se}$ mimics a fast beta-decay, permitting then to move upward to nuclei with masses $A < 100$. This $\nu$p-process is a primary process, could explain the observational results promoting the LEPP and can also fill in the light p-isotopes which encountered difficulties in the classical p-process picture [16]. The question is whether, as a function of elapsed time after the initiation of the explosion shock, the neutrino wind changes from proton-rich to neutron-rich and thus would permit to originate an r-process. Thus, both of the latter two discussed processes are really related to the supernova explosion mechanism itself, while the classical p-process requires only the existence of a supernova shock wave. In the following sections we will therfore proceed from the more simple to the more problematic cases in the sequence discussing first the p-process, then the $\nu$p-process and finally the r-process, after first giving a short overview on the status of the core collapse supernova explosion mechanism.

## 2. Core Collapse Supernova Explosions

The problem of core collapse supernova explosions is an old one and the attempt to understand the mechanism has been ongoing for more than 40 years, linking it to massive stars and the collapse of the Fe-core after having passed all nuclear burning stages. Since the sixties the explosion mechanism has been related to neutrino emission from the hot collapsed core, interrupted by a period when it was speculated that the strength of the bounce at nuclear densities could permit shock waves with sufficient energies to lead to prompt explosions [6]. However, this became questionable when previously neglected neutrino scattering processes were introduced (e.g. neutrino-electron scattering), which permitted to replace lost low energy neutrinos, leading to a continuous energy leakage and to the death of the prompt shock within 10 ms after bounce. Since then, and with the first neutrino detection from a core collapse supernova (SN1987A), the hope has been that further improvement would lead to successful explosions via energy deposition through neutrino and anti-neutrino captures on neutrons and protons. Two different paths were explored. 1. Convective instabilities, but with still simplified neutrino transport, causing either (a) convective transport in the core and leading to higher neutrino luminosities [45] or (b) higher energy deposition efficiencies in convective regions [34]. 2. Improved neutrino transport schemes, leading to higher neutrino luminosities via the full solution of the Boltzmann transport equation for neutrino scattering and neutrino reactions [56]. There has been substantial progress in 1D-3D supernova explosion simulations over recent years [51, 89, 39, 90, 9, 8, 27, 10, 3, 40, 52, 24, 20, 21] and a solution (in 3D) seems close. However, a fundamental understanding and robust predictions are still missing. Related to the explosion is also the so-called neutrino wind, emitted for seconds after the successful shock wave generation [70, 26] and considered also as a possible source of the r-process to produce the heaviest elements via neutron captures [84, 100, 71, 88, 93, 85, 89]. Neutrino emission, from the hot and dense matter in the proto-neutron star and from infalling matter, and its time and spectral characteristics [20, 21] lies at the focus of the supenova mechanism and related nucleosynthesis and influences also neutrino nucleosynthesis in the outer mass zones [33].

Given this situation, at present the self-consistent prediction of supernova nucleosynthesis yields seems not possible. However, supernova nucleosynthesis has a long tradition [101, 87, 61, 13, 62]. All of the past predictions relied on an artificially introduced explosion, either via a piston or a thermal bomb introduced into the progenitor star model. In this approach, the mass cut between the ejecta and the remnant does not emerge from the simulation and is freely chosen, guided by constraints on $^{56}$Ni ejecta and/or entropy jumps. While the approach of artificially introduced explosions makes sense and is fully correct for the outer stellar layers (see section on the classical p-process), it clearly is incorrect for the innermost ejected layers which should be directly related to the physical processes causing the explosion. This affects the Fe-group composition and the $\nu$p- and r-process. Here we will make use of 1D approximations [26, 97] and free parameter studies [18, 19]. The relevant thermonuclear reaction rates to be employed in such calculations have been provided by experiment [12, 2] or theory [98, 76, 29]. The weak interaction rates stem from phenomenological approaches or shell model calculations [28, 50], beta-decay properties from experiment or QRPA predictions [58], beta-delayed or neutron-induced fission predictions are related to mass models and fission barriers [86, 53, 65, 67, 30], neutrino-induced reactions make use of RPA calculations [46, 54].

The resulting nucleosynthesis ejecta have to be confronted with observations related to galactic chemical evolution. Cool low-mass stars have an evolution time comparable to the lifetime of the Galaxy, and, at the present epoch, we can observe both young and very old objects among them. The study of chemical abundances in cool stars allows to determine the history of chemical enrichment of galactic matter because their atmospheres preserve much of the chemical composition of the gas out of which the star formed. Core collapse supernovae dominated nucleosynthesis in the early Galaxy, before the onset of type Ia supernova explosions and the main s-process. Detailed chemical analysis of the most metal-poor stars can, therefore, provide insight into the synthesis of the first heavy elements [82, 11, 37, 38]. Several studies [35, 82, 38, 15] have presented arguments supporting constant relative ratios of r-process element abundances during the history of the Galaxy for the elements with $Z = 56 - 70$. This suggests that a unique r-process exists in nature, at least for heavy elements, while the lighter r-process and possibly p-process elements might be produced in different supernovae with varying amounts, including also $\nu$p-process nuclei observed as part of the LEPP [91].

## 3. The p-Process

A number of proton-rich (p-)isotopes of naturally occurring stable nuclei cannot be produced by neutron captures along the line of stability. The currently most favored production mechanism for those 35 p-isotopes between Se and Hg is photodisintegration of intermediate and heavy elements at high temperatures in late evolution stages of massive stars [99, 79]. However, not all p-nuclides can be produced satisfactorily, yet. A well-known deficiency in the model is the underproduction of the Mo-Ru region, but the region $151<A<167$ is also underproduced, even in recent calculations [77, 4, 74, 16]. There exist deficiencies in astrophysical modeling and the employed nuclear physics. Recent investigations have shown that there are still considerable uncertainties in the description of nuclear properties governing the relevant photodisintegration rates. This has triggered a number of experimental efforts to directly or indirectly determine reaction rates and nuclear properties for the $\gamma$-process [78]. Here it is important to investigate the the sensitivity of the location of the $\gamma$-process path with respect to reaction rate uncertainties.

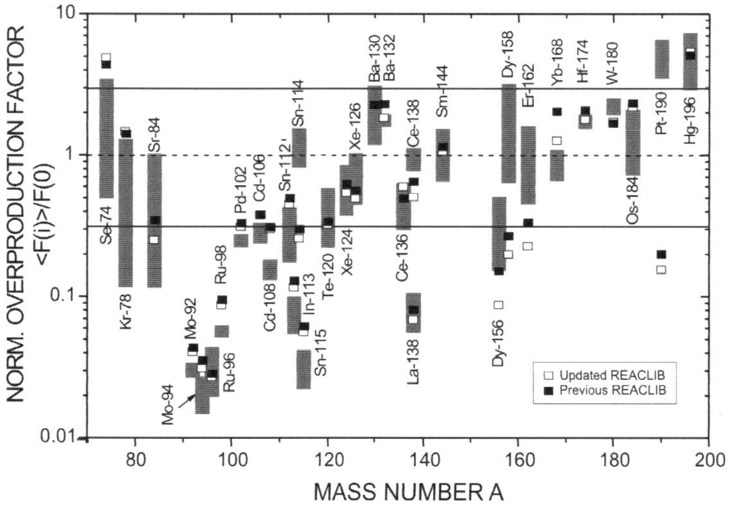


**Figure 1.** Normalized overproduction factors of p-process nuclei derived with the [74] (open squares) and [16] (full squares) reaction library. In addition, the results from a range of stellar models (10-25$M_\odot$) from [80] are given for comparison. A value equal to unity corresponds to relative solar abundances.

Concerning the astrophysical modeling, only a range of temperatures has to be considered which are related to the explosive Ne/O-burning zones of a supernova explosion, i.e. 2-3$\times 10^9$K. The $\gamma$-process starts with the photodisintegration of stable seed nuclei that are present in the stellar plasma. During the photodisintegration period, neutron, proton, and alpha-emission channels compete with eachother and with beta-decays further away from stability. In general, the process, acting like "spallation" of pre-existing nucei commences with a sequence of $(\gamma, n)$-reactions, moves the abundances to the proton-rich side. At some point in a chain of isotopes, $(\gamma, p)$ and/or $(\gamma, \alpha)$-reactions become faster than neutron emissions, and the flow branches and feeds other isotopic chains. At late times photodisintegrations become less effective, when decreasing temperatures shift the branching points and make beta-decays more important. Finally the remaining unstable nuclei decay back to stability. The branchings established by the dominance of proton and/or $\alpha$-emission over neutron emission are crucial in determining the radioactive progenitors of the stable p-nuclei and depend on the ratios of the involved reaction rates. Numerous experimental and theoretical efforts have been undertaken to improve the reaction input, especially with respect to open questions in optical potentials for alpha particles and protons [32, 43, 44, 102].

Applications of p-process network calculations to the temperature profiles of initiated explosions have been performed [80, 74, 16]. Here we present the results of a 25$M_\odot$ mass model [16] with two reaction rate libraries without and with inclusion of all experimental improvements, existing at that point. It is noticed that the nuclear uncertainties cannot change the underproduction of especially the light p-nuclei. Another process seems to be required to supply these missing abundances.

## 4. The $\nu$p-Process

Neutron-deficient nuclei can be produced by two astrophysical nucleosynthesis processes: the rp-process in X-ray bursts (which, however, does not eject matter into the interstellar medium

[92, 81, 22] and the recently discovered $\nu$p-process in core collapse supernovae [26, 70, 95]. The $\nu$p-process occurs in explosive environments when proton-rich matter is ejected under the influence of strong neutrino fluxes. This includes the inner ejecta of core-collapse supernova [8, 90, 52] and possible ejecta from black hole accretion disks in the collapsar model of gamma-ray bursts [83]. The matter in these ejecta is heated to temperatures well above $10^{10}$ and becomes fully dissociated into protons and neutrons. The ratio of protons to neutrons is mainly determined by neutrino and antineutrino absorptions on neutrons and protons, respectively. Similar neutrino and antineutrino energy spectra and fluxes produce proton-dominated matter due to the n-p mass difference. When the matter expands and cools, the free neutrons and protons combine into $\alpha$-particles. Later, at temperatures around $5\times10^9$K, alpha-particles assemble into heavier nuclei via unstable intermediate nuclei, e.g. the triple-$\alpha$ reaction via unstable $^8$Be, but - depending on the entropy and the expansion of matter - only a fraction of those form iron-group nuclei (alpha-rich freeze-out). In case of a proton-rich environment, there are also still free protons available at the time of the alpha freeze-out. Once the temperature drops to about $2\times10^9$K, the composition of the ejecta consists mostly of $^4$He, protons, and iron group nuclei with N≈Z (mainly $^{56}$Ni) in order of decreasing abundance. Without neutrinos, synthesis of nuclei beyond the iron peak becomes very inefficient due to bottleneck (mainly even-even $N = Z$) nuclei with long beta-decay half-lives and small proton-capture cross sections. However, the matter is subject to a large neutrino/antineutrino flux from the proto-neutron star.

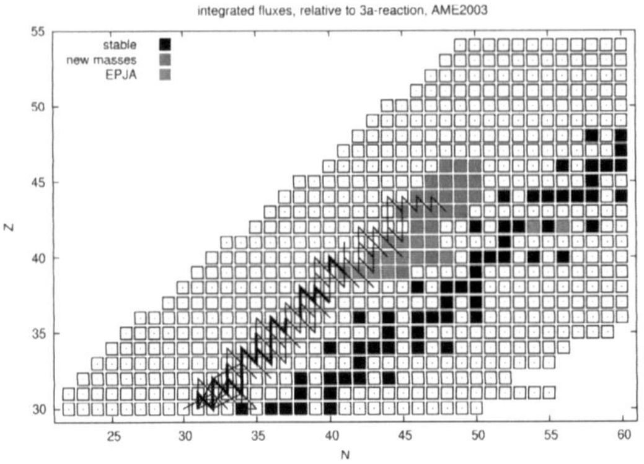


**Figure 2.** $\nu$p-process path employing AME2003 [5] and latest mass measurements [97].

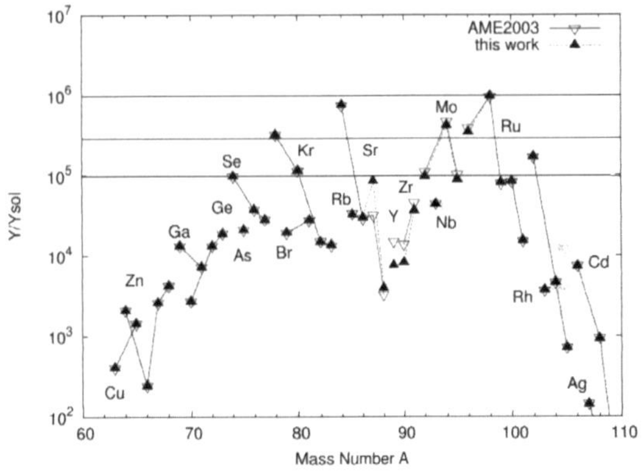


**Figure 3.** Final abundances normalized to solar after decay for two sets of thermonuclear reaction rates/masses. Matter up to $A = 100$ can be produced easily.

While neutrons are bound in neutron-deficient $N = Z$ nuclei and neutrino captures on these nuclei are negligible due to energetics, antineutrinos are readily captured both on free protons and on heavy nuclei on a timescale of a few seconds. As protons are more abundant than heavy nuclei, antineutrino captures occur predominantly on protons, leading to residual neutron densities of $10^{14}10^{15}$ cm$^3$ for several seconds. These neutrons are easily captured by heavy neutron-deficient nuclei, for example $^{64}$Ge, inducing $(n,p)$ reactions with time scales much shorter than the beta-decay half-life. This permits further proton captures and allows the nucleosynthesis flow to continue to heavier nuclei. The $\nu$p-process [26] is this sequence of $(p,\gamma)$-reactions, followed by $(n,p)$-reactions or beta-decays, where the neutrons are supplied by antineutrino captures on free protons. Here we show $\nu$p-process nucleosynthesis for the explosion of a 15M$_\odot$ star [39], also utilized in [70, 23], which synthesizes efficiently nuclei with

$A > 90$. Two sets of astrophysical reaction rates were used in the reaction network, both based on theoretical rates from the NON-SMOKER code [76], but once with with the latest excited state information and masses from the AME2003 compilation [5] and another set also with the latest mass measurements [42, 97].

Fig. 3 shows the final abundances normalized to solar abundances after decay to stability for these two sets of thermonuclear reaction rates. Only nuclei produced in the p-rich ejecta are shown. As is clearly seen, there is no difference in the yields for the two different sets of rates except for a few nuclei in the mass range $85 < A < 95$, namely $^{87,88}$Sr, $^{89}$Y, and $^{90,91}$Zr. This can be directly traced back to the large change in the mass of $^{88}$Tc ($\Delta M$= 1031 keV). This change in mass leads to an increase in the reaction rate for $^{88}\mathrm{Tc}(\gamma,p)^{87}\mathrm{Mo}$ at the relevant temperatures and therefore a relative suppression of the opposite capture rate. Fig. 2 shows the time-integrated reaction flows relative to the triple-alpha-reaction employing the masses from AME2003 only and the masses including the latest measurements, respectively. The total flow reaching $^{94}$Pd is very similar in both cases. These results show that the $\nu$p-process can easily produce the light p-nuclei of Mo and Ru, which are deficient in p-process calculations. Further processing depends on the expansion (speed) of matter and the overlying mass of ejecta.

## 5. The r-Process

A rapid neutron-capture process (r-process) in an explosive environment is traditionally believed to be responsible for the nucleosynthesis of about half of the heavy elements above Fe. While in recent years the high entropy (neutrino) wind (HEW) of core-collapse supernovae has been considered to be one of the most promising sites, hydrodynamical simulations still encounter difficulties to reproduce the astrophysical conditions under which this process occurs. The classical "waiting-point" approximation, with the basic assumptions of an Fe-group seed, an $(n,\gamma)-(\gamma,n)$-equilibrium for constant neutron densities $n_n$ at a chosen temperature $T$ over a process duration $\tau$, and an instantaneous freezeout, has helped to gain improved insight into the systematics of an r-process in terms of its dependence on nuclear-physics input and astrophysical conditions [14, 47, 48]. Taking a specific seed nucleus, the solar r-process pattern peaks can be reproduced by a variation/superposition of neutron number densities $n_n$ and durations $\tau$. Whether the solar r-process abundances $N_{r,\odot} \simeq N_\odot - N_{s,\odot}$ are fully reproduced in each astrophysical event, i.e., whether each such event encounters the full superposition of conditions required, is a matter of debate [96, 69, 82, 37, 73, 18]. In realistic astrophysical environments with time variations in $n_n$ and $T$, it has to be investigated whether at all and for which time duration $\tau$ the supposed $(n,\gamma)-(\gamma,n)$-equilibrium of the classical approach will hold and how freeze-out effects change this behavior. In general, late neutron captures may alter the final abundance distribution. In this case neutron capture reactions will be important. Also $\beta$-delayed neutrons can play a role in forming and displacing the peaks after freeze-out.

### *5.1. The High Entropy Neutrino Wind*

For many years since [100, 84, 71] the high entropy wind has been considered as the most promising (realistic?) environment, expelled from newly formed (hot) neutron stars in core-collapse supernovae, which continue to release neutrinos after the supernova shock wave is launched. These neutrinos interact with matter of the outermost proto-neutron star layers which are heated and ejected in a continuous wind. The late neutrino flux also leads to moderately neutron-rich matter [71] via interactions with neutrons and protons and causes matter ejection with high entropies. Problems were encountered to attain entropies sufficiently high in order to produce the heaviest r-process nuclei [88, 93, 85]. Recent hydrodynamic simulations for core-collapse supernovae support the idea that these entropy constraints can be fulfilled in the late phase (after the initial explosion) when a reverse shock is forming [27, 3, 10, 40, 66].

The question is whether such high entropies occur at times with sufficiently high temperatures when an r-process is still underway [49]. Exploratory calculations to obtain the necessary conditions for an r-process in expanding high-entropy matter have been undertaken by a number of groups [36, 55, 63, 93, 85, 94, 103, 49]. Our recent investigations focussed (a) on the effects of varying nuclear physics input [mass models FRDM (Finite Range Droplett Model [57], ETFSI-1 (Extended Thomas-Fermi with Strutinsky Integral) [1], ETFS-Q with quenching of shell closures far from stability [68], the mass formula by Duflo & Zuker (DUFLO-ZUKER) [17] and HFB-17 (a rencent Hartree-Fock-Bogoliubov approach) [31]] and (b) the detailed understanding of the nuclear flow through the chart of nuclides, testing equilibria, freeze-out and delayed neutron capture. To investigate these effects we have applied a full network containing up to 6500 nuclei and the corresponding nuclear masses, cross sections and $\beta$-decay properties.

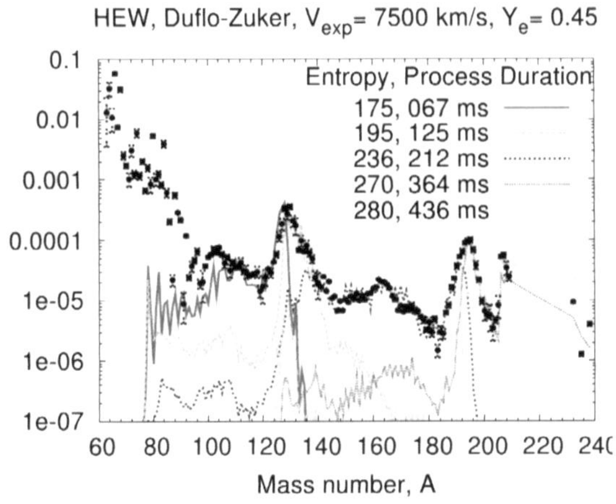


**Figure 4.** High entropy neutrino wind results for the mass model Duflo-Zucker [17], expansion parameters and proton/nucleon ratio $Y_e$ as given in the label, for a variation in extropies per baryon and $k_B$.

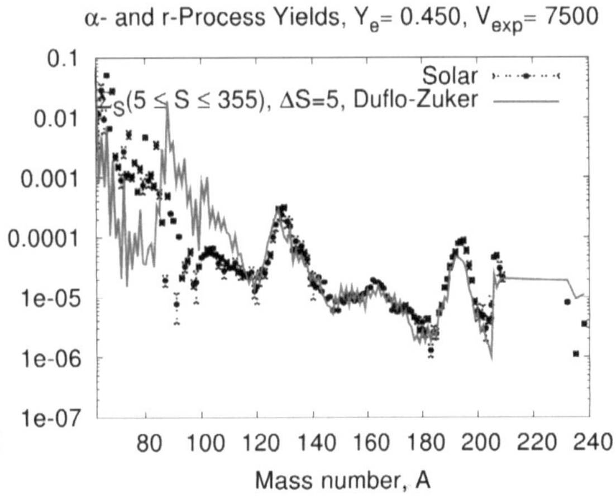


**Figure 5.** A superposition of entropies (assuming equal mass ejecta per entropy interval). It is obvious that the light r-process abundances are not reprocuced correctly for such $Y_e$'s.

The calculations presented here are based on trajectories for densities and temperatures originating from expansions of a complete parameter study in terms of entropy $S$, electron fraction $Y_e$ and expansion velocity $V_{exp}$, the latter being related to the expansion timescale $\tau_{exp}$ [25, 19]. Here we only show the results utilizing the Duflo-Zuker mass model (a) for a range of entropies and (b) a superposition of entropies with weights corresponding to equal mass ejecta per entropy interval. We assume that in the late phases of the neutrino wind of a deleptonized neutron star conditions with $Y_e < 0.5$ prevail (see discussion in section 6).

*5.2. Strong r-Processes with Fission Cycling*

Either higher entropies than utilized in the previous subsection or conditions with intrinsically high neutron densities (like expanding neutron star matter with $Y_e \approx 0.1 - 0.2$) can lead to neutron/seed ratios which are sufficiently high to reach fissionable nuclei in the r-process. The fission fragments can again capture neutrons and produce fissionable nuclei, leading to an r-process with fission recycling [75, 54]. This requires reliable fission barriers (and fission fragment distributions) to test the possibility for the production of superheavy elements. It was shown recently that neutron-induced fission is more important in r-process nucleosynthesis than beta-delayed fission [64, 54]. Thus, the need to provide a compilation of neutron-induced fission rates is obvious and has been performed recently [65, 30, 67]. Comparison of rates obtained with different sets for mass and fission barrier predictions give a measure of the uncertainties involved. In order to explore a range of synthesis conditions realistically, our aim is to provide a range of reaction rates within the presently existing nuclear uncertainties.

We show here results of [67], based on extensions of [76] and [65] for the full region $84 \leq Z \leq 110$, in order to provide the necessary input for nucleosynthesis studies under high neutron densities. Results for neutron-induced fission are given in Fig. 6 for a variety of fission barrier predictions (ETFSI [53], TF [60], FRDM Moller08 [59], and HFB [30]) in comparison to experiment. One realizes that there exist still large uncertainties in fission barrier predictions, even near stability where experimental information is available for comparison. This situation worsens towards heavier nuclei and far from stability. As shown in Fig. 7, it can be seen that even for the most recent advances, $(n, f)$ cross sections obtained with theoretically predicted barriers seem to diverge from experimental values for the heaviest nuclei.

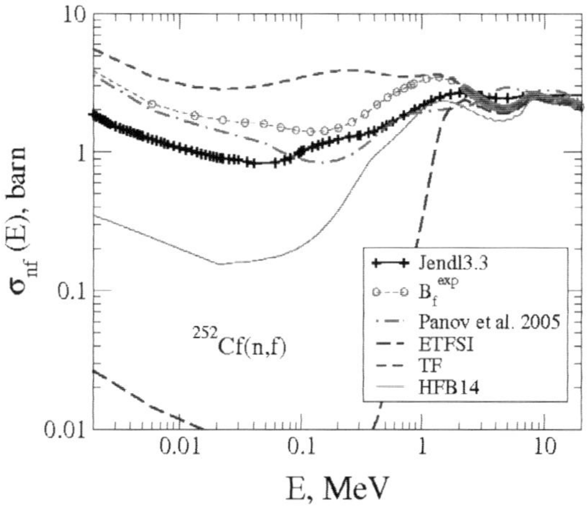


**Figure 6.** $^{252}$Cf$(n, f)$ cross section from our calculations with experimental fission barriers $B_f^{exp}$, ETFSI, TF, and HFB predictions as well as an older version of our code (Panov 2005) in comparison to experiment (JENDL3.3).

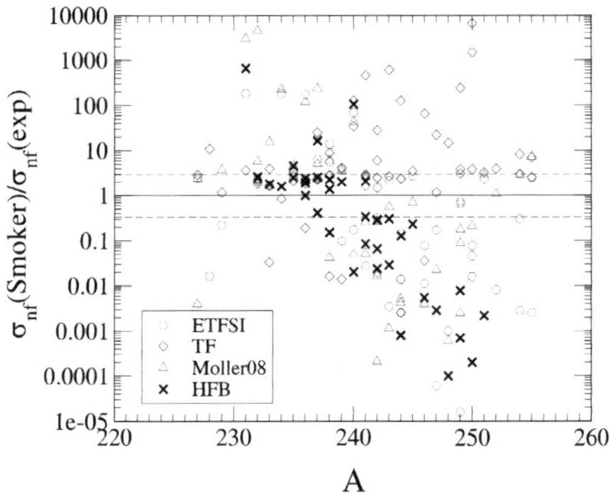


**Figure 7.** Ratio of $(n, f)$-cross section predictions with different fission barrier sets in comparison to experiment. In some cases the ratios seem to diverge towards heavier mass numbers $A$.

## 6. Conclusions

In the preceding sections we have discussed the status of three processes, the (classical) p- and r-process and the recently discovered $\nu$p-process. All of them seem to be related to massive stars, and very probably to core collapse supernova events. The classical p-process is identified with moderate photodisintegration processing, acting like spallation on previously existing heavy nuclei in the outer shells of explosive Ne/O burning. This is a secondary process, requiring existing heavy nuclei from previous stellar populations. In recent years the nuclear physics basis of this process has been quite firmly established and can explain the abundance of the proton-rich p-isotopes, amounting to typically 1% of the elemental abundance. This is different for the light p-isotopes, where this fraction increases up to 10% and all attempts to explain them by a classical p-process fail. The newly discovered $\nu$p-process can just fill this gap and seems also be able to explain the observed LEPP ("light" element primary process) abundances in low metallicity stars. The ejecta innermost core collapse supernova ejecta of the early neutrino wind are proton-rich, of primary origin and produce nuclei up to $A = 100$. The origin of the r-process stays a problem. A moderately neutron-rich, high entropy neutrino wind has been identified with the site of the r-process for many years now. Recent core collapse studies, however, indicate the neutrino wind to be proton-rich for many seconds. A major question is how this turns to be neutron-rich in late phases, what physics causes this change (the nuclear EoS or neutrino properties?) and how very high entropies can be attained to produce also the heaviest nuclei. Present observations indicate that in most cases the latter is not taking place, causing only a weak r-process. Whether either high entropies are only attained in exceptional cases or other origins of low entropy, highly neutron-rich matter is the origin of the main r-process has to be explored, in parallel with the still remaining challenges of nuclear physics far from stability.

## References

[1] Aboussir Y, Pearson J M, Dutta A K, Tondeur F 1995 *At. Data Nucl. Data Tables* **61** 127
[2] Angulo C et al. 1999 *Nucl. Phys.* A **656** 3
[3] Arcones A, Janka H-T, Scheck L 2007 *A&A* **467** 1227
[4] Arnould M, Goriely S 2003 *Phys. Rep.* **384** 1
[5] Audi G, Wapstra A H, Thibault C. 2003 *Nucl. Phys.* A **729** 337
[6] Bethe H A 1990 *Rev. Mod. Phys.* **62** 801
[7] Buras R et al. 2003 *Phys. Rev. Lett.* **90** 241101
[8] Buras R, Janka H-T, Rampp M, Kifonidis K 2006 *A&A* **457** 281
[9] Burrows A, Livne E, Dessart L, Ott C D, Murphy J 2006 *Ap. J.* **640** 878
[10] Burrows A, Dessart L, Ott, C D, Livne E 2007 *Phys. Rep.* **442** 23
[11] Cayrel R et al. 2004 *A&A* **416** 1117
[12] Caughlan G R, Fowler W A 1988 *At. Data Nucl. Data Tables* **40** 283
[13] Chieffi A, Limongi M. 2002 *Ap. J.* **577** 281
[14] Cowan J J, Thielemann F-K, Truran J W 1991 *Phys. Rep.* **208** 267
[15] Cowan J J, Sneden C 2006 *Nature* **440** 1151
[16] Dillmann I, Rauscher T, Heil M, Käppeler F, Rapp, W, Thielemann F-K 2008 *J. Phys.* G **35** 014029
[17] Duflo J, Zuker A P 1995 *Phys. Rev.* C **52** 23
[18] Farouqi K et al. 2009 *Ap. J.* **694** L49
[19] Farouqi K, Kratz K-L, Pfeiffer B, Rauscher T, Thielemann F-K, Truran J W 2009, submitted to *Ap. J.*
[20] Fischer T, Whitehouse S C, Mezzacappa A, Thielemann F-K, Liebendörfer M 2009 *A&A* **499** 1
[21] Fischer T et al. 2009, submitted to *A&A*
[22] Fisker J L, Schatz H, Thielemann F-K 2008 *Ap. J. Suppl.* **174** 261
[23] Fisker J L, Hoffman R D, Pruet J 2009 *Ap. J.* **690** L135
[24] Foglizzo T 2009 *Ap. J.* **694** 820
[25] Freiburghaus C et al. 1999 *Ap. J.* **516** 381
[26] Fröhlich C. et al. 2006 *Phys. Rev. Lett.***96** 142502
[27] Fryer C L, Young P A, Hungerford A L 2006 *Ap. J.* **650** 1028
[28] Fuller G M, Fowler W A, Newman M J 1985 *Ap. J.* **293** 1
[29] Goriely S, Hilaire S, Koning A J 2008 *A&A* **487** 767
[30] Goriely S, Hilaire S, Koning A J, Sin M, Capote R 2009 *Phys. Rev.* C **79** 024612
[31] Goriely S, Chamel N, Pearson J M 2009 *Phys. Rev. Lett.* **102** 152503
[32] Gyürky G et al. 2006 *Phys. Rev.* C **74** 025805
[33] Heger A et al. 2005 *Phys. Lett.* B **606** 258
[34] Herant M, Benz W, Hix W R, Fryer C L, Colgate S A 1994 *Ap. J.* **435** 339
[35] Hill V et al. 2002 *A&A* **387** 560
[36] Hoffman R D, Woosley S E, Qian Y-Z 1997 *Ap. J.* **482** 951
[37] Honda S, Aoki W, Ishimaru Y, Wanajo S 2006 *Ap. J.* **643** 1180
[38] Ivans I I, Simmerer J, Sneden C et al. 2006 *Ap. J.* **645** 613
[39] Janka H-T, Buras R, Rampp M 2003 *Nucl. Phys.* A **718** 269
[40] Janka H-T, Langanke K, Marek A, Martinez-Pinedo G, Müller B 2007 *Phys. Rep.* **442** 38
[41] Käppeler F, Gallino R, Busso M, Picchio G, Raiteri C M 1990 *Ap. J.* **354** 630
[42] Kankainen A et al. 2006 *Eur. Phys. J.* A **29** 271
[43] Kiss G G et al. 2007 *Phys. Rev.* C **76** 055807
[44] Kiss G G, Rauscher T, Gyürky G, Simon A, Fülöp Z, Somorjai E 2008 *Phys. Rev. Lett.* **101** 191101
[45] Keil W, Janka H-T, Mller E. 1996 *Ap. J.* **473** L111

[46] Kolbe E, Langanke K, Martinez-Pinedo G, Vogel P 2003 *J. Phys.* G **29** 2569
[47] Kratz K-L, Bitouzet J-P, Thielemann F-K, Möller P, Pfeiffer B 1993 *Ap. J.* **403** 216
[48] Kratz K-L, Farouqi K, Pfeiffer B, Truran JW, Sneden C, Cowan J J 2007 *Ap. J.* **662** 39
[49] Kuroda T, Wanajo S, Nomoto K 2008 *Ap. J.* **672** 1068
[50] Langanke K, Martínez-Pinedo G 2003 *Rev. Mod. Phys.* **75** 819
[51] Liebendörfer M et al. 2003 *Nucl. Phys.* A **719** 144
[52] Liebendörfer M et al. 2008 *J. Phys.* G **35** 014056
[53] Mamdouh A, Pearson J M, Rayet M, Tondeur F 1998 *Nucl. Phys.* A **644** 389
[54] Martinez-Pinedo G et al. 2007 *Prog. Part. Nucl. Phys.* **59** 199
[55] Meyer B S, Brown J 1997 *Ap. J.* **112** 199
[56] Mezzacappa A, Bruenn S W 1993 *Ap. J.* **405** 669; **410** 740
[57] Möller P, Nix J R, Myers W D, Swiatecki W J 1995 *At. Data Nucl. Data Tables* **59** 185
[58] Moller P, Nix J R, Kratz K-L 1997 *At. Data Nucl. Data Tables* **66** 131
[59] Möller P, Sierk A J, Ichikawa T et al. 2008 submitted to *Phys. Rev.* C
[60] Myers W D, Swiatecki W J 1999 *Phys. Rev.* C **60** 014606
[61] Nakamura T et al. 2001 *Ap. J.* **555** 880
[62] Nomoto K. et al. 2006 *Nucl. Phys.* A **777** 424
[63] Otsuki K, Tagoshi H, Kajino T, Wanajo S 2000 *Ap. J.* **533** 424
[64] Panov I V, Thielemann F-K 2003 *Nucl. Phys.* A **718** 647
[65] Panov I V et al. 2005 *Nucl. Phys.* A **747** 633
[66] Panov I V, Janka H-T 2009 *A&A* **494** 829
[67] Panov I V et al. 2009 submitted to *A&A*
[68] Pearson J M, Nayak R C, Goriely S 1996 *Phys. Lett.* B **387** 455
[69] Pfeiffer B, Kratz K-L, Thielemann F-K, Walters W B 2001 *Nucl. Phys.* A **693** 282
[70] Pruet J et al. 2006 *Ap. J.* **644** 1028
[71] Qian Y-Z, Woosley S E 1996 *Ap. J.* **471** 331
[72] Qian Y-Z, Vogel P, Wasserburg G J 1998 *Ap. J.* **494** 285
[73] Qian Y-Z, Wasserburg G J 2007 *Phys. Rep.* **442** 237
[74] Rapp W, Görres J, Wiescher M, Schatz H, and Käppeler F 2006 *Ap. J.* **653** 474
[75] Rauscher T, Applegate J, Cowan J, Thielemann F-K, Wiescher M 1994 *Ap. J.* **429** 499
[76] Rauscher T, Thielemann F-K 2000 *At. Data Nucl. Data Tables* **75** 1; **79** 47; **88** 1
[77] Rauscher T, Heger A, Homan R D, Woosley S E 2002 *Ap. J.* **576** 323
[78] Rauscher T 2006 *Phys. Rev.* C **73** 015804
[79] Rayet M, Prantzos N, Arnould M 1990 *A&A* **227** 271
[80] Rayet M, Arnould M, Hashimoto M, Prantzos N, Nomoto K 1995 *A&A* **298** 517
[81] Schatz H et al. 1998 *Phys. Rep.* **294** 167
[82] Sneden C, Cowan J J, Lawler J E et al 2003 *Ap. J.* **591** 936
[83] Surman R, McLaughlin G C, Hix W R 2006 *Ap. J.* **643** 1057
[84] Takahashi K, Witti J, Janka H-T 1994 *A&A* **286** 857
[85] Terasawa M, Sumiyoshi K, Yamada S, Suzuki H, Kajino T 2002 *Ap. J.* **578** L137
[86] Thielemann F-K, Metzinger J, Klapdor H V 1983 *Z. Phys.* A **309** 301
[87] Thielemann F-K, Nomoto K, Hashimoto M 1996 *Ap. J.* **460** 408
[88] Thompson T A, Burrows A, Meyer B S 2001 *Ap. J.* **562** 887
[89] Thompson T A, Burrows A, Pinto P A 2003 *Ap. J.* **592** 434
[90] Thompson T A, Quataert E, Burrows A 2005 *Ap. J.* **620** 861
[91] Travaglio C, Gallino R, Arnone E, Cowan J J, Jordan F, Sneden C 2004 *Ap. J.* **601** 864
[92] Wallace R K, Woosley S E 1981 *Ap. J. Suppl.* **45** 389
[93] Wanajo S, Kajino T, Mathews G J, Otsuki K 2001 *Ap. J.* **554** 578
[94] Wanajo S, Goriely S, Samyn M, Itoh N 2004 *Ap. J.* **606** 1057
[95] Wanajo S 2006 *Ap. J.* **647** 1323
[96] Wasserburg G J, Busso M, Gallino R 1996 *Ap. J.* **466** L109
[97] Weber C et al. 2008 *Phys. Rev.* C **78** 054310
[98] Woosley S E, Fowler W A, Holmes J A, Zimmerman B A 1978 *At. Data Nucl. Data Tables* **22** 371
[99] Woosley S E, Howard W 1978 *Ap. J. Suppl.* **36** 285
[100] Woosley S E, Wilson J R, Mathews G J, Hoffman R D, Meyer B S 1994 *Ap. J.* **433** 229
[101] Woosley S E, Weaver T A 1995 *Ap. J. Suppl.* **101** 181
[102] Yalçın C et al. 2009 *Phys. Rev.* C **79** 065801
[103] Yoshida T, Terasawa M, Kajino T, Sumiyoshi K 2004 *Ap. J.* **600** 204

Journal of Physics: Conference Series **202** (2010) 012007 doi:10.1088/1742-6596/202/1/012007

# Nucleosynthesis in neutrino-driven winds: influence of the nuclear physics input

**Almudena Arcones & Gabriel Martínez-Pinedo**

Institut für Kernphysik,TU Darmstadt, Schlossgartenstr. 9, D-64289 Darmstadt, Germany
GSI Helmholtzzentrum für Schwerionenforschung GmbH, Planckstr. 1,D-64291 Darmstadt, Germany

E-mail: a.arcones@gsi.de

**Abstract.** We have performed hydrodynamical simulations of the long-time evolution of proto-neutron stars to study the nucleosynthesis using the resulting wind trajectories. Although the conditions found in the present wind models are not favourable for the production of heavy elements, a small enhancement of the entropy results in the production of r-process elements with $A \approx 195$. This allows us to explore the sensitivity of their production to the hydrodynamical evolution (wind termination shock) and nuclear physics input used.

## 1. Introduction

Half of the heavy elements are produced by rapid neutron capture (known as r-process). In the last years different scenarios have been proposed, but many questions remain still open (see [1] for a review). There are two conditions for a successful r-process: 1) the number of neutrons per seed nuclei has to be high ($\approx 100$). 2) Once the astrophysical site has been identified, nucleosynthesis calculations based on simulations have to reproduce observations. In order to get the ratio between two elements as observed in ultra metal poor stars, in addition to high neutron-to-seed ratio, one needs to find the right evolution of matter along the NZ plane, even after r-process freezes out when matter moves back to stability. The question then emerges: where in the Universe can such conditions be realized?

Our results presented here show that the final abundances depend strongly on dynamical evolution and on the variation of the nuclear physics input. In particular, we have explored the influence of various theoretical mass models with consistent sets of neutron capture rates.

## 2. Neutrino-driven wind

There are several possible scenarios where r-process takes place [1]. Based on galactic chemical evolution [2] core-collapse supernovae and the subsequent neutrino-driven winds remain the most promising scenario. Therefore, we concentrate our study on neutrino-driven winds [3], which are baryonic outflows from proto-neutron stars formed in core-collapse supernova explosions. This supersonic outflow expands through slow-moving matter, which was ejected at the beginning of the explosion and leads to a wind termination shock or reverse shock. This shock has a great influence on the dynamical evolution of the outflow, and therefore on the nucleosynthesis.

The conditions for r-process in neutrino-driven winds are well understood [4] but requires extreme high entropy, fast expansion or very low electron fraction. In recent simulations [3], that we use for our study, the entropies are too low or the electron fraction too high to get a high enough neutron-to-seed ratio. Therefore it is still an open question whether there is some physical aspects missing in the long-time evolution of the ejecta or if the r-process takes place in some other astrophysical site. More investigations of the neutrino-driven wind and other astrophysical environments are necessary to understand where heavy elements are created. However, the available simulations are a useful basis to explore the influence of the nuclear physics input on nucleosynthesis, when the entropy is increased by a small factor. This simulates a dynamical evolution of the matter that will be similar also to other astrophysical environments where the r-process may occur.

## 3. Nucleosynthesis studies

We take the temperature and density evolution from trajectories of hydrodynamical simulations [3] as input to the nucleosynthesis network combined with a given electron fraction. In addition, the density is reduced by a factor of two in order to increase the entropy. As discussed, this allows us to study the later r-process phase.

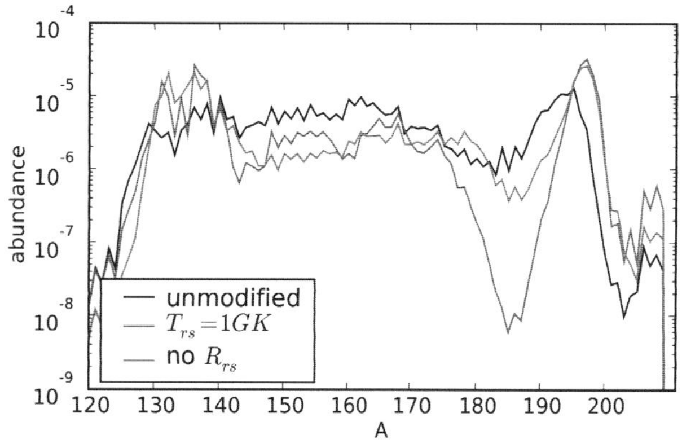


**Figure 1.** Final abundances for the same trajectory (i.e., same initial neutron-to-seed ratio) but different position of the reverse shock. For the black line, the position of the reverse shock is not changed from the simulation [3]. For the green line, the reverse shock has been moved to a temperature of 1 GK. The red line represents a case without reverse shock, where we have taken the expansion to be adiabatic at the temperature of the former reverse shock. In all cases the entropy has been increased by a factor of two compared to the value from the simulation [3]. The mass model used here is ETFSI-Q [12]. Other mass models will be explored in Fig. 2.

### *3.1. Dependence on late-time dynamical evolution and the reverse shock*

The r-process starts when the ejecta reach temperatures of around 1 GK and is very sensitive to its subsequent evolution, that depends on the interaction of the wind with the supernova ejecta. This interaction produces a wind termination shock which is not a steady-state phenomenon, therefore, it can only be studied within a full hydrodynamical simulation. However, there are some possibilities to parametrize the behaviour of the reverse shock using an artificial outer boundary with constant pressure [6, 7], temperature [8, 9], or density [10, 11]. We have used data directly from our simulations [3] and subsequently changed the position of the reverse shock in a consistent way (for details see Ref. [5]).

We find that, when the wind termination shock occurs at high temperature ($\sim$ 1 GK), the r-process takes place in (n,$\gamma$)-($\gamma$,n) equilibrium, as in the classical r-process. In this case the neutron separation energies determine the abundances along an isotopic chain. On the other hand, when the reverse shock is at low temperatures ($<$ 0.5 GK) photo-dissociation becomes negligible and there is a competition between neutron capture and beta decay. Consequently the relevant nuclear physics input will depend on the dynamical evolution of the outflow. This translates into different r-process paths and differences in the final abundances, as shown in Fig. 1.

*3.2. Dependence on mass models*

Here we describe only the case where the evolution takes place at high temperatures and therefore in (n,$\gamma$)-($\gamma$,n) equilibrium. Other cases and more details can be found in Ref. [5]. In Fig. 2 we present the final abundances obtained for the same trajectory but different mass models. We find that the region before the third peak is remarkable different but also the abundances between peaks vary considerably.

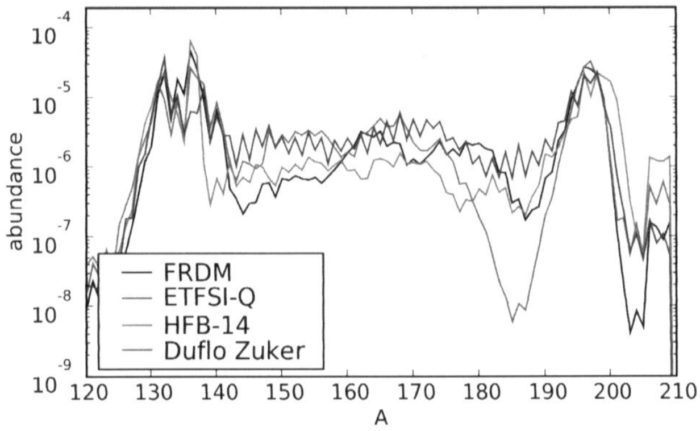


**Figure 2.** Final abundances obtained from different mass models. All cases are calculated with the same trajectory which reaches (n,$\gamma$)-($\gamma$,n) equilibrium, because the reverse shock is at 1 GK.

In order to assess the impact of nuclear masses, we compare the abundances obtained from FRDM to ETFSI-Q at freeze-out ($Y_n/Y_{\rm seed} = 1$) and when all decays have occurred. In Fig. 3 we observe remarkable odd-even effects following the behavior of the neutron separation energies. However, the final abundances in Fig. 4 are smoother similar to solar abundances.

In the long-time evolution there is a competition between beta decay and neutron capture and we have found that neutron captures still play an important role when matter moves back to stability, even when neutron densities and neutron-to-seed ratios are low ($N_n \approx 10^{17}{\rm cm}^{-3}$ and $Y_n/Y_{\rm seed} \approx 10^{-5}$). Neutron captures can fill holes, move peaks to higher mass number and reduce odd-even effects in the abundances. Moreover, the masses also enter in the neutron capture cross sections, and this can explain the differences in Fig. 4 between the two mass models.

## 4. Conclusions

The late-time evolution of the ejecta, also after freeze-out of the r-process, is very important to determine details in the final abundances. Therefore, we performed nucleosynthesis studies in neutrino-driven winds by means of long-time hydrodynamical simulations of core-collapse supernova explosions. The conditions found in the simulations (low wind entropies and/or high electron fraction) do not allow the formation of heavy elements. However, an artificial increase of the entropy by a factor around two is enough to reach A=195 and allow us to explore the sensitivity of the wind termination shock and the nuclear physics input.

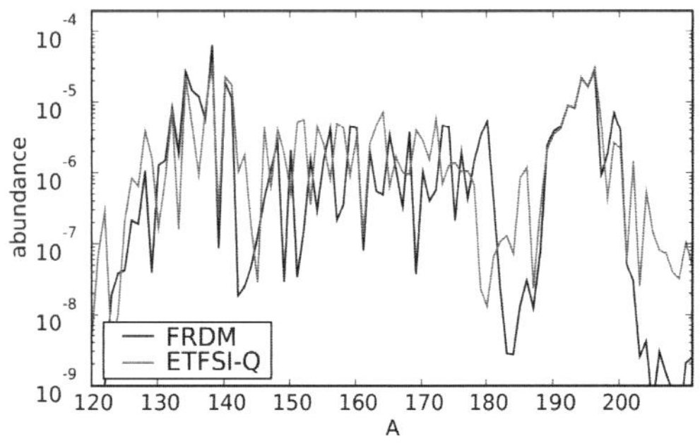


**Figure 3.** Abundances for $Y_n/Y_{seed} = 1$.

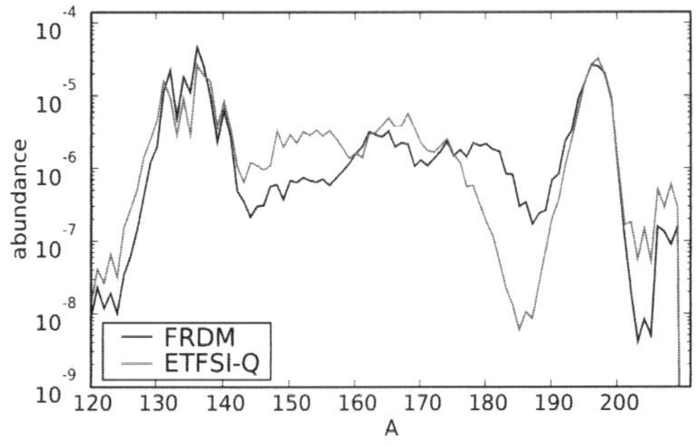


**Figure 4.** Final abundances after all decays.

Depending on whether the evolution takes place at high or low temperatures the relevant nuclear physics input will change. When (n,$\gamma$)-($\gamma$,n)equilibrium is valid the masses and beta decays determine the abundances. In the low temperature case, beta decays and neutron capture are crucial. In both cases, the final abundance distribution is determined by a competition between neutron capture and beta decays, when matter moves back to stability.

### *4.1. Acknowledgments*

A. Arcones acknowledges support by the Deutsche Forschungsgemeinschaft through contract SFB 634.

## References

[1] Arnould M, Goriely S and Takahashi K 2007 *Phys. Repts.* **450** 97–213
[2] Ishimaru Y and Wanajo S 1999 *ApJL* **511** L33–L36
[3] Arcones A, Janka H T and Scheck L 2007 *Astron.& Astroph.* **467** 1227–1248
[4] Otsuki K, Tagoshi H, Kajino T and Wanajo S 2000 *ApJ* **533** 424–439
[5] Arcones A and Martínez-Pinedo G (in prep.)
[6] Sumiyoshi K, Suzuki H, Otsuki K, Terasawa M and Yamada S 2000 *PASJ* **52** 601–611
[7] Terasawa M, Sumiyoshi K, Yamada S, Suzuki H and Kajino T 2002 *ApJ* **578** L137–L140
[8] Wanajo S, Itoh N, Ishimaru Y, Nozawa S and Beers T C 2002 *ApJ* **577** 853–865
[9] Wanajo S 2007 *ApJL* **666** L77–L80
[10] Kuroda T, Wanajo S and Nomoto K 2008 *ApJ* **672** 1068–1078
[11] Panov I V and Janka H T 2009 *Astron.& Astroph.* **494** 829–844
[12] Pearson J. M., Nayak R. C. and Goriely S. 1996 *Phys. Lett. B* **387** 455–459

Journal of Physics: Conference Series **202** (2010) 012008 doi:10.1088/1742-6596/202/1/012008

# Network calculations for r-process nucleosynthesis

**I Petermann[1,2], G Martínez-Pinedo[1], A Arcones[2,1], W R Hix[3], A Kelić[1], K Langanke[1,2], I Panov[6], T Rauscher[4], K-H Schmidt[1], F-K Thielemann[4], N Zinner[5]**

[1] GSI Helmholtzzentrum für Schwerionenforschung, Darmstadt, Germany
[2] Institut für Kernphysik, TU Darmstadt, Germany
[3] Physics Division, Oak Ridge National Laboratory, Oak Ridge, TN37831-6374, USA
[4] Department für Physik und Astronomie, Universität Basel, Switzerland
[5] Department of Physics, Harvard University, Cambridge, MA 02138
[6] Institute for Theor. and Exp. Physics, B Cheremushkinskaya St 25, 117218 Moscow, Russia

**Abstract.** The r-process is known to be responsible for the synthesis of about half of the elements heavier than iron, nevertheless its astrophysical site has not yet been clearly ascertained, but observations indicate that at least two possible sites should contribute to the solar system abundance of r-process elements. The r-process being responsible for the production of elements heavier than $Z = 56$ operates rather robustly always resulting in a similar relative abundance pattern. From the nuclear-physics point of view the r-process requires the knowledge of a large number of reaction rates involving exotic nuclei that are not accessible by experiment and data have to be provided by theoretical predictions. We have developed for the first time a complete database of reaction rates that in addition to neutron-capture rates and $\beta$-decay half-lives includes the dominant reactions that can induce fission (neutron-capture, $\beta$-decay and spontaneous fission) and the corresponding fission yields. In addition, we have implemented these reaction rates in a fully implicit reaction network. The influence of the nuclear physics input constituted in the reaction rates based on the two mass models FRDM and ETFSI and on the astrophysical conditions simulating a cold or hot environment are examined.

## 1. Introduction

The r-process is a series of rapid neutron-capture reactions and $\beta$-decays in explosive scenarios with high neutron densities. Its astrophysical site remains uncertain, but observations indicate the contribution of at least two sites, namely core collapse supernovae from stars of different masses and neutron-star mergers. Observed abundances of r-process elements with $Z > 56$ in metal-poor stars follow a very robust pattern, that is consistent with scaled solar abundances[1]. Our aim is to explore whether or not fission can provide an explanation for the robustness.

## 2. Fission in the r-process

R-process simulations require in addition to neutron captures and $\beta$-decays a full set of fission rates [2], supplemented by all corresponding fission yields. We have considered neutron-induced fission, computed using the SMOKER code [3], $\beta$-delayed fission and spontaneous fission. For all of them fission yields have been computed using the statistical code ABLA [4], taking the neutrons evaporated explicitly into account. The resulting rates have been implemented in a fully implicit network, that includes approximately 7000 nuclei from $n, p$ up to $^{300}$Ds. The set of differential equations is linearized and solved using the Newton-Raphson method [5]. The

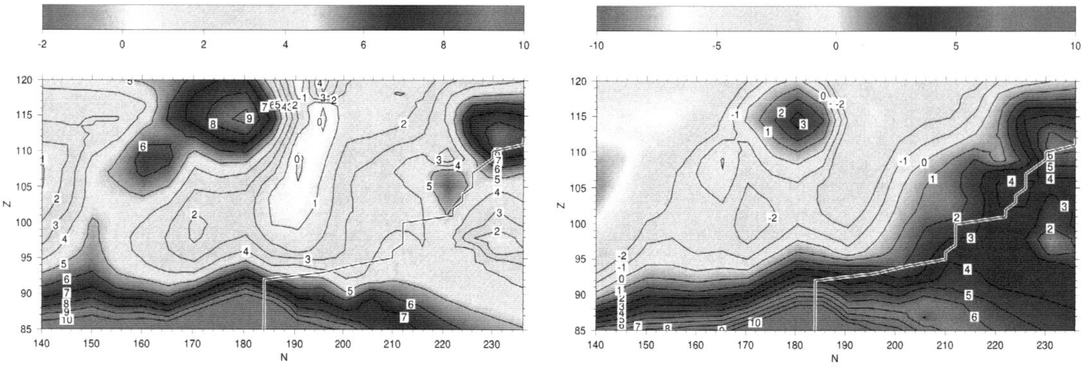


**Figure 1.** Region of the nuclear chart where fission takes place during the r-process. In the left panel the contour lines show the Myers-Swiatecki fission barrier heights ($B_f$) in MeV. The right panel shows contour lines of the quantity $B_f - S_n$, with $S_n$ the neutron separation energy taken from the FRDM mass model. The unlabeled line shows the location of the neutron dripline.

inclusion of fission maintains the sparseness of the matrix and allows to use sparse matrix solvers, in our case the Parallel Sparse Direct Linear Solver (PARDISO) was used [6].

Fig. 1 shows the region where fission takes place during the r-process for the combination of Myers-Swiatecki [7] fission barrier heights and the neutron separation energies based on the FRDM mass model. Once the matter-flow breaks out of the magic neutron number $N = 184$ the fission barriers decrease drastically, and fission becomes possible.

## 3. Network calculations

We have performed network calculations using trajectories obtained from model M15-l1-r1 from the hydrodynamical simulations of ref. [8]. By increasing the entropy obtained in these simulations, large enough neutron-to-seed ratios could be achieved for a successful r-process.

Besides the total fission rate it is also instructive to specify the dominating fission channel. Fig. 2 shows the evolution of the fission rates for the different channels. They are computed determining the instantaneous increase in the total abundance of r-process nuclei, having $A \geq 5$, due to each fission channel. Clearly neutron induced fission is the dominating channel with $\beta$-delayed fission having a slight influence. The end of the r-process can be seen around 1.3 s by the sudden drop in the neutron-to-seed ratio. This is followed by a phase where all heavy fissioning nuclei decay over a timescale of several hundreds of seconds. During this phase neutron induced fission still dominates due to the neutrons produced by fission.

For an examination of the possible build-up of superheavy elements a comparison of the largest rates including $\beta$-delayed fission, spontaneous fission, alpha and $\beta^-$ decay for FRDM with Myers-Swiatecki barriers and ETFSI masses and barriers [9] is shown in Fig.3. For FRDM an impassable barrier made up by spontaneous fission rates prevents a progressive motion of the nucleosynthesis front towards (super)heavy elements. For the case of the ETFSI mass model the area of spontaneous fission constituting the largest rate shows a clear split with a small passage permitting to reach heavier elements.

The lowermost point in the N-Z-plane constitutes the last element that can via $\beta$-decay contribute to a build-up of heavier elements while nuclei beyond that point will immediately decay via spontaneous fission in lighter fragments that are not available for an augmentation of heavy elements. This point lies at higher proton and neutron numbers for the ETFSI mass model, which can serve as a possible explanation for the observation of a slightly larger amount of heavy elements in this mass model compared to FRDM as shown in Fig. 4 in a direct comparison

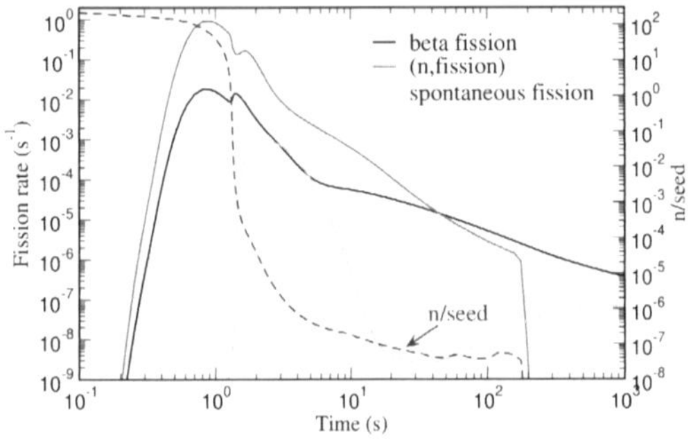


**Figure 2.** Evolution of the fission rate for the different channels for a calculation with initial neutron-to-seed ratio 295. Time zero corresponds to the beginning of the r-process phase after alpha-rich freeze-out. The evolution of the neutron-to-seed ratio is also shown (dashed line).

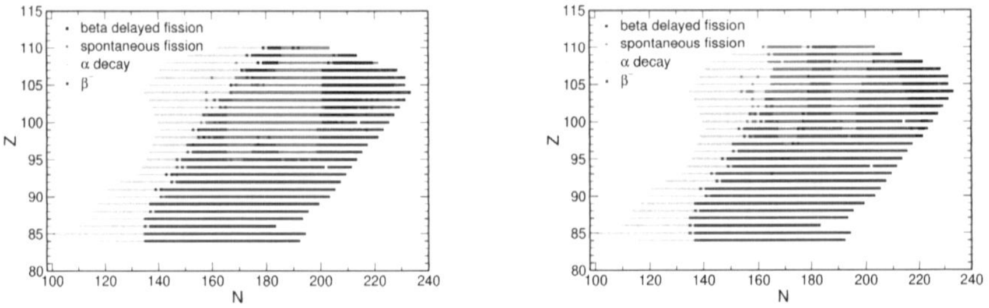


**Figure 3.** The largest rates for FRDM (left) and ETFSI (right).

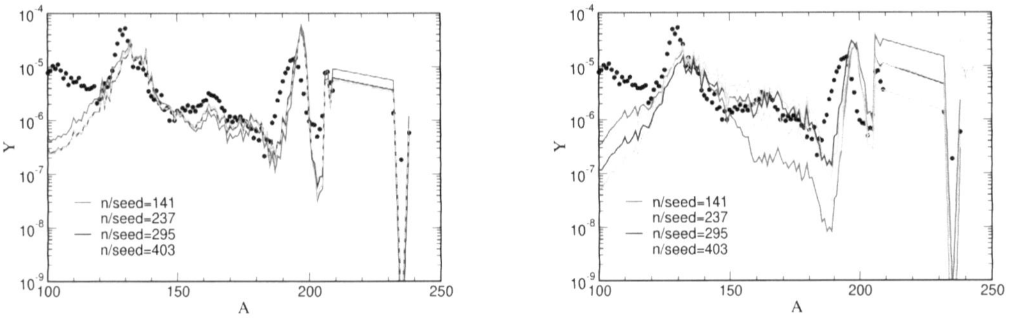


**Figure 4.** R-process abundances after a time of 13 Gy resulting from four calculations with different initial neutron-to-seed ratios for the two mass models FRDM and ETFSI. The black dots show the r-process solar distribution.

of the abundances at a time obtained after a time of 13 Gy, in each case for four initial neutron-to-seed ratios. In addition only small differences comparing the four abundance curves in the case of FRDM are existent, while the convergence behaviour in the case of ETFSI is observable, but much less distinct. In the case of a neutron-to-seed ratio of 141 a clear divergence compared to the ones for n/seed=237, 295 can be seen. Besides the impact of the nuclear physics input an analysis of the effect of the astrophysical conditions on the final abundance pattern was made. Therefor calculations were performed in a low-temperature environment, where the neutron emission by photodisintegration is insignificant and in a hot environment, where an $(n,\gamma)$-$(\gamma,n)$ equilibrium is maintained for an essential period of time. Starting with an initial neutron-to-seed ratio of 295 a comparison for n/seed=100 and n/seed=1, constituting the termination

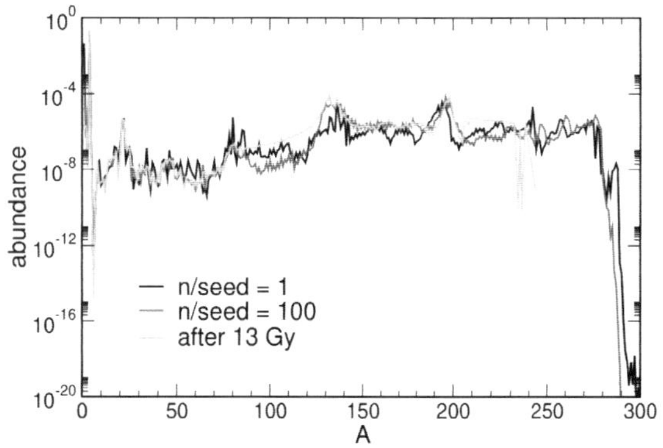


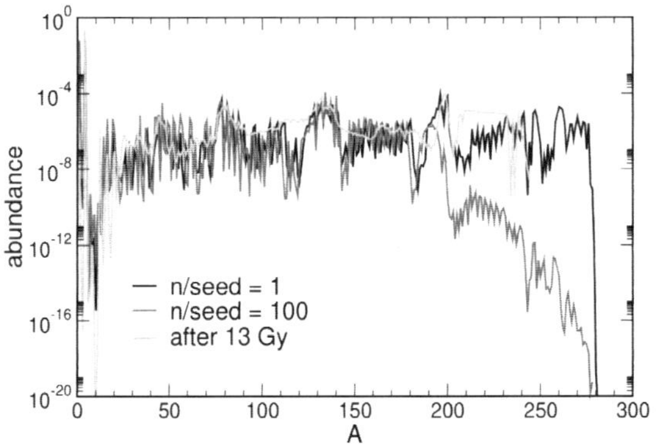


**Figure 5.** Abundances for neutron-to-seed ratios 100 and 1 for cold (left) and hot (right) astrophysical conditions with FRDM mass model and initial neutron-to-seed ratio 295.

point of the r-process, is shown for the two cases in Fig.5. For conditions of a cold r-process (left) the maximum massnumber A$\sim$290 is already reached at early times (n/seed=100) allowing for a longer period of time where fission can take place supplementing nuclei in the medium-heavy massrange made available for an onward fission cycling process. This is in contrast to the case of a hot r-process where at a neutron-to-seed ratio of 100 a significant abundance is only existent up to a massnumber of A=200, compared to a an A(max) of 280. A pronounced pattern in the abundance distribution indicates the sensitivity to the underlying structure of the neutron-separation energies. Despite the distinct differences at intermediate timesteps the final abundance distribution at a time of 13 Gy is comparable, the peaks at A=130 and 195 are comparable well reproduced, nevertheless variations are seen in the smaller mass range.

## 4. Summary

The influence of fission in the r-process has been examined using a full set of reaction rates that include all possible fission channels, supplemented with the corresponding fission yield distributions. In our calculations neutron induced fission turned out to be the dominating channel, which is in agreement with results obtained in ref. [11] in studies of r-process in neutron star mergers. Furthermore, calculations with different initial neutron-to-seed ratios give very similar final abundance distributions for calculations based on the mass model FRDM, while a convergence is less obvious in the case using the ETFSI mass model. Trajectories simulating a cold or hot environment give comparable results in the final abundance pattern, while the strong influence of neutron separation energies is seen in the case of a hot r-process where the (n,$\gamma$)-($\gamma$,n) equilibrium is maintained for a longer timescale.

## References

[1] J. J. Cowan and C. Sneden, Nature **440**, 1151 (2006).
[2] J. J. Cowan, F.-K. Thielemann, J. W. Truran, Phys. Rep. **208** 267-394 (1991).
[3] I. Panov et al., Nucl.Phys. A **747**, 633-654 (2005). I. Panov, *et al.*, in preparation.
[4] J.-J. Gaimard and K.-H. Schmidt, Nucl. Phys. A **531**, 709 (1991).
[5] W. R. Hix, F.-K. Thielemann, Journal of Computational and Applied Mathematics **109**, 321-351 (1999).
[6] O. Schenk and K. Gärtner, Journal of Future Generation Computer Systems **20**, 475 (2004).
[7] W. D. Myers and W. J. Swiatecki, Phys. Rev. C **60**, 014606 (1999).
[8] A. Arcones, H.-Th. Janka, L. Scheck, A&A, **467**, 1227 (2007).
[9] A. Mamdouh, J. M. Pearson, M. Rayet, F. Tondeur, Nucl. Phys. A **644**, 389 (1998).
[10] P. Moller, J. R. Nix and W. J. Swiatecki, Atomic Data Nucl. Data Tables **59** 185 (1995).
[11] I. V. Panov and F.-K. Thielemann, Astronomy Letters **30**, 647 (2004).

# Nucleosynthesis of proton-rich nuclei. Experimental results on the *rp*-process

**D Galaviz[1,2,3,*], A M Amthor[2,3,4], D Bazin[2], A. D. Becerril[2,3,5], B. A. Brown[2,5], A. A. Chen[6], A. Cole[2,3], J. M. Cook[2,3], T. Elliot[2,3,5], A. Estrade[2,3,5], Zs. Fülöp[7], A. Gade[2], T. Glasmacher[2,5], A. Heger[3,8], M. E. Howard[9], R. Kessler[10], G. Lorusso[2,3,5], M. Matos[2,3], F. Montes[2,3,5], W. Müller[2], J. Pereira[2,3], H. Schatz[2,3,5], B. Sherrill[2,3,5], F. Schertz[10], Y. Shimbara[2,3], E. Smith[9], K. Smith[2,3,5], A. Tamii[11], A. Stolz[2], D. Weisshaar[2], M. Wallace[2,5], M. Wiescher[12], and R. G. T. Zegers[2,3,5]**

[1] Centro de Física Nuclear da Universidade de Lisboa, 1070-325 Lisbon, Portugal
[2] National Superconducting Cyclotron Laboratory, Michigan State University, East Lansing, Michigan 48824, USA
[3] Joint Institute for Nuclear Astrophysics, Michigan State University, East Lansing, Michigan 48824, USA
[4] Grand Accelerateur National d'Ions Lourds, Bd Henri Becquerel, BP 55027, F-14076 Caen Cedex 5, France
[5] Department of Physics and Astronomy, Michigan State University, East Lansing, Michigan 48824, USA
[6] Department of Physics and Astronomy, McMaster University, Hamilton, ON L8S 4M1, Canada
[7] ATOMKI, H-4001 Debrecen, POB. 51, Hungary
[8] School of Physics and Astronomy, University of Minnesota, Twin Cities, Minnesota 55455-0149, USA
[9] Department of Physics, Ohio State University, Columbus, Ohio 43210, USA
[10] Institut für Kernchemie, University of Mainz, D-55128 Mainz, Germany
[11] Research Center for Nuclear Physics, Osaka University, Ibaraki, Osaka 567-0047, Japan
[12] Department of Physics, University of Notre Dame, Notre Dame, Indiana 46556, USA

E-mail: galaviz@cii.fc.ul.pt

**Abstract.** We report in this study the nuclear properties of proton-rich isotopes located along the *rp*-process path. The experiments have recently been performed at the National Superconducting Cyclotron Laboratory (NSCL) at Michigan State University. The level properties above the proton separation energy of the nuclei $^{30}$S, $^{36}$K and $^{37}$Ca were measured with precision of $< 10$ keV. This will allow a reduction in the determination of the astrophysical (p,$\gamma$) reaction rate under *rp*-process conditions.

## 1. Introduction

Over the past two decades, the study of extra-solar X-ray objects has entered a new era with the launching of precision X-ray observatories, XMM-Newton and Chandra X-ray observatory being the two latest examples operating since 1999. These and other observatories have provided and continue to produce data with a high degree of sensitivity and resolution, leading to new discoveries such as super-bursts or ms-oscillations, and therefore challenging physicists to describe these novel phenomena. Among the observed objects, Type I X-ray bursts are characterized as thermonuclear explosions on the surface of accreting neutron stars in a binary system. After accumulation of hydrogen and helium from the envelope of the companion star on the surface of the neutron star, the burst is powered by the rapid proton capture process (*rp*-process) [1, 2], which consist of a series of ($\alpha$,p) and (p,$\gamma$) reactions, and subsequent $\beta^+$-decays, creating proton-rich isotopes and releasing energy on the order of $10^{40}$ ergs over a period between 10 and 100 seconds. For further details on X-ray sources, refer to the recent review of Schatz and Rehm [3].

Nuclear physics is a key input necessary to the modeling of the evolution of the burst. Nuclear masses, $\beta$-decay half lives, and ($\alpha$,p) and (p,$\gamma$) reaction rates determine quantities such as the time scale, light output and energy release in the burst, as well as the reaction path and the final abundance distribution of elements synthesized during the explosion. As for the determination of the (p,$\gamma$) reaction rates, in most of the cases at masses $A \leq 40$, the reaction is dominated by a resonant contribution from one or a few levels close to the particle threshold. A precise knowledge of the nuclear structure and masses is essential to provide the most accurate possible proton-capture reaction rates in this region of nuclei, whose reactions power the rise of the burst.

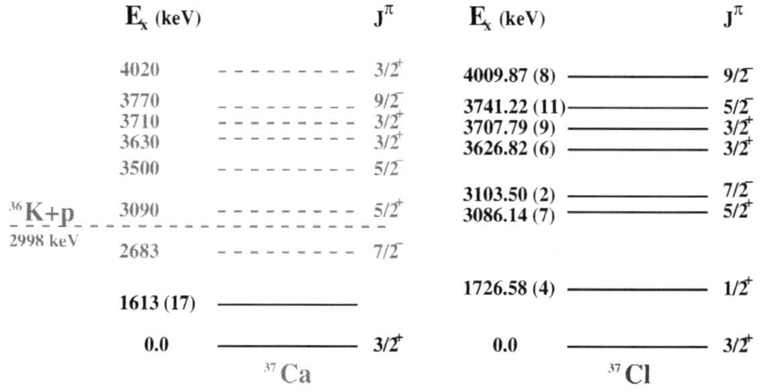


**Figure 1.** (Color online) Knowledge on the level scheme for the A = 37 mirror nuclei $^{37}$Ca (left) and $^{37}$Cl (right) at the time the experiments were performed [4], represented by solid lines. The value of the proton separation energy in $^{37}$Ca [5] recently measured is also shown. Only the first excited state in $^{37}$Ca had been previously measured [6]. The dashed lines in the case of $^{37}$Ca are taken directly from the mirror states in $^{37}$Cl with a Coulomb shift estimated for the negative parity states from the fp-shell [7]

As an example, the present knowledge on the level scheme of the nucleus $^{37}$Ca is shown in figure 1. Those states located close to and above the proton separation energy will contribute significantly to the proton capture reaction rate at burst temperatures. Typical uncertainties from shell model calculations of the order of 100 keV could easily translate into orders of magnitude uncertainty in the proton capture reaction rate [8]. A precise determination of the level structure of nuclei involved in the *rp*-process would reduce the uncertainty in the calculation of (p,$\gamma$) reaction rates. In order to achieve this goal, several experiments have recently been performed at the National Superconducting Cyclotron Laboratory (NSCL) at Michigan State University aiming to improve the knowledge of the level structure of proton-rich isotopes located along the *rp*-process path. The experimental procedure followed was similar to that applied to the study of the properties of the isotope $^{33}$Ar [8].

This contribution presents the status of the experimental results from measurements on the isotopes $^{30}$S, $^{36}$K and $^{37}$Ca using a specific experimental technique at the NSCL. The detailed discussion of the results obtained will be the topic of upcoming publications [9, 10] and is outside the scope of this manuscript. The possible reduction of the uncertainties in the determination of the reaction rate and their implications in X-ray burst modeling are considered as well.

## 2. Experimental Procedure

The experiments were performed at the Coupled Cyclotron Facility (CCF) at the NSCL. Two different settings were chosen for the investigation of the nuclei of interest. For the case of $^{30}$S, a $^{36}$Ar primary beam was accelerated up to 140 MeV/u and fragmented on a $^{9}$Be target of 1151 mg/cm$^2$ thickness at the entrance of the A1900 fragment separator [11]. The secondary beam emerging from the A1900 consisted predominantly of $^{31}$S (74.7 %), but also $^{32}$Cl (11.6 %) and $^{30}$P (12.6 %). The beam was guided towards the entrance of the vertical S800 spectrometer [12] with an energy of approximately 71 MeV/u, and further down to the target area of the S800. A polypropylene secondary target was used to accomplish the one neutron removal process on the $^{31}$S secondary beam to obtain $^{30}$S*. The photon-decay of the produced $^{30}$S nuclei was detected by 17 detectors of the Segmented Germanium Array (SeGA) [13], distributed over two rings under 90 and 37 degrees respectively to the beam direction. The measured gamma rays were detected in coincidence with the recoil nuclei at the focal plane of the S800.

In the study of the level properties of $^{36}$K and $^{37}$Ca, a primary beam of $^{40}$Ca was accelerated to 140 MeV/u at the CCF. A secondary beam containing $^{38}$Ca (61 %), $^{37}$K (28 %), $^{36}$Ar (10 %) and $^{35}$Cl (1 %) was delivered to the S800 experimental area with a setup similar to that previously described. The photons from the decay of the produced $^{36}$K and $^{37}$Ca nuclei were again measured using the SeGA array, and detected in coincidence with the recoil nuclei at the S800 focal plane.

Several nuclear species were produced from the interaction of the secondary beams with the reaction target. The situation is partially shown in figure 2 for the particular cases of $^{31}$S (left panel) and $^{38}$Ca (right panel). Both figures show $N - Z$ in the abscissa, which has been derived from the time needed by the particles to travel from the beginning of the S800 spectrometer to its final focal plane. The ordinate axis gives $Z$, which roughly corresponds to the energy lost by the particles at an ionization chamber located at the S800 focal plane. The high energy loss and timing resolutions enabled an easy separation of the isotope of interest for further analysis.

## 3. Present status of the analysis

The photons emitted by the nuclei under study were measured by the detectors from the SeGA. The velocity of the beam was over 1/3 of the speed of light and thus the Doppler shift between the particle and the laboratory frame was significant. Details of the reconstruction analysis will be published elsewhere [9, 10]. The resulting spectra for the three nuclei under consideration in this work, $^{30}$S, $^{36}$K and $^{37}$Ca are displayed in figure 3.

All measured transitions in $^{37}$Ca, with the exception of the first excited state at 1.61 MeV, are signatures of states that have been detected for the first time, and they will be the topic of a dedicated paper [10]. Similarly, the improved precision of the measured levels in $^{36}$K and its effect in the uncertainties determining the $^{35}$Ar(p,$\gamma$)$^{36}$K reaction rate will be extensively discussed elsewhere [10]. The focus of this paper will be the motivation and preliminary results obtained for the the nucleus $^{30}$S.

Interest in $^{30}$S has increased enormously in recent years as it is considered a possible *waiting-point* on the *rp*-process [14, 15]. This statement is made on the (p,$\gamma$)($\gamma$,p) equilibrium between $^{30}$S and $^{31}$Cl [16], and remains under consideration based on the spectroscopic properties of $^{30}$S and the rate of the $^{30}$S($\alpha$,p) reaction [17], which is a possible alternative to the proton capture.

The main source of uncertainty in the $^{29}$P(p,$\gamma$)$^{30}$S reaction rate is the location of two resonances above the proton threshold ($S_p$=4400 (3) keV [18]), predicted by Iliadis *et al.* [19] at

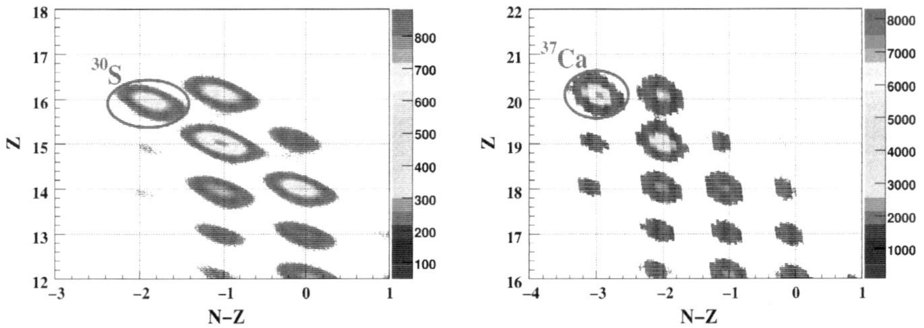


**Figure 2.** (Color online) Particle identification plots. In both cases, the time-of-flight information on the X-axis has been transformed into $N - Z$, whereas the energy loss of the particles is expressed in $Z$ units along the Y-axis. The reaction products from the $^{31}$S beam are shown in the left panel, while the right panel presents the nuclei observed from the reactions on the $^{38}$Ca beam. The quality of the separation in both cases permitted a clear identification of the nuclei of interest for further analysis.

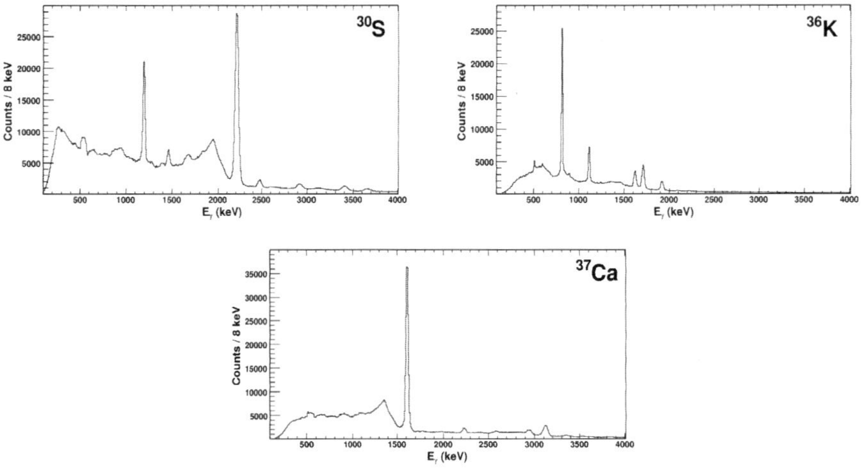


**Figure 3.** Doppler corrected photon spectrum measured in coincidence with recoil $^{30}$S (upper left), $^{36}$K (upper right) and $^{37}$Ca (bottom). The most dominant transition in all cases corresponded to the decay from the first excited state in the studied nucleus.

4733 ($3^+$) and 4888 ($2^+$) keV. Typical uncertainties in the determination of the energy levels of the order of 100 keV translate into an uncertainty of up to several orders of magnitude in the determination of the $^{29}$P(p,$\gamma$)$^{30}$S reaction rate. The reaction rate for the considered reaction in the relevant temperature range for X-ray bursts is shown in figure 4. The rate derived using this spectroscopic information is given by the light-green area.

Several measurements performed over the years [20, 21, 22, 23, 24] mapped the level structure of $^{30}$S using different experimental techniques, from transfer reactions to beta-decay studies. None of these attempts detected the existence of the two mentioned levels just above the particle threshold. Only recently, Bardayan *et al.* [25] presented the measurement of the $3^+$

state at an energy of 4704 (5) keV by means of the $^{32}$S(p,t)$^{30}$S reaction. That measurement resulted in a reduction of the proton capture rate uncertainties at low temperatures, which is also the relevant temperature range for *novae* nucleosynthesis. The proton capture rate based on Bardayan's result is shown by the red area in figure 4. At higher temperatures however, the contribution from the resonant capture via the $2^+$ state dominates the process and uncertainties increase again.

In our experiment we detected all known transitions up to 5.1 MeV using a single experimental approach. Additionally, we observed two novel gamma lines at energies that are identified as candidates of the de-excitation of the missing excited states close above the proton threshold. The analysis of the data is in a preliminary stage. Final results and details of the analysis will be presented in a dedicated manuscript [9].

## 4. Implications on X-ray burst modeling

The reaction rate for the $^{29}$P(p,$\gamma$)$^{30}$S reaction calculated based on the preliminary spectroscopic information derived from our measurement is shown in figure 4 as the black curve. The uncertainties shown in the figure are exclusively due to the errors in the determination of the resonance energies. The preliminary reaction rate presented in this work shows an uncertainty of roughly a factor of 2 over the entire temperature range.

In order to estimate the impact of the new rate in the X-ray burst modeling, calculations using the one-dimensional multi-zone KEPLER model [26] are underway. Preliminary results show that the reduction of the uncertainties and the deduced rate for the $^{29}$P(p,$\gamma$)$^{30}$S reaction do not have a remarkable impact on the evolution of the burst within this model.

The model of Fisker *et al.* [14] showed a double-peak structure (not observed in the results of the KEPLER code) with origin in the *waiting points* $^{30}$S and $^{34}$Ar. A possibly higher rate would increase the production rate of $^{30}$S and could even further pronounce this effect. This is presently under investigation. It becomes clear that studies evaluating the impact of the uncertainties of proton capture reaction rates during *rp*-process nucleosynthesis are necessary in order to address the nuclear physics needs of the models and improve the planning of future experiments.

## 5. Summary

This report summarizes selected results obtained in experiments recently performed at the NSCL on nuclei located along the *rp*-process. The level structure of $^{30}$S, $^{36}$K and $^{37}$Ca around the proton separation energy was determined with high precision. Based on the obtained data, we expect a reduction of the uncertainties in the determination of proton capture reaction rates to roughly a factor of 2 in the temperature range relevant for *rp*-process nucleosynthesis.

### *5.1. Acknowledgments*

This work has been supported by US National Science Foundation under grants PHY-0216783 (JINA), PHY-0606007 (NSCL) and PHY-0758099.

## 6. References

[1] Wallace R K and Woosley S E 1981 *Astroph. J. Suppl.* **45** 389
[2] Schatz H, Aprahamian A, Görres J *et al.* 1998 *Phys. Rep.* **294** 167
[3] Schatz H and Rehm K 2006 *Nucl. Phys. A* **777** 601
[4] http://www.nndc.bnl.gov/ensdf
[5] Ringle R, Sun T, Bollen G *et al.* 2007 *Phys. Rev. C* **75** 055503
[6] Benenson W, Kashy E and Proctor I D 1973 *Phys. Rev. C* **8** 210
[7] Brown B A private communication
[8] Clement R R C, Bazin D, Benenson W *et al.* 2004 *Phys. Rev. Lett.* **92** 172502
[9] Galaviz D, Amthor M, Bazin D *et al.* 2009 *in preparation*

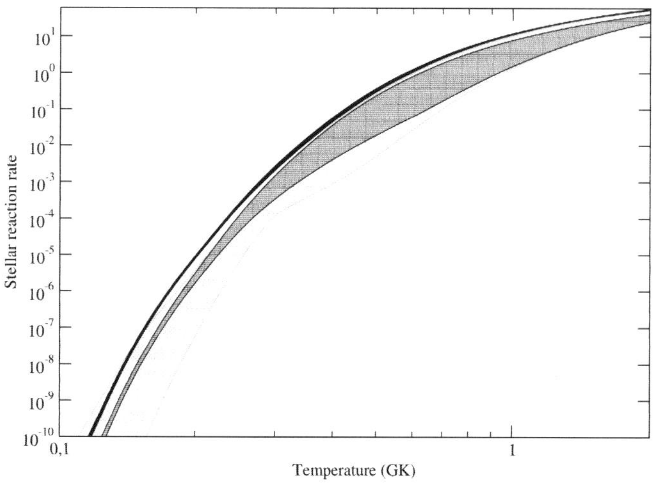


**Figure 4.** (Color online) Reaction rate for the $^{29}$P(p,$\gamma$)$^{30}$S reaction at the energy range relevant for X-ray bursts. The rate derived using the spectroscopic information of Iliadis *et al.* [19] is shown by the light-green area. The result from Bardayan *et al.* [25] is shown in red. The preliminary result based on the information obtained in our measurements is shown by the black area. The uncertainties in the determination of the rate are based exclusively in the error of the resonance energies.

[10] Amthor A M, Galaviz D, Bazin D *et al.* 2009 *in preparation*
[11] Morrissey D J, Sherrill B, Steiner M *et al.* 2003 *Nucl. Instrum. Meth.* **B204** 90
[12] Bazin D, Caggiano J A, Sherrill B M *et al.* 2003 *Nucl. Instrum. Meth.* **B204** 629
[13] Mueller W F, Church J A, Glasmacher T *et al.* 2001 *Nucl. Instrum. Meth.* **A466** 492
[14] Fisker J L, Thielemann F K, and Wiescher M 2004 *Astrophys. J. Lett.* **608** L61
[15] Fisker J L, Schatz H, and Thielemann F K 2008 *Astrophys. J. Suppl.* **174** 261
[16] Wrede C, Caggiano J A, Clark J A *et al.* 2009 *Phys. Rev. C* **79** 045808
[17] Kahl D, Chen A A, Binh D N *et al.* 2009 *PoS (NIC X)* 171
[18] Audi G and Wapstra A H 1995 *Nucl. Phys.* **A595** 409
[19] Iliadis C, DÁuria J, Starrfield S *et al.* 2001 *Astrophys. J. Suppl.* **134** 151
[20] Paddock R A 1972 *Phys. Rev. C* **5** 485
[21] Caraça J M G, Gill R D, Cox A J and Rose H J 1972 *Nucl. Phys.* **A193** 1
[22] Kuhlmann E, Albrecht W and Hoffmann A 1973 *Nucl. Phys.* **A213** 82
[23] Yokota H, Fujioka K, Ichimaru K *et al.* 1982 *Nucl. Phys.* **A383** 298
[24] Fynbo H, Axelsson L, Äystö J *et al.* 1999 *Phys. Rev. C* **59** 2275
[25] Bardayan D W, Blackmon J C, Fitzgerald R P *et al.* 2007 *Phys. Rev. C* **76** 045803
[26] Woosley S E, Heger A, Cumming A *et al.* 2004 *Astrophys. J. Suppl.* **151** 75

Journal of Physics: Conference Series **202** (2010) 012010 doi:10.1088/1742-6596/202/1/012010

# Studies of astrophysically interesting nucleus $^{23}$Al

**A Saastamoinen[1], L Trache[2], A Banu[2], M A Bentley[3], T Davinson[4], T Eronen[1], J C Hardy[2], V E Iacob[2], D Jenkins[3], A Jokinen[1], M McCleskey[2], B Roeder[2], G Tabacaru[2,*], R E Tribble[2], P J Woods[4], J Äystö[1] and the IGISOL/MARS groups**

[1] Department of Physics, University of Jyväskylä, P.O.Box 35(YFL) FI-40014, Finland
[2] Cyclotron Institute, Texas A&M University, College Station, TX-77843, USA
[3] Department of Physics, University of York, Heslington, York, YO10 5DD, UK
[4] School of Physics and Astronomy, University of Edinburgh, Edinburgh, EH9 3JZ, UK
[*] On leave from "Natl. Inst. for Physics and Nucl. Eng. - Horia Hulubei", Bucharest, Romania

E-mail: ajsaasta@jyu.fi

**Abstract.** We have studied the $\beta$-delayed proton decay of $^{23}$Al with a novel detector setup at the focal plane of the MARS separator at the Texas A&M University to resolve existing controversies about the proton branching of the IAS in $^{23}$Mg and to determine the absolute proton branchings by combining our results to the latest $\beta\gamma$-decay data. We have made also a high precision mass measurement of the ground state of $^{23}$Al to establish more accurate proton separation energy of $^{23}$Al. Here the description of the used techniques along with preliminary results of the experiments are given.

## 1. Introduction

Classical novae are relatively common events in our galaxy with a rate of a few per year detected. The present understanding is that novae occur in interacting binary systems where hydrogen-rich material accretes on a white dwarf from its low-mass main-sequence companion. At some point in the accretion the hydrogen-rich matter compresses leading to a thermonuclear runaway [1]. Understanding the dynamics of the nova outbursts and the nucleosynthesis fueling it is crucial in understanding the chemical evolution of the galaxy.

The key parameters in determining the astrophysical reaction rates dominated by resonant proton capture, as is the case with many sd-shell nuclei, are the energies and decay widths of the associated nuclear states. Accurate determination of these energies can be achieved through traditional decay spectroscopy, accompanied with high-precision mass measurements of the ground states by using Penning trap mass spectrometery. One of the key reactions that possibly deplete $^{22}$Na produced in the so called NeNa-cycle and for which the reaction rates are known only with large uncertainties is the radiative proton capture $^{22}$Na(p,$\gamma$)$^{23}$Mg [2, 3]. This reaction rate is dominated by the capture through low-energy proton resonances which correspond to excited states in $^{23}$Mg nucleus above the proton separation threshold.

The relevant states in $^{23}$Mg can be studied via $\beta$-decay of $^{23}$Al, populating the excited states of $^{23}$Mg that are decaying by both proton and $\gamma$-emission. Earlier works on the $\beta$-decay of $^{23}$Al show contradicting results for the lowest states above the $^{22}$Na+p threshold [4, 5] The scope of the present work is to solve this controversy and to deduce the absolute proton branchings from the excited states of $^{23}$Mg by combining our data to existing decay data [3].

## 2. $\beta$-decay of $^{23}$Al

The $\beta$-decay of $^{23}$Al was studied at the Cyclotron Institute of the Texas A&M University. In this experiment the $^{23}$Al beam was produced in inverse-kinematics reaction $^{1}$H($^{24}$Mg,$^{23}$Al)2n by bombarding a hydrogen gas target with $^{24}$Mg beam at 48 MeV/u. The recoil products were separated with the Momentum Achromat Recoil Separator (MARS) [6], resulting a beam of $^{23}$Al with typical intensity of 4000 pps and purity of better than 95%.

The beam was taken into the detector setup, illustrated in Fig 1, consisting of a 65$\mu$m thick Double-Sided Silicon Strip Detector (DSSSD) with 16+16 $3 \times 50$ mm$^2$ strips, a 1 mm thick Si-pad detector and a high-purity germanium detector (HPGe). The beam implantation depth was controlled by using a rotatable 300 $\mu$m Al degrader, allowing us to tune the beam into the center of the DSSSD. The beam was pulsed with an implantation period of 1 second and a decay period of 1 second. The data was collected only during the decay part of the cycle.

The particle detectors were calibrated online with beams of $^{20}$Na, $^{21}$Mg, $^{22}$Mg and the germanium detector with $^{24}$Al. Both, the DSSSD and the HPGe were gated with the $\beta$-spectrum from the Si-pad detector. As the DSSSD used had fairly large pixel size, the $\beta$-response extends up to several hundred keV even with a pure source. One way to extract out the meaningful data is to measure the actual $\beta$-response from the detector by using an implanted source that does not emit any other charged particles and try to deduct this contribution from the interesting data. In this case the $\beta$-response was measured with $^{22}$Mg. The measured $\beta$-spectrum was smoothed to get rid of statistical fluctuations and then scaled so that it matched the $^{23}$Al spectrum at 150 keV. The resulting spectrum is illustrated in figure 2.

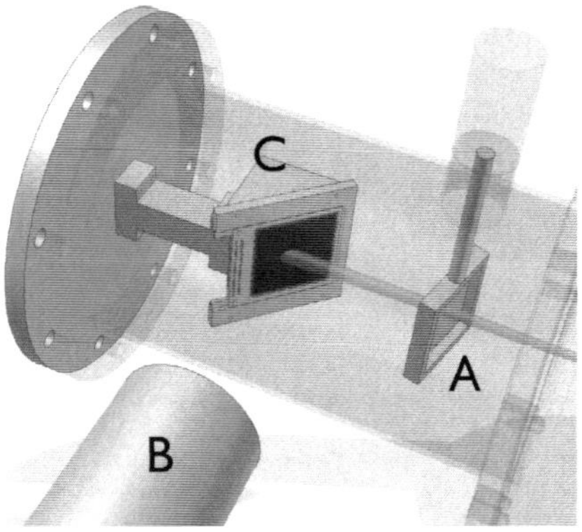


**Figure 1.** A sketch of the setup used in this experiment. The beam enters the setup from right through the rotatable aluminum degrader (A) and is implanted into the DSSSD backed with a Si-pad detector (C). The HPGe detector (B) was located outside the chamber at 90° to the beam.

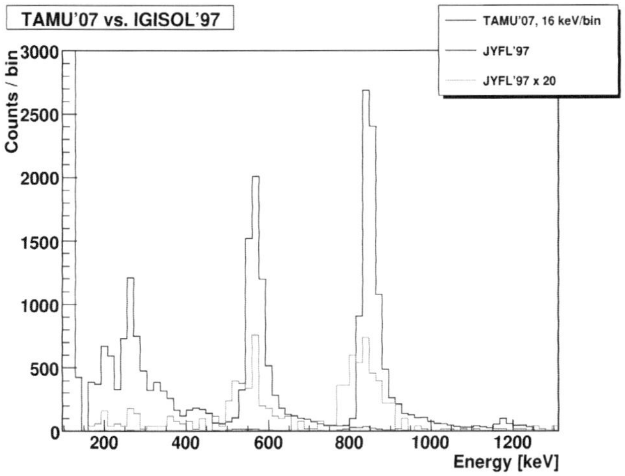


**Figure 2.** The background-reduced total decay energy spectrum from this work (black) compared with the proton spectrum from Ref. [5]. The old Jyväskylä data is multiplied by factor of 20 to make it more visible (red).

Even from the raw spectrum, it is clear that there is no anomalously large proton branch from the IAS in $^{23}$Mg, as reported in Ref. [4]. From the background reduced spectrum, illustrated in Fig. 2, it is clear that our result agree with Ref. [5]. From these proton and $\gamma$-data one can extract the relative proton intensities and then by using the existing absolute $\gamma$-branchings [3] one can assign the absolute proton branchings from the states studied.

## 3. Mass of $^{23}$Al

A high-precision mass measurement of $^{23}$Al was conducted by using the JYFLTRAP setup at the IGISOL facility in the Accelerator Laboratory of the University of Jyväskylä. The $^{23}$Al beam was produced by using Ion Guide Isotope Separator On Line (IGISOL) method [7]. From the IGISOL facility, ions having the same $A/q = 23$ were sent into a gas-filled radiofrequency quadrupole (RFQ) cooler-buncher [8] to prepare the samples for injection into the JYFLTRAP Penning trap setup consisting of two identical, cylindrical traps inside the same superconducting 7 T magnet [9]. The first trap is filled with low-pressure helium gas and works as a purification trap with a mass resolving power of a few $\times 10^5$. The second trap is a precision trap where the mass of the ion is determined by using the time-of-flight ion cyclotron resonance (TOF-ICR) method [10, 11]. The absolute mass of the ion of interest is determined from the ratio of the measured cyclotron frequencies of the sample and a well known reference case: $m = (\nu_{c,ref}/\nu_c) \cdot (m_{ref} - m_e) + m_e$. Figure 3 illustrates typical TOF resonance curves from this experiment.

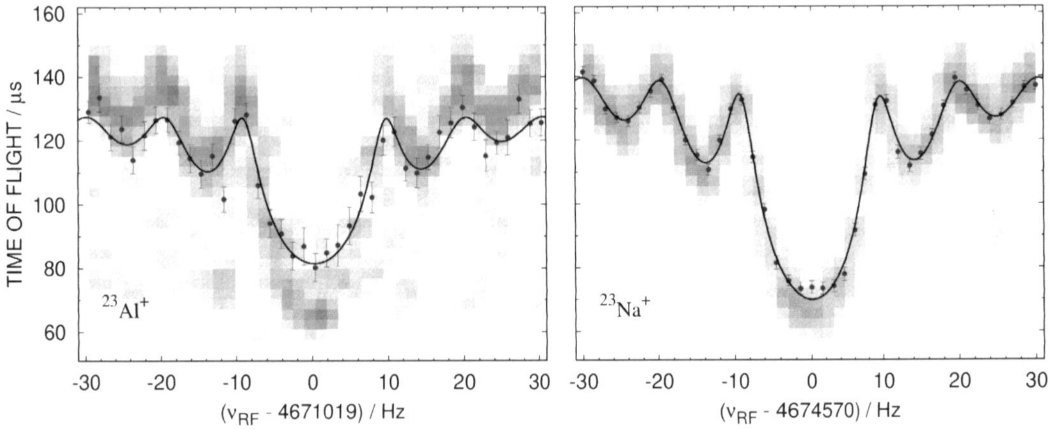


**Figure 3.** Typical cyclotron resonance curves of $^{23}$Al$^+$ and $^{23}$Na$^+$. The RF-excitation time was 100 ms.

In this experiment $^{23}$Na was used as reference for determining the masses of $^{23}$Al and $^{23}$Mg. The full analysis and the implication for the Isobaric Multiplet Mass Equation are discussed in a separate article [17]. One can calculate a new value for the $^{23}$Al proton separation energy, $S_p(^{23}\mathrm{Al}) = 141.11(43)$ keV, by combining our result for the mass excess of $^{23}$Al, 6748.07(34) keV, and the mass excesses of $^{22}$Mg, -399.79(25) keV, [18] and $^1$H, 7288.97050(11) keV, [12].

## 4. Conclusions and outlook

From the $\beta$-decay data, it is clear that there is no exceptionally large proton branching from the IAS in $^{23}$Mg and thus strong isospin mixing as proposed by Tighe et al. in Ref. [4]. Our decay data allows us to extract the proton branchings from the excited states of $^{23}$Mg that are close to $^{22}$Na+p threshold and thus crucial to the $^{22}$Na(p,$\gamma$)$^{23}$Mg. reaction rate. The detailed analysis is to be finished and published in near future. The same setup has been improved and optimized for further experiments, which include $\beta$-decay of $^{31}$Cl and similar studies under planning.

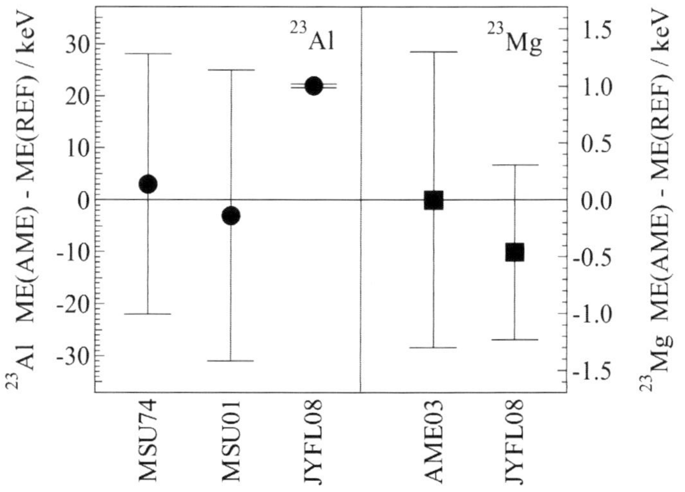


**Figure 4.** Differences between the different measurements and the latest Atomic Mass Evaluation (AME03 [12]) value. MSU74 refers here to [13], MSU01 refers to [14] and JYFL08 is from the work presented in this paper. The AME value for $^{23}$Mg is a combination of two spectrometric studies [15, 16].

The $S_p$ value resulting from our mass measurement is higher than the previous value indicating a reduced halo nature. This new value has influence on the calculated astrophysical S-factor for the proton capture reaction $^{22}$Mg(p,$\gamma$)$^{23}$Al and its corresponding reaction rate in the stellar enviroments. It also shows $^{23}$Al to be more resilent to destruction through photodissociation, making this isotope a more important player in the reaction networks in the explosive H-burning processes, like novae and X-ray bursts.

## Acknowledgements
This work has been supported by the EU 6th Framework programme "Integrating Infrastructure Initiative - Transnational Access", Contract Number: 506065 (EURONS, JRA TRAPSPEC) and by the Academy of Finland under the Finnish Centre of Excellence Programme 2006-2011 (Project No. 213503, Nuclear and Accelerator Based Physics Programme at JYFL) and by the US DOE under grant DE-FG02-93ER40773. A.S. acknowledges the support from the Jenny and Antti Wihuri foundation.

## References
[1] J Jose *et al* 2006 *Nucl. Phys* A **777** 550
[2] D G Jenkins *et al* 2004 *Phys. Rev. Lett.* **92** 031101
[3] V E Iacob *et al* 2006 *Phys. Rev.* C **74** 045810
[4] R J Tighe *et al* 1995 *Phys. Rev.* C **52** R2298
[5] K Peräjärvi *et al* 2000 *Phys. Lett.* B **492** 1
[6] R E Tribble *et al* 2002 *Nucl. Phys.* A **701** 278c
[7] J Äystö 2001 *Nucl. Phys.* A **693** 477
[8] A Nieminen *et al* 2002 *Phys. Rev. Lett.* **88** 094801
[9] V S Kolhinen *et al* 2004 *Nucl. Instr. Meth.* A **528** 776
[10] G Gräff *et al* 1980 *Z. Phys.* A **297** 35
[11] M König *et al* 1995 *Int. J. Mass Spectrom. Ion Processes* **142** 95
[12] Audi G, Wapstra A H and Thibault C 2003 *Nucl. Phys.* A **729** 337
[13] W Benenson *et al* 1975 *Phys. Lett.* B **58** 46
[14] J A Caggiano *et al* 2001 *Phys. Rev.* C **64** 025802
[15] J M Freeman *et al* 1962 *Nucl. Phys.* **38** 89
[16] J A Nolen *et al* 1974 *Nucl. Instr. Meth.* **115** 189
[17] A Saastamoinen *et al* *Submitted to Phys. Rev. C*
[18] A Parikh *et al* 2005 *Phys. Rev.* C **71** 055804

Journal of Physics: Conference Series **202** (2010) 012011 doi:10.1088/1742-6596/202/1/012011

# $^{96}$Ru(p,$\gamma$)$^{97}$Rh measurement at the GSI storage ring

**Q. Zhong[1,2], T. Aumann[1], S. Bishop[3], K. Blaum[9], K. Boretzky[1], F. Bosch[1], H. Bräuning[1], C. Brandau[1,3], T. Davinson[4], I. Dillmann[3], O. Ershova[1,5], H. Geissel[1], G. Gyürky[6], M. Heil[1], F. Käppeler[7], A. Kelić[1], C. Kozhuharov[1], C. Langer[1,5], T. Le Bleis[1,5,10], Y. A. Litvinov[1,9], G. Lotay[3], J. Marganiec[1], N. Petridis[5], R. Plag[1,5], U. Popp[1], R. Reifarth[1,5], B. Riese[1], C. Rigollet[8], C. Scheidenberger[1], H. Simon[1], T. Stöhlker[1], T. Szücs[6], G. Weber[1], H. Weick[1], D. F. A. Winters[1], N. Winters[1], P. J. Woods[4]**

[1] GSI Helmholtzzentrum für Schwerionenforschung GmbH, Darmstadt, 64291, Germany
[2] China Institute of Atomic Energy (CIAE) 102413 Beijing China
[3] Technische Universität München, 85748 Garching, Germany
[4] University of Edinburgh, Edinburgh, UK
[5] J.W. Goethe Universität, Frankfurt a.M, 60438, Germany
[6] Institute of Nuclear Research of the Hungarian Academy of Sciences, Hungary
[7] Forschungszentrum Karlsruhe, Institut für Kernphysik, Karlsruhe, Germany
[8] Kernfysisch Versneller Instituut , Netherlands
[9] Max-Planck-Institut für Kernphysik, 69117 Heidelberg, Germany
[10] Institut Pluridisciplinaire Hubert Curien, Strasbourg, France

E-mail: `r.reifarth@gsi.de (R. Reifarth)`

**Abstract.** A pioneering experiment was recently performed at the Experimental Storage Ring (ESR) at GSI. Fully stripped ions of $^{96}$Ru were injected into the storage ring and slowed down to a few MeV per nucleon. The $^{97}$Rh ions from the $^{96}$Ru(p,$\gamma$) reaction at a newly developed hydrogen jet target were detected with Double Sided Silicon Strip Detectors (DSSSD) mounted inside a pocket. The experiment and the status of the analysis at a beam energy of 11 MeV per nucleon will be presented.

## 1. Introduction

The nucleosynthesis of elements beyond iron is dominated by neutron captures in the s and r processes [1]. However, 32 stable, proton-rich nuclides between $^{74}$Se and $^{196}$Hg cannot be formed during those processes, because they are shielded from the s-process flow and r-process $\beta$-decay chains. These nuclei are thought to be produced in the p process, where proton-rich nuclei are made by sequences of photodisintegrations and proton captures reactions and following $\beta^+$ decays on existing r- and s-seed nuclei. Rate predictions for the (p,$\gamma$) and ($\alpha$,$\gamma$) reactions are very difficult since only a handful of experimental data for stable isotopes have been determined in the Gamow window of the p process so far. The presently favored sites for the p process are the explosively burning O/Ne layers in supernovae (SN) II and the explosive carbon burning in the core of SN Ia. The p-process models for SN II and for SN Ia are capable of reproducing the p abundances within a factor of three [2, 3, 4]. However, the light p nuclei $^{92,94}$Mo, and $^{96,98}$Ru are underproduced by more than a factor of 10 [3]. Cross-section measurements on (p,$\gamma$) and

($\alpha$,$\gamma$) reactions in the astrophysically interesting energy range are already very challenging on stable nuclei. Only a minute part of the nuclei involved in p-process networks, however, is stable. The majority of the isotopes crucial for the final p-abundance are unstable. The most promising approach to determine the desired reaction rates is to produce the isotopes at Radioactive Ion Beam facilities and to investigate the reactions in inverse kinematics.

In order to prove the principle of this approach, an experiment was performed at the Experimental Storage Ring (ESR) at the GSI Helmholtzzentrum für Schwerionenforschung GmbH. Fully stripped ions of $^{96}$Ru were injected into the ESR, slowed down to a few MeV per nucleon, interacted with a hydrogen target, and the reaction products were detected with particle detectors.

## 2. Experiment

The experiment was performed at the ESR with a circumference of 108.36 m, which has two 9.5 m long straight experimental sections, see Figure 1. In one of the experimental sections an electron cooler is installed while the other one hosts a gas target [5, 6].

In order to produce fully stripped ions, $^{96}$Ru was first accelerated to 100 MeV/u and then stripped using a 11 mg/cm$^2$ carbon foil. Afterwards the fully stripped ions were injected into the ESR, and slowed down to energies of 9, 10, and 11 MeV/u. Electron cooling was applied before and after the slowing down phase. After slowing down about $5 \cdot 10^6$ $^{96}$Ru$^{44+}$ ions were stored in the ESR circulating at a frequency of $\approx$500 kHz. The thickness of the newly developed cryogenically cooled liquid microjet hydrogen target [6] was $\approx 10^{13}$/cm$^2$ resulting in a luminosity of $2.5 \cdot 10^{25}$/cm$^2$/s.

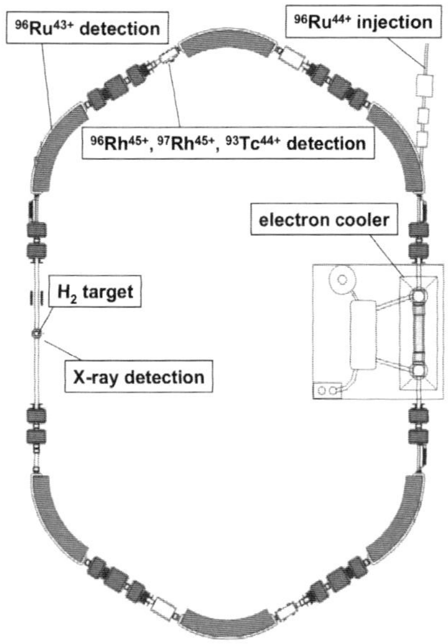


**Figure 1.** Layout of the experimental storage ring at GSI. The experimental section on the left contains the hydrogen jet target and the electron cooler device is mounted on the right. The ring has a circumference of 108 m and a maximum bending power of 10 Tm.

The most important interactions of fully stripped $^{96}$Ru$^{44+}$ ions with hydrogen are electron pick-up (atomic process) and proton-induced nuclear reactions like (p,$\gamma$), (p,$\alpha$), and (p,n). Depending on the exact mechanism, the electron pick-up results in $^{96}$Ru$^{43+}$ ions and possibly X-rays following the deexcitation of the atomic shell. Multi-Wire Proportional Counters (MWPC) were installed after the $H_2$-target on the outer side of the ESR in order to detect the $^{96}$Ru$^{43+}$ ions. Germanium X-ray detectors were mounted close to the target at angles of 35° and 90° to detect the emitted X-rays.

The products of nuclear reactions, in particular the $^{97}$Rh$^{45+}$ ions following the proton capture reaction under investigation, were detected with two Double Sided Silicon Strip Detectors

(DSSSD) mounted in a pocket on the inside of the ESR. Each detector had 16 stripes in X- and Y-direction. The pocket had a stainless steel window of 25 $\mu$m. The thickness of this window restricted the possible energies of the $^{96}$Ru ions to energies above $\approx$ 9 MeV, since the freshly produced $^{97}$Rh ions are otherwise not able to penetrate the window and could therefore not be detected with the DSSSDs. This means that we were not able to measure inside or close to the astrophysically interesting Gamow window around 3 MeV with this proof-of-principle experiment.

## 3. Preliminary result and discussion

The (p,$\gamma$) cross section was measured relative to the well known electron recombination cross section at 9, 10, and 11 MeV per nucleon. The advantage of this approach is that the -otherwise difficult to determine- overlap between beam and target cancels out. The number of electron-pick-up events was determined using the X-ray detectors close to the target as well as the MWPC downstream of the target.

If K-REC is the radiative electron-capture process to the K-Shell, $N_{(p,\gamma)}$ and $N_{K-REC}$ the number of detected (p,$\gamma$) and K-REC events, $\varepsilon_{(p,\gamma)}$ and $\varepsilon_{K-REC}$ the respective detection efficiencies, $\sigma_{(p,\gamma)}$ and $\sigma_{K-REC}$ the respective cross sections, one finds:

$$\sigma_{(p,\gamma)} = \sigma_{K-REC} \frac{\varepsilon_{K-REC}}{\varepsilon_{(p,\gamma)}} \frac{N_{(p,\gamma)}}{N_{K-REC}} . \tag{1}$$

The cross section $\sigma_{K-REC}$ can be calculated based on theory with an uncertainty in the range of 5% [7]. K-REC is detected using the X-ray detectors and their efficiencies were calibrated using a standard X-ray source.

The detection efficiency for (p,$\gamma$) events using the DSSSDs is approximately 100%, because all ions hit the detectors. Since the measurement had to be performed above 9 MeV in the CM-frame, not only events from proton capture occur in the DSSSDs. The important components in the position spectrum of the DSSSDs are (p,$\gamma$), (p,$\alpha$), and (p,n) reactions. An additional background component is caused by elastic scattering of $^{96}$Ru$^{44+}$ ions on protons, see Figure 2. Figure 2 does not consider all details of the acceptance of the storage ring, which will modify the distribution in the DSSSDs.

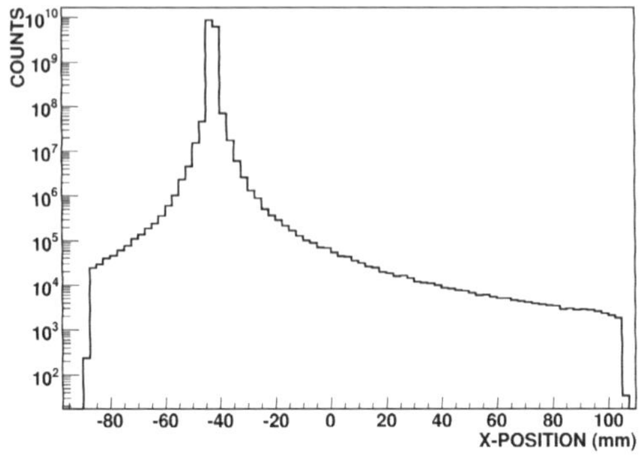


**Figure 2.** The x-distribution at the position of the DSSSD particle detectors as expected from elastic scattering of $^{96}$Ru$^{44+}$ ions on protons predicted using the Monte Carlo code MOCADI [8]. The position of the primary beam is at -40 mm.

The shape of the different nuclear components can be predicted using the code LISE++ [9], see left part of Figure 3. The right part of Figure 3 shows a fit, where besides (p,$\gamma$) only (p,$\alpha$) and the elastic scattering were considered. The (p,n) component was neglected for this preliminary analysis, because the exact shape depends strongly on the exact geometry of the beam line components and its contribution to the (p,$\gamma$) peak is expected to be small. The final analysis will contain all components. Because of this assumption, we report here an upper limit of the (p,$\gamma$) cross section. So far only first results at 11 MeV per nucleon are available. The derived upper limit of the $^{96}$Ru(p,$\gamma$)$^{97}$Rh cross section is 4.0 mb, which is comparable to the predictions from the NON-SMOKER code [10].

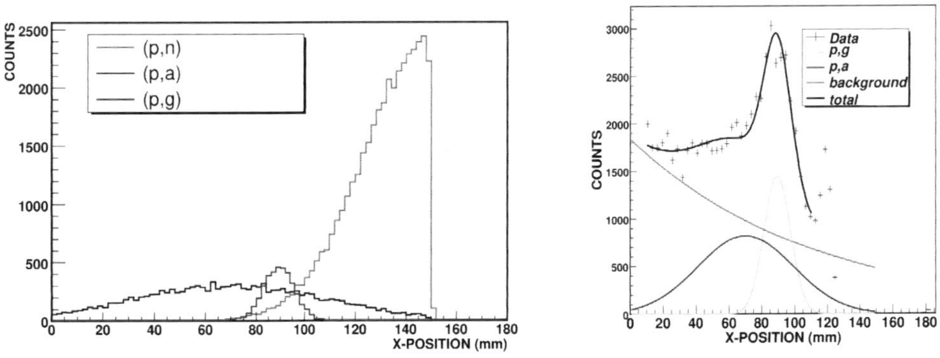


**Figure 3.** The left part shows the LISE++ [9] shape predictions of the different nuclear components. The right part shows a fit of the experimental data neglecting the (p,n) component. The position of the primary beam is at -40 mm.

## 4. Summary and outlook

A pioneering experiment was performed at the Experimental Storage Ring (ESR) at GSI. Fully stripped ions of $^{96}$Ru were injected into the storage ring and slowed down to 9-11 MeV per nucleon. The $^{97}$Rh ions following the $^{96}$Ru(p,$\gamma$) reaction at a cryogenically cooled liquid microjet hydrogen target were detected with Double Sided Silicon Strip Detectors (DSSSD) mounted inside a pocket. We could prove the feasibility of this approach and clearly detected events originating from proton captures on $^{96}$Ru. The analysis is still in progress and we could therefore only report an upper limit on the cross section at 11 MeV, which is slightly above the prediction from the NON-SMOKER code. The detailed analysis will eventually include a simulation of all beam line components and reaction channels. It will also be extended down to 10 and 9 MeV. In the future, such experiments have to be performed with particle detectors inside the ultra-high vacuum of the ESR in order to measure in the Gamov window. Additionally a demonstration experiment for $(\alpha, \gamma)$ reactions is planned.

## Acknowledgments

This work is supported by the HGF Young Investigators Projects VH-NG-327 and VH-NG-331 and the ERC grant no. 203175.

## References

[1] Burbidge E, Burbidge G, Fowler W and Hoyle F 1957 *Rev. Mod. Phys.* **29** 547
[2] Prantzos N, Hashimoto M, Rayet M and Arnould M 1990 *Astron. Astrophys.* **238** 455
[3] Rayet M, Arnould M, Hashimoto M, Prantzos N and Nomoto K 1995 *Astron. Astrophys.* **298** 517 – 527
[4] Howard W, Meyer B and Woosley S 1991 *Ap. J.* **373** L5
[5] Reich H, Bourgeois W, Franzke B, Kritzer A and Varentsov V 1997 *Nuclear Physics A* **626** 417
[6] Kühnel M, Petridis N, Winters D, Popp U, Dörner R, Stöhlker T and Grisenti R 2009 *NIM A* **602** 311
[7] Eichler J and Stöhlker T 2007 *Physics Reports* **439** 1–99
[8] Iwasa N, Geissel H, Münzenberg G, Scheidenberger C, Schwab T and Wollnik H 1997 *NIM B* **126** 284
[9] Tarasov O and Bazin D 2008 *Nucl. Inst. Meth. B* **266** 4657
[10] Rauscher T and Thielemann F K 2000 *Atomic Data Nucl. Data Tables* **75** 1

# First approach to half-life measurements around N=126

**Teresa Kurtukian-Nieto**
Centre d'Études Nucléaires de Bordeaux Gradignan - Université Bordeaux 1 - UMR 5797 CNRS/IN2P3, Chemin du Solarium, BP 120, 33175 Gradignan, France
E-mail: kurtukia@cenbg.in2p3.fr

**Abstract.** The present piece of work reviews recent progress on $\beta$-decay half-life measurements around the neutron shell N=126. The production of heavy neutron-rich nuclei is discussed, including the present status of the experimental data available for this region of the chart of nuclides. The impact of the theoretical calculations used to model the properties of nuclei involved in the r-process nucleosynthesis is discussed.

## 1. Introduction

The half-life, a fundamental property of radioactive nuclei, plays a decisive role in several astrophysical processes, being the r-process nucleosynthesis [1] one remarkable example. In the r-process, the abundance pattern presents pronounced peaks around the mass numbers A$\sim$ 80, 130 and 195 that clearly reflect the influence of the nuclear structure at the magic neutron numbers N= 50, 82 and 126. As the half-lives of nuclei at magic neutron numbers are relatively long, they determine the dynamical time scale of the r-process and build-up the abundance peaks at the waiting points.

For the vast majority of nuclei on or close to the r-process path, no experimental information is available. This is remarkably critical for the region around the neutron shell N=126, associated to the waiting point A=195, which has remained over the years as a practically unexplored territory, since the nuclei involved are so neutron-rich that their production in the laboratory becomes a real challenge. These nuclei represent the last bottleneck for proceeding into the actinide region, and become an important normalisation point for the r-process models, for determining the end-point and time scale of the r-process, and the production of trans-bismuth elements.

While the synthesis in the laboratory of all such nuclei for half-life measurements will be out of reach, the study of the ones which are closer to stability, and therefore expected to be the major contributors to the overall timescale of the r-process, as the key r-nuclei $^{192}Dy$, $^{193}Ho$, $^{194}Eu$, $^{195}Tm$, $^{196}Yb$ and $^{197}Lu$, will probably be within reach in the future. In fact, a first approach for measuring half-lives around the neutron shell N=126 has been performed at GSI [2] based on the event-by-event particle identification and implantation-$\beta$-decay time correlation of a mixed beam produced by cold fragmentation [3]. This work represents a first step towards the study of the r-process waiting-point nuclei.

In the following sections, a review of the different experimental approaches for producing and studying nuclei around the neutron shell N=126 will be presented, including the most recent experimental half-life data available. Also the comparison between the data and the different theoretical calculations used to model the nuclear properties of the r-nuclei will be discussed, including their implications on the r-process nucleosynthesis.

## 2. Production of heavy neutron-rich nuclei

In order to study experimentally this region of the chart of nuclides, the first challenge we have to face is the production of such nuclei in the laboratory. In general there exists two complementary methods to produce and separate heavy neutron-rich nuclei: the in-flight [4] technique and the isotope-separation on-line (ISOL) technique [5]. Both methods have extensively been used for $\beta$-decay studies of r-process-relevant nuclei at different regions of the chart of the nuclides.

The ISOL technique suffers from specific difficulties for heavy neutron-rich nuclei. Elements between Hf ($Z = 72$) and Pt ($Z = 78$) are difficult to extract from the production target due to their low volatility, and for neutron-rich isotopes of the adjacent heavier elements, the ISOL technique suffers from an overwhelming isobaric contamination of isobars with higher $Z$, which are produced with much higher cross sections. Background suppression can be achieved by combining resonant-laser ionisation and the specific condition that the isotope of interest is longer lived than the unwanted species. But these conditions are not generally applicable.

The in-flight separation is universal, offering the advantage of fast ($< 1\mu s$), highly efficient, chemistry-independent separation. This technique allows to select the desired region on the chart of nuclides and to perform experiments focusing either on one nucleus or on a cocktail beam, by the adequate tuning of the magnetic fields of the separator. The nuclei to be studied are then implanted in a detection setup, and the contaminants due to the small fraction of inevitable secondary reactions, produced during the stopping process, are eliminated by a further range selection at the implantation stage. As a consequence, the clean particle identification for each implanted nucleus permits the direct correlation of an individual fragment with its subsequent $\beta$ decay on an event-by-event basis.

## 3. Experimental approaches to measure $\beta$-decay half-lives

The experimental approach of choice for measuring half-lives will depend on the lifetimes themselves and on the production of the nuclei investigated. These approaches can be subdivided mainly into two groups: Activation method and delayed-coincidence method.

- Activation method
  In the first case, it is feasible to produce a large number of nuclei, in a given time period, which is short compared to their half-life. The experiment thus consists of an activation phase and a recording phase. These phases might be repeated several times.
  There exists some variants of the activation-type experiment. First, the decays are registered; secondly, the number of nuclei still present in the probe is determined as a function of time. In both cases, the spectrum follows an exponential function.
  For very long half-lives, the decay rate is approximately constant during the experiment, and the value of the half-life can be obtained directly from the ratio of the number of detected decays per time unit over the total number of nuclei in the sample.
  In all these cases, the background may consist of nuclei with other half-lives (contaminations, daughter nuclei) or a constant rate of parasitic events.
- Delayed-coincidence method
  In this second case, the experiment is characterised by short half-lives and/or low production rates. If the rate of produced nuclei is lower than the inverse of its half-life, it is convenient to record the time of production and look for the time of the consecutive decay. These time differences are sorted into a spectrum.
  The data can be recorded as follows:

- All decay-like events, after any produced nucleus,
- all decay-like events, after the last produced nucleus,
- only the first decay-like event, after the last produced nucleus.

Several methods have been developed to extract half-lives of radioactive species from experimental data and to determine their statistical uncertainties. A review on these methods is given in Ref. [2]. In general, if the average production rate is constant, there is no time structure in the production rate, and the standard analytical methods can be used to extract the half-lives. However, if the source intensity is not constant, that is, if it is modulated with a periodic function, the recorded time-correlation spectra do not directly reflect the decay properties of the nuclei involved but also show signatures of the time structure of the beam. In this case, the standard analytical methods can not be used and a new numerical method has been developed in Ref. [2] to deal with the determination of the half-lives from such spectra with a complex time-dependent structure.

The statistical analysis to deduce the decay constant $\lambda$ can be a complex task since the radioactive decays can only be observed in a limited time range, and in addition, events of other species which decay with different decay constants or background events may be mixed in. Also if daughter nuclei produced in the primary decay are also radioactive, an even more complex situation appears.

## 4. Status of half-lives around the neutron-shell N=126

Along the N=126 neutron shell, south of lead, only five isotones have been identified, and the half-life of only three of them has been already measured. Figure 1 shows the present status of the chart of nuclides in this region, and Table 1 shows a summary of the experimental half-life information available for these nuclei. Colour boxes in the figure indicate the known half-lives [6], while empty boxes mark nuclei identified with no experimental half-life information available.

**Table 1.** Isotones along N=126 south of $^{208}Pb$

| Isotone | $T_{1/2}$ | Reference |
|---|---|---|
| $^{207}Tl$ | 4.77(2) min | [7] |
| $^{206}Hg$ | 8.32(7) min | [8] |
| $^{205}Au$ | 31(2) s | [9] |
| $^{204}Pt$ | — | [10] |
| $^{203}Ir$ | — | [10] |

The half-life given for $^{207}Tl$ is the weighted average of 4.76(2) min [11], 4.77(5) min [12], 4.79(2) min [13] and 4.77(3) min [14]. For $^{206}Hg$, the value listed in Table 1, is the weighted average of 8.5(1) min [15], 8.1(4) min [16] and 8.15 (10) min [17]. The half-life of $^{205}Au$, 31(2) s, is taken from [18]. This last one is the most recent experimental half-life information at the neutron shell N=126, and dates from 1994. As can be seen, few progress has been obtained over the last years in studying the properties of nuclei in this region. The reason for this situation is that the production in the laboratory of heavy neutron-rich nuclei is a real challenge. However, a recent experiment performed at GSI [10] in 2003 allowed for the production and identification of $^{204}Pt$ and $^{203}Ir$ by means of cold fragmentation of a $^{208}Pb$ beam at 1 GeV/A, opening a way

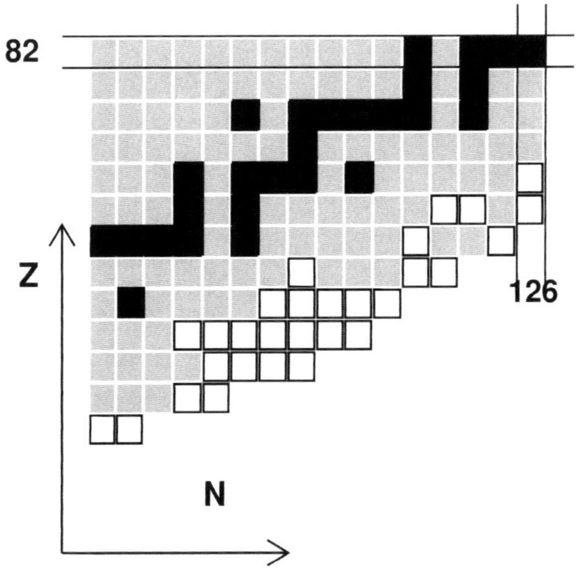


**Figure 1.** Chart of the nuclides around N = 126. The colour indicates the known half-lives [6], while empty boxes mark nuclei identified with no experimental half-life information available.

towards the study of heavy nuclei around the shell N=126 (see Figure 2). This work allowed the production and identification of more than 190 heavy neutron-rich nuclei, 25 of which were synthesised for the first time. Figure 3 shows the region of the chart of nuclides covered in this experiment. The solid line represents the limits of the chart of nuclides before that work, and the dashed line, the limits of the known half-lives. Also in this work the half-life of 7 nuclei was measured for the first time. The values are listed in Table 2.

**Table 2.** $\beta$-decay half-lives measured in [19] compared with different model calculations. See text for details.

| Nuclide | $T_{1/2}$ (s) [19] | DF3+QRPA (s)[20] | FRDM+QRPA (s) [21] |
|---|---|---|---|
| $^{202}$Ir | $11 \pm 3$ | 9.8 | 68.4 |
| $^{199}$Ir | $6^{+5}_{-4}$ | 46.7 | 370.6 |
| $^{198}$Ir | $8 \pm 2$ | 19.1 | 377.1 |
| $^{200}$Os | $6^{+4}_{-3}$ | 6.9 | 187.1 |
| $^{199}$Os | $5^{+4}_{-2}$ | 6.6 | 106.8 |
| $^{196}$Re | $3^{+1}_{-2}$ | 1.4 | 3.6 |
| $^{195}$Re | $6 \pm 1$ | 8.5 | 3.3 |
| $^{194}$Re | $1 \pm 0.5$ | 2.1 | 70.8 |

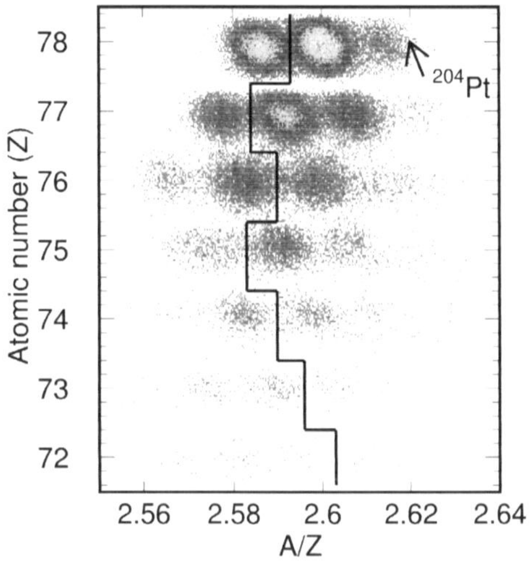


**Figure 2.** Two-dimensional cluster plot of Z vs. A/Z, including the identification of $^{204}Pt$ [19]

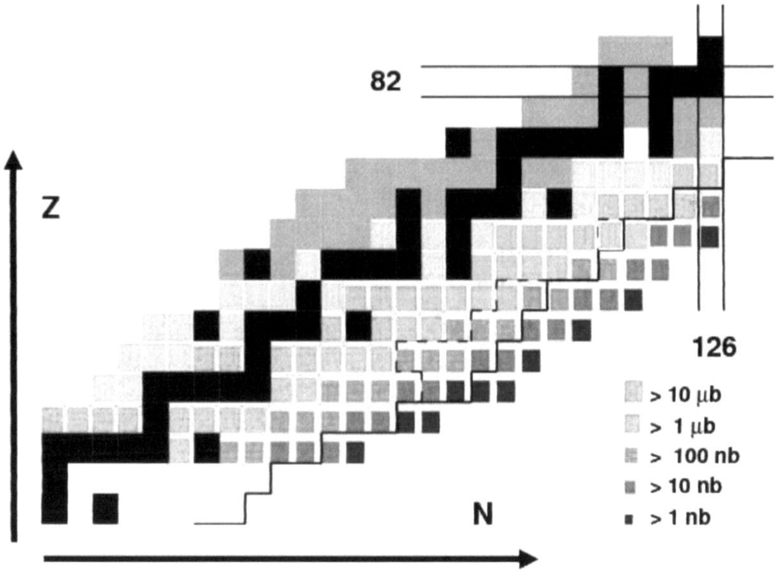


**Figure 3.** Fragments residues in the reaction $^{208}Pb$ (1 A GeV) + Be. [19]. The solid line represents the limits of the chart of nuclides before this work. The dashed line shows the limit of the known half-lives.

## 5. Nuclear properties of the nuclei at N=126 and their implication in the r-process

As pointed out before, there is no experimental data for r-nuclei at $N = 126$ on the r-process path ($Z \approx 65 - 72$). As a consequence, r-process simulations still rely on theoretical models for the $\beta$-decay half-lives. When comparing the experimental data available close to N=126 with the model calculations available in this region, it can be seen, as shown in Table 2, that the DF3+QRPA model agrees, in general, rather well with the experimental half-lives. On the other hand, the FRDM+QRPA half-lives are for most nuclei more than one order of magnitude longer than the data.

The DF3+QRPA model is the only one that includes Gamow-Teller (GT) and first-forbidden (FF) transitions on the same footing. Therefore, the good agreement of the experimental half-lives with this model gives strong evidence for important contributions of FF transitions to the total half-lives, arising from $\nu 3p_{1/2} \rightarrow \pi 3d_{3/2}$ and $\nu 1i_{13/2} \rightarrow \pi 2h_{11/2}$ neutron-proton transitions, and hence it seems that pure GT estimates of the half-lives of the $N = 126$ waiting points are insufficient. Since the differences between these two models qualitatively persist up to the $N = 126$ waiting-point nuclei around $Z = 65$ as shown in Ref. [22], the r-process matter flow to the heavier fissioning nuclei [23] may be faster than currently expected.

## 6. Summary and outlook

In summary, half-lives of nuclei approaching the r-process waiting point at $N = 126$ have been measured for the first time after many years of no progress in this region of the chart of nuclides. This work may be considered as the starting point of a new era in exploring the properties of heavy neutron-rich nuclides. It relies on the production of the nuclides of interest by cold fragmentation and the application of a novel very efficient analysis method.

The prospects for extending this experimental approach to more neutron-rich isotopes of elements below lead and reaching r-nuclei such as $^{192}Dy$, $^{193}Ho$, $^{194}Eu$, $^{195}Tm$, $^{196}Yb$ and $^{197}Lu$, are very promising, in particular when higher beam intensities become available in the new-generation facilities around the world as GSI-FAIR[24], NSCL/FRIB[25] at MSU and RIBF/RIKEN[26].

## References

[1] E.M. Burbidge *et al.* 1957 *Rev. Mod. Phys.* **29**, 547.
[2] T. Kurtukian-Nieto *et al.* 2008 *Nuc. Instr. and Meth. A* **589**, 472.
[3] J. Benlliure, *et al.* 1999 *Nucl. Phys. A* **660**, 87.
[4] D. Morrissey, B. M. Sherrill, 2004 *Lect. Notes Phys.* **651**, 113.
[5] P. Van Duppen, 2006 *Lect. Notes Phys.* **700**, 37.
[6] National Nuclear Data Center. http://www.nndc.bnl.gov/chart.
[7] M. J. Martin 1993 *Nuclear Data Sheets* **70**, 315.
[8] F.G. Kondev 2008 *Nuclear Data Sheets* **109**, 1527.
[9] F.G. Kondev 2004 *Nuclear Data Sheets* **101**, 521.
[10] T. Kurtukian-Nieto *et al.* 2006 Progress in the investigation of nuclei approaching the r-process waiting point A = 195. *Proceedings of Science* POS(NIC-IX)008.
[11] M. Curie *et al.* 1931 *Rev. Modern Phys.* **3**, 427
[12] K. Fajans, A. F. Voigt 1940 *Phys. Rev.* **58**, 177.
[13] B. W. Sargent, L. Yaffe, A. P. Gray. 1953 *J. Phys* **31**, 253.
[14] J. M. Trischuck, E. Kankeleit. 1967 *Nucl. Phys. A* **90**, 33.
[15] P. Kauranen. 1962 *Ann. Acad. Sci. Fennicae, Ser. A VI* **96**.
[16] G. K. Wolf, F. Lux, H.-J. Born. 1964. *Radiochim. Acta* **3**, 206.
[17] G. K. Wolf. 1968. *Nucl. Phys. A* **116**, 387.
[18] Ch.Wennemann *et al.* 1994 *Z. Phys. A* **347** 185.
[19] T. Kurtukian-Nieto, Ph.D. Thesis, University of Santiago de Compostela, 2006. $http : //www - wnt.gsi.de/charms/DoktorArbeiten/Teresa - Kurtukian/PhD_{T}hesis_{T}KN.pdf$
[20] T. Kurtukian-Nieto *et al.* 2009 *Nucl. Phys. A* **827** 587.
[21] P. Möller, B. Pfeiffer, and K.-L. Kratz, 2003 *Phys. Rev.* **C 67**, 055802.
[22] I. N. Borzov *et al.*, 2008 *Nucl. Phys. A* **814**, 159.
[23] G. Martínez-Pinedo *et al.*, 2007 *Prog. Part. Nucl. Phys.* **59**, 199.
[24] http://www.gsi.de/fair
[25] http://www.frib.msu.edu
[26] http://www.rarf.riken.go.jp/ribf

# Differences between stellar and laboratory reaction cross sections

**Thomas Rauscher**

Department of Physics, University of Basel, 4056 Basel, Switzerland

E-mail: Thomas.Rauscher@unibas.ch

**Abstract.** Nuclear reactions proceed differently in stellar plasmas than in the laboratory due to the thermal effects in the plasma. On one hand, a target nucleus is bombarded by projectiles distributed in energy with a distribution defined by the plasma temperature. The mostly relevant energies are low by nuclear physics standards and thus require an improved description of low-energy properties, such as optical potentials, required for the calculation of reaction cross sections. Recent studies of low-energy cross sections suggest the necessity of a modification of the proton optical potential. On the other hand, target nuclei are in thermal equilibrium with the plasma and this modifies their reaction cross sections. It is generally expected that this modification is larger for endothermic reactions. We show that there is a large number of exceptions to this rule.

## 1. Introduction

Stellar plasmas are characterized by their composition and temperature. The energy distribution of nuclei according to the temperature is given by a Maxwell-Boltzmann (MB) function. This function peaks at comparatively low energy and thus the astrophysically relevant interaction energies of nuclei are low by nuclear physics standards. This is a challenge for experimental and theoretical nuclear physics, especially when dealing with charged-particle induced reactions. The thermalization of nuclei is very fast, even compared to reaction timescales. This populates excited states and reactions can proceed from the ground state and the excited states, contrary to reactions in the laboratory which only include targets in the ground state. The resulting change in the effective cross section is called stellar enhancement and usually given by the ratio of the stellar cross section to the ground state one, $f = \sigma^*/\sigma^{\mathrm{g.s.}}$. The stellar enhancement factor $f$ can only be studied theoretically.

## 2. Coulomb suppression of the stellar enhancement factor

Stellar reaction rates are obtained by folding *stellar* cross sections with the MB distribution of the projectiles. It can be shown [1, 2] that *stellar* rates obey reciprocity whereas rates based on laboratory cross sections do not. Since excited states in both the target and the final nucleus contribute, it is easy to see that more transitions beyond the g.s. transitions are involved in the final nucleus of an exothermic reaction compared to its target nucleus [2, 3, 4] and thus $f_{\mathrm{endo}} > f_{\mathrm{exo}}$. When studying an astrophysically relevant reaction in the laboratory, $f$ should be as close as possible to Unity. Therefore, it seems advantageous to study exothermic reactions.

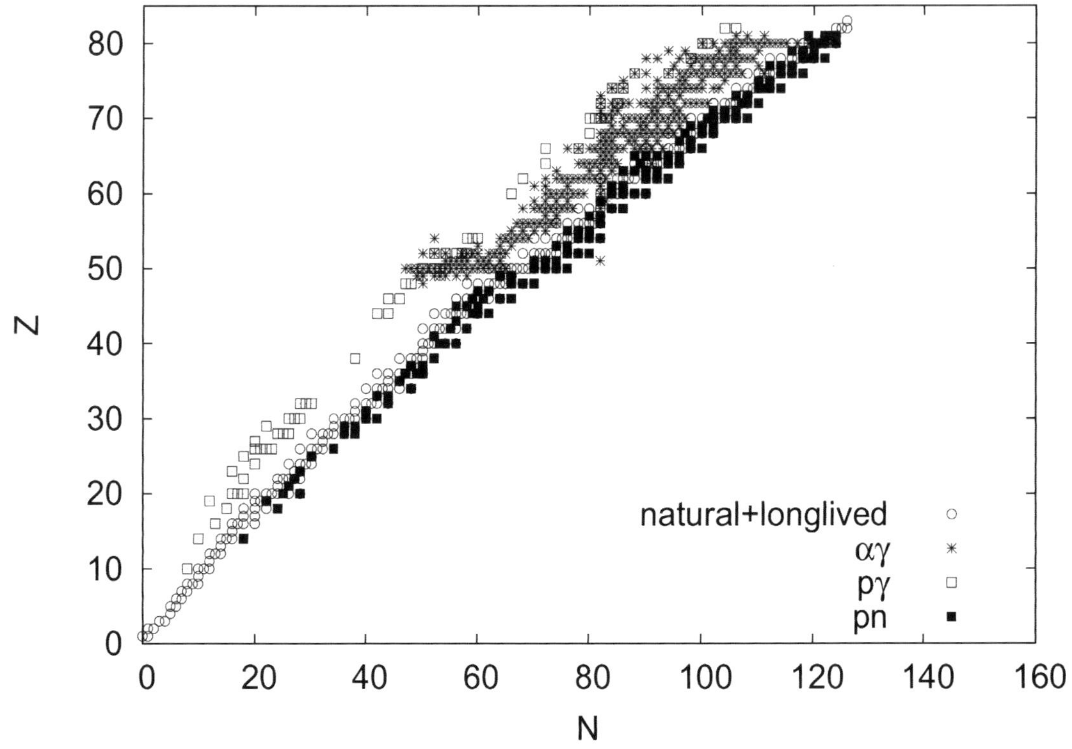


**Figure 1.** Targets for selected reactions with $f_{\mathrm{endo}} < f_{\mathrm{exo}}$. Natural and longlived nuclides are shown to guide the eye.

Closer inspections reveals [3, 4] that the above rule is not appropriate in all cases. If transitions from excited state are systematically suppressed, there can be cases for which $f_{\mathrm{endo}} < f_{\mathrm{exo}}$ although the endothermic reaction in principle may allow more transitions from excited states. Charged particle compound reactions provide such a suppression mechanism. With increasing excitation energy, the relative energy of the transitions between excited states and the compound state is decreasing. With a sufficiently high Coulomb barrier, the transitions from excited states are efficiently suppressed. We found [3, 4] that the stellar enhancement factor of the endothermic direction $f_{\mathrm{endo}}$ is lower than the one for the exothermic direction $f_{\mathrm{exo}}$ for more than 1200 reactions. This includes some reactions at or close to stability relevant in the p-process and reactions on highly unstable nuclei in the r-, rp-, and $\nu$p-process paths. Examples are shown in Fig. 1, detailed lists of reactions are given in [4].

## 3. The low-energy optical potential for protons

Measuring low-energy cross sections for charged-particle reactions is problematic due to the Coulomb barrier causing the astrophysically relevant cross sections to be tiny. Even more problematic is the standard way to obtain information on optical potentials through elastic scattering experiments. The scattering cross section at low-energy becomes indistinguishable from Rutherford scattering. Accordingly, the database for reactions involving protons and $\alpha$s at astrophysically relevant energies is incomplete.

Proton capture rates at $2-4$ GK are important in the $\gamma$-process which is thought to produce the majority of p-nuclei. An important difference to cross sections at higher energies is in the

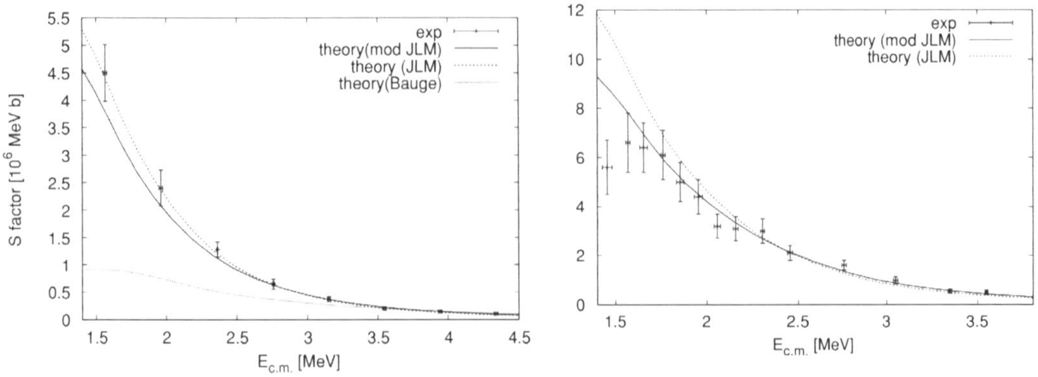


**Figure 2.** Astrophysical S-factors for (p,$\gamma$) reactions on $^{70}$Ge (left) and $^{74}$Se (right; all data are from [5]) compared with theoretical values obtained with different optical potentials: "standard" potential (JLM, [6, 7]), our new potential (mod JLM), and the potential by [9] (Bauge).

fact that the cross sections of the p-process are mostly sensitive to the particle widths instead of the $\gamma$-widths. This is due to the Coulomb barrier, reducing the charged particle width at low energy to values below those of the $\gamma$-width. A series of (p,$\gamma$) and (p,n) reactions was measured at $\gamma$-process energies recently (see, e.g., [5, 3] and references therein). The latter reactions are especially useful for testing the proton potential because the neutron width will (almost always) be larger than the proton width at all energies.

Several optical potentials were tested against the data and it was found that a modification of a widely used microscopic potential reproduces the data best. This is the potential by [6] with low-energy modifications by [7] (JLM) which were especially provided for applications in nuclear astrophysics. This microscopic potential is derived by applying the Brueckner-Hartree-Fock approximation with Reid's hard core nucleon-nucleon interaction and adopting a local density approximation.

Despite of overall good agreement, systematic deviations at low energy are still found (see, e.g., Fig. 2, more examples are shown in [5]). By variation of the optical potential we found that an increase by 70% in the strength of the imaginary part considerably improved the reproduction of the data in Hauser-Feshbach calculations (denoted by "mod JLM" in the figures). This increased absorption is allowed within the previous parameterization because the isoscalar and especially the isovector component of the imaginary part is not well constrained at low energies, as also noted by [6, 7, 8, 9, 10].

Also shown are results obtained with another recent, Lane-consistent new parameterization of the JLM potential (Bauge [9]). It yields worse agreement at low energy but it does neither include the additional modifications of [7], nor can it constrain the imaginary isovector part well at low energy [8, 9, 10] because it was fitted to data at higher energy.

Simultaneously with the (p,$\gamma$) data, also (p,n) data is described well by our newly modified potential, as shown in [5]. It also works well when comparing to even more recent data (see Fig. 3) which was derived independently [3, 4]).

Required input to the calculation of the optical potential is the nuclear density distribution. Fig. 3 also shows the dependence of the results when employing a droplet model density [11] and one from an Hartree-Fock-Bogolyubov model [12]. For the reactions considered here, the droplet description yields better agreement to the data in both absolute scale and energy dependence of the theoretical S-factor. Therefore, all other shown results make use of this description if not explicitly mentioned otherwise. For comparison, Fig. 3 also shows the results when employing the optical potentials of [8, 9] with both densities. In the original work, HFB densities were

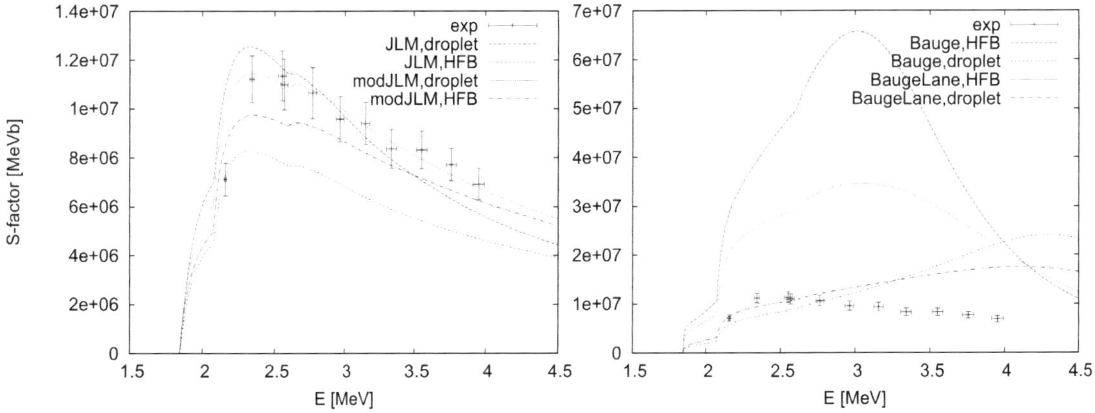


**Figure 3.** Astrophysical S-factors of $^{85}$Rb(p,n)$^{85}$Sr (exp. data from [3]) compared with theory using different optical potentials and nuclear densities. Shown are results with nuclear density from a droplet model [11] and from a HFB model with Skyrme interaction [12], applied in the calculation of the "standard" potential [6, 7] and our new modified version of this potential (left panel) as well as the potential of [9] (Bauge, right panel).

employed [8, 9].

Summarizing, it was found that an improved reproduction of low-energy proton-induced data can be achieved by utilizing the potential of [6, 7] with a 70% increase in the strength of the imaginary part. This is not to say that the potential of [6, 7] should generally be changed but rather that there is room for a possible energy-dependent modification of the imaginary potential strength, acting at low proton energies. This can either be achieved by adapting (refitting) the parameterization of [6] or by externally applying an enhancement factor of the imaginary part which decreases with increasing energy.

## References

[1] Fowler W A 1974 *Quart. J. Roy. Astron. Soc.* **15** 82
[2] Holmes J A, Woosley S E, Fowler W A and Zimmerman B A 1976 *At. Data Nucl. Data Tables* **18** 305
[3] Kiss G G, Rauscher T, Gyürky Gy, Simon A, Fülöp Zs. and Somorjai E 2008 *Phys. Rev. Lett.* **101** 191101
[4] Rauscher T, Kiss G G, Gyürky Gy, Simon A, Fülöp Zs. and Somorjai E 2009 *Phys. Rev. C* submitted
[5] Kiss G G, Gyürky Gy, Elekes Z, Fülöp Zs, Somorjai E, Rauscher T and Wiescher M 2007 *Phys. Rev. C* **76** 055807
[6] Jeukenne J P, Lejeune A and Mahaux C 1977 *Phys. Rev. C* **16** 80
[7] Lejeune A 1980 *Phys. Rev. C* **21** 1107
[8] Bauge E, Delaroche J P and Girod M 1998 *Phys. Rev. C* **58** 1118
[9] Bauge E, Delaroche J P and Girod M 2001 *Phys. Rev. C* **63** 024607
[10] Goriely S and Delaroche J P 2007 *Phys. Lett.* **B 653** 178
[11] von Groote H, Hilf E R and Takahashi K 1976 *At. Data Nucl. Data Tables* **17** 418
[12] Samyn M, Goriely S and Pearson J M 2003 *Nucl. Phys.* **A718** 653

Journal of Physics: Conference Series **202** (2010) 012014 doi:10.1088/1742-6596/202/1/012014

# Photoactivation of the p-nucleus $^{92}$Mo with bremsstrahlung at ELBE

**M Erhard[1,3], E Grosse[1,2], A R Junghans[1], J Klug[1,4], C Nair[1], G Rusev[1,5], K D Schilling[1], R Schwengner[1] and A Wagner[1]**

E-mail: erhard@pd.infn.it

[1] Institut f. Strahlenphysik, Forschungszentrum Dresden-Rossendorf, 01314 Dresden, Germany

[2] Institut f. Kern- u. Teilchenphysik, Technische Universität Dresden, 01062 Dresden, Germany

**Abstract.** In nuclear network calculations especially the p-nucleus $^{92}$Mo is frequently underproduced. Since experimental data of the photodisintegration of $^{92}$Mo so far do not exist, it was necessary to measure the reaction yields with the photoactivation method using the bremsstrahlung facility ELBE at FZ Dresden-Rossendorf. Also the reaction $^{100}$Mo($\gamma$,n) could be tested while irradiating Mo samples of natural composition. The photon fluence was measured with nuclear resonance fluorescence (NRF) of strong and well-known transitions in $^{11}$B. As a normalization standard the photoactivation reaction $^{197}$Au($\gamma$,n) was used. For the 65 s half-life of $^{91m}$Mo a fast pneumatic delivery was used. The photoactivation yields of the ($\gamma$,p) and ($\gamma$,n) reactions are compared to yield integrals of the bremsstrahlung spectral shape folded with the cross sections derived from Hauser-Feshbach nuclear model calculations using the TALYS program and cross sections from earlier experiments.

## 1. Introduction

There are 35 proton-rich nuclei heavier than iron that cannot be produced by slow or rapid neutron capture since they are blocked by stable isotopes. Originally they were thought to be produced by proton capture but due to the high Coulomb barrier there is no site were temperatures are high enough. In asymptotic giant branch (AGB) stars on the other hand, with temperatures of 2 to 3 GK there are enough photons with energies above the separation energies that can cause photodisintegration reactions. Astrophysical network calculations try to predict elemental and isotopic abundances [1]. Especially the lightest Mo and Ru isotopes are inherently underproduced in p-process networks. Most of the photodisintegration cross sections are not yet measured at astrophysically relevant energies. Therefore network calculations have to resort to Hauser-Feshbach model calculations which may be imprecise.

## 2. Experimental setup and analysis

Photodisintegration reactions were investigated using the photoactivation method: at the superconducting Electron Linear accelerator with high Brilliance and low Emittance (ELBE) targets of natural isotopic composition are irradiated with intense bremsstrahlung of up to $10^9$ (MeV cm$^2$ s)$^{-1}$ and endpoint energies up to 20 MeV. Afterwards the decay of the radioactive isotopes is measured by $\gamma$-ray spectroscopy with a well-shielded high purity (HP)Ge detector.

[3] Pres. add.: Istituto Nazionale di Fisica Nucleare, Sezione di Padova, Via Marzolo 8, 35131 Padova (PD), Italy

[4] Pres. add.: Vattenfall Power Consultant AB, Jämtlandsgatan 99, P.O. Box 527, 16216 Stockholm, Sweden

[5] Pres. add.: Dept. of Physics, Duke Univ., Triangle Universities Nuclear Laboratory, Durham, NC 27708, USA

The activation yield corresponds to the convolution of the bremsstrahlung spectral shape and the real photodisintegration cross section: $Y_{\rm act}(E_0) = \int_{S_x}^{E_0} \sigma_{\gamma,x}(E) \cdot \Phi_\gamma(E, E_0)\,{\rm d}E$. At ELBE photoactivation can be studied at two experimental sites:

(A) In bremsstrahlung produced [2] by the electron beam impinging on a 7 $\mu$m thin Nb radiator. The setup is located behind a 2.6 m long collimator in a separate cave optimized for photon scattering studies. The photon fluence distribution $N_\gamma(E_i)$ at photon energy $E_i$ scattered off $^{11}$B is measured with 4 HPGe detectors, each two of them at polar angles of 90° and 127°, respectively. To obtain the photon fluence from a certain transition, the number of counts in the photopeak has to be divided by the number of $^{11}$B target nuclei, the transition strength, the angular correlation coefficient, the full-energy peak efficiency, and the dead-time and pile-up correction factor. Also resonant self absorption in the $^{11}$B target was found to be important and was corrected. The absolute efficiency was simulated with the Monte-Carlo Code GEANT3 [3] and normalized with calibrated $\gamma$-ray sources.

(B) In the bremsstrahlung behind the thick graphite electron beam dump. Here the photon flux is 50 to 150 times higher. Since a direct flux measurement is not possible there, the experimental cross section of the $^{197}$Au($\gamma$,n) reaction [4] was used determine the the absolute value of the photon fluence. The reaction $^{197}$Au($\gamma$,n) has been investigated as a standard for photoactivation, see [5] and references therein. Mo targets have been activated at both sites with different bremsstrahlung spectral shape to reduce systematic uncertainties. At the target site A only ($\gamma$,n) and ($\gamma$,p) reactions with higher cross sections and half-lives $T_{1/2} \gtrsim 10\,$min were accessible, while at the site B all photodisintegration reactions of $^{92}$Mo including ($\gamma$,$\alpha$) have been investigated. The reaction $^{92}$Mo($\gamma$,n)$^{91{\rm m}}$Mo ($T_{1/2} = 65\,$s) was accessible by a fast pneumatic delivery [6]. The bremsstrahlung distribution was simulated with the Monte-Carlo N-Particle code MCNP4C2 [7] in realistic geometry for the thin target radiator (site A) and for the thick graphite target (B) [8].

The endpoint energy of the bremsstrahlung $E_{\gamma,\rm max} \approx 2 \cdot E_{\rm p} + S_{\rm p,n}({\rm D})$ was determined using the deuteron break-up reaction D($\gamma$,p)n. It corresponds approx. to the deuteron separation energy $S_{\rm p,n}$(D) plus two times the kinetic energy of the proton $E_{\rm p}$, measured with 4 Si detectors perpendicular to the beam and at 45° to the normal of a 40 $\mu$m thick foil of deuterated polyethylene $[{\rm CD}_2]_{\rm n}$. The proton spectra were Monte-Carlo simulated taking into account the experimental geometry, reaction kinematics and energy loss inside the target foil. By comparison of the measured spectra with the simulated spectra the endpoint energy was determined [5]. The total error of the endpoint energy amounts to $\Delta E_{\rm p} \approx 40\,$keV.

The photoactivation yield is calculated by the number of counts in the photopeak of the decay measurement by taking into account the number of target nuclei, the photon emission probability, the full-energy peak efficiency of the setup with well-shielded HPGe detectors using GEANT3 simulations normalized by measurements with calibrated sources, corrections for the decay during and in between irradiation and measurement, and eventually summing corrections as well as dead-time and pile-up corrections. The detailed analysis is described in [5].

## 3. Results and Conclusion

At ELBE we measured for the first time not only the $^{92}$Mo($\gamma$,n), but also the ($\gamma$,p) reaction at astrophysically relevant energies. In fig. 1 the activation yields of the photodisintegration reactions on $^{92}$Mo are displayed for irradiation site A and B. They are compared with the yield integrals of H-F model calculations done with the TALYS code based on different nuclear models. The results confirm essentially H-F model calculations of the photodisintegration reactions ($\gamma$,n) and ($\gamma$,p). The underprediction of $^{92}$Mo cannot be explained by imprecise cross sections. The agreement of the measured activation yield for $^{92}$Mo($\gamma$,p)$^{91{\rm m}}$Nb with a simulated yield integral using thin and thick target bremsstrahlung spectra as described above, shows that systematic uncertainties related to the bremsstrahlung photon fluence are small.

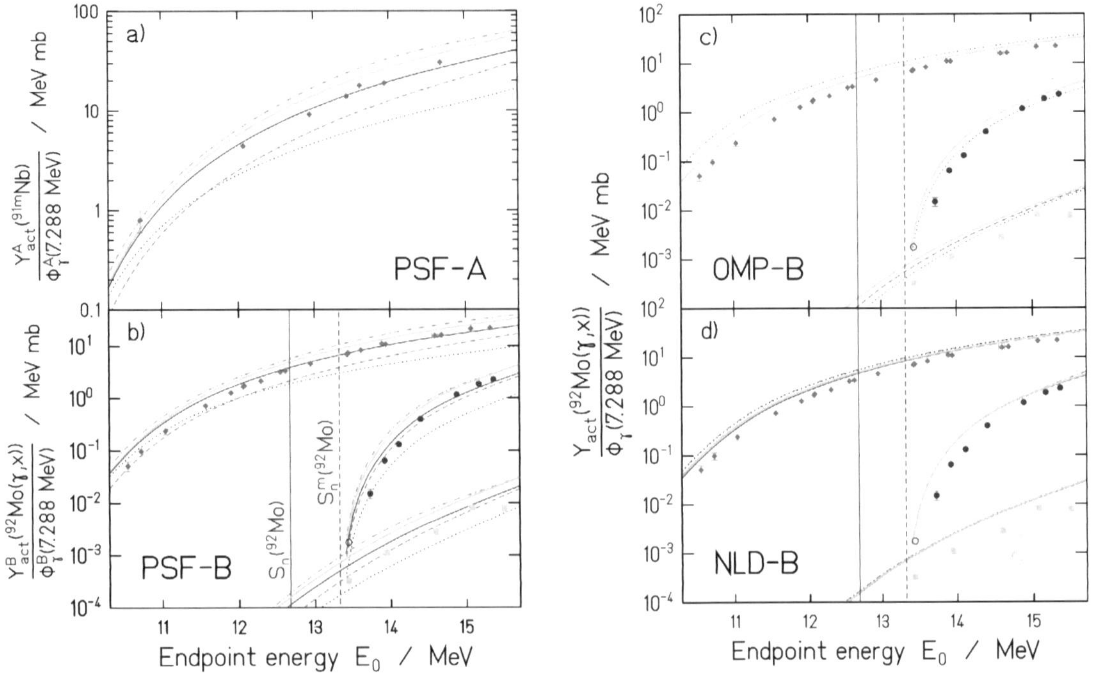


**Figure 1.** Photoactivation yields of $^{92}$Mo($\gamma$,p) are shown as red diamonds, ($\gamma$,n) as blue points, and ($\gamma$,$\alpha$) as green squares, normalized to the photon fluence at 7.3 MeV. At activation site A (panel a)) the photon fluence is normalized with photons scattered off $^{11}$B and at the irradiation site B (panel b) to d)) with the activation yield and yield integral of $^{197}$Au($\gamma$,n). The experimental data are compared with yield integrals based on different nuclear models calculated with TALYS [9]. In the two left panels the effect of different photon strength functions (PSF) is shown (details see text). Different optical model potentials (OMP) are displayed in panel b) and the variation of nuclear level density (NLD) models is given in panel d).

The PSF shows the biggest impact on the yield integral curve. The thin blue solid curves use a standardized Lorentzian with a single resonance for the shape of the giant dipole resonance (GDR) with parameters given by Dietrich and Berman [10]. A novel parametrization of the PSF [11, 6] describes photon scattering data, photoneutron data and averaged resonance neutron capture data (ARC) quite well for all nuclei above $A > 80$. There the PSF is parametrized as function of the ground state deformation parameters taking into account triaxial deformation with up to three Lorentzians were determined. The corresponding yields are shown here as thick green solid lines. The PSF by Kopecky and Uhl [12] is based on a generalized Lorentzian with a $\gamma$-ray energy dependent damping correction (red dashed line). The PSF derived from the Hartree-Fock-BCS model (black dotted curve) and the Hartree-Fock-Bogolyubov model [13] (magenta dash-dotted line) also available in TALYS do not describe our experimental yields very well.

TALYS results using different OMPs do not deviate very much by each other, as can be seen in fig. 1, panel c). The choice of optical model should influence the transmission coefficients for neutron and charged particle emission. TALYS uses by default a phenomenological parametrization of the spherical potential in the optical model of Koning and Delaroche [14] (thick green solid line) which can be adjusted with the $\alpha$-OMP by McFadden and Satchler [15] (red dashed curve). As an alternative the semimicroscopical OMP of Jeukenne, Lejeune, and Mahaux [16] is implemented in TALYS (blue dotted curve). Especially below the neutron separation energy the standard OMP fits best. Five different models of the NLD can be chosen in TALYS and

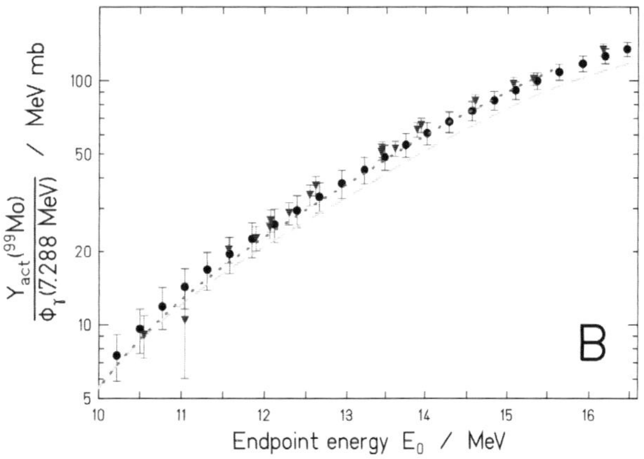


**Figure 2.** Photoactivation yields of $^{100}$Mo($\gamma$,n) irradiated at the target site B, normalized to the photon fluence at 7.3 MeV (blue triangles), compared with yield integrals of photonuclear data of Beil *et al* [18] (black points) corrected by a factor of 0.85 suggested by Berman *et al* [19], and HF-models calculated with TALYS (green dashed line) and NON-SMOKER [20] (red dotted curve).

are described in detail in the article of Koning *et al* [17] (and references therein). By default TALYS uses the constant-temperature model at low energies and the Fermi-Gas Model at high energies (thick green solid line). The deviations of the NLD in the yield curves from each other is less than a factor 1.5 over the measured energy region.

The activation yield of the $^{100}$Mo($\gamma$,n)$^{99}$Mo reaction could be compared with the yield integral of the photoneutron measurements of positron annihilation in flight from Saclay [18] (fig. 2). Applying a correction factor of 0.85, suggested by Berman *et al* [19], they are in excellent agreement with the ELBE data. Also shown are H-F model calculations with TALYS (GDR parametrization of [11]) and the photodisintegration cross section of the NON-SMOKER code [20] which is in very good agreement, too.

We thank P. Michel and the ELBE-crew for providing stable, high-intensity electron beams and A. Hartmann and W. Schulze for continuous valuable technical support.

## References

[1] Arnould A and Goriely S 2003 *Phys. Rep.* **384** 1–84
[2] Schwengner R *et al* 2005 *Nucl. Instr. Meth. Phys. Res.* A **555** 211–9
[3] 1993 *GEANT – Detector Description and Simulation Tool* CERN Program Library Long Writeup W5013, CERN Geneva, CH `http://wwwasd.web.cern.ch/wwwasd/geant/index.html`
[4] Veyssiere A, Beil H, Bergere R, Carlos P and Lepretre A 1970 *Nucl. Phys.* **A159** 561–76
[5] Nair C *et al* 2008 *Phys. Rev.* C **78** 055802
[6] Erhard M *et al* 2009 *submitted to Phys. Rev.* C
[7] 2000 *MCNP—a general Monte Carlo N-particle transport code* ed Briesmeister J F, Los Alamos National Laboratory Report LA13709
[8] Erhard M *et al* 2006 *Int. Symp. on Nuclear Astrophysics - Nuclei in the Cosmos - IX* CERN, Proc. of Science 028 PoS(NIC-IX)056
[9] Koning A J, Hilaire S and Duijvestijn M 2005 *TALYS: Comprehensive Nuclear Reaction Modeling, Proc. of the Int. Conf. on Nuclear Data for Science and Technology - ND2004. Sep. 26 – Oct. 1, 2004. Santa Fe, New Mexico (USA)* ed Haight R C *et al* AIP Conf. Proc. vol 769, 1154–9
[10] Dietrich S S and Berman B L 1988 *At. Data Nucl. Data Tables* **38** 199–338
[11] Junghans A R, Rusev G, Schwengner R, Wagner A and Grosse E 2008 *Phys. Lett.* B **670** 200–4
[12] Kopecky J and Uhl M 1990 *Phys. Rev.* C **41** 1941–55
[13] Belgya T *et al* 2006 *Handbook for calculations of nuclear reaction data, RIPL-2: Reference Input Parameter Library-2* (Vienna: IAEA) IAEA-TECDOC-1506 `http://www-nds.iaea.org/RIPL-2/`
[14] Koning A J and Delaroche J P 2003 *Nucl. Phys.* **A713** 231–310
[15] McFadden L and Satchler G R 1966 *Nucl. Phys.* **84** 177–200
[16] Jeukenne J-P, Lejeune A and Mahaux C 1977 *Phys. Rev.* C **16** 80–96
[17] Koning A J, Hilaire S and Goriely S 2008 *Nucl. Phys.* **A810** 13–76
[18] Beil H, Bergère R, Carlos P, Leprêtre A, De Miniac A and Veyssière A 1974 *Nucl. Phys.* **A227** 427–49
[19] Berman B L *et al* 1987 *Phys. Rev.* C **36** 1286–92
[20] Rauscher T and Thielemann F-K 2004 *At. Data Nucl. Data Tables* **88** 1–81

# S-factor measurement of the $^{13}C(p,\gamma)^{14}N$ reaction in reverse kinematics

**G. Genard[1], P. Descouvemont[2] and G. Terwagne[1]**

[1]Research Centre in Physics of Matter and Radiation (PMR), Laboratoire d'Analyses par Réactions Nucléaires (LARN), University of Namur (FUNDP), Rue de Bruxelles, 61, B-5000 Namur, Belgium

[2]Physique Nucléaire Théorique et Physique Mathématique, C.P. 229, Université Libre de Bruxelles, B-1050 Brussels, Belgium

email: ggenard@fundp.ac.be

**Abstract**. We measure the S-factor of the $^{13}C(p,\gamma)^{14}N$ reaction in reverse kinematics for energies ranging from 561 down to 225 keV with a low background experimental setup. The results are compared with previous measurements and an *R*-matrix treatment is applied to the data in order to obtain the properties of the 511 keV resonance that dominates the cross section at low energies.

## 1. Introduction

The aim of the nuclear astrophysics is to measure the cross section of the nuclear reactions occurring in the core of the stars, providing thereafter the reaction rates. This work has to be done in optimal conditions because of the very low intensities encountered. In particular, if one is interested in a radiative capture reaction, it consists mainly in decreasing the effect of all background sources that can interact in the photon detector. With this idea, we developed an experimental setup for low background measurements and decided to study proton induced reactions by working in reverse kinematics, suppressing therefore the usual beam induced noise [1]. The first one we studied is the $^{13}C(p,\gamma)^{14}N$ reaction that exhibits some astrophysical interest. Indeed, it appears in the CNO cycle and is also, in AGB stars, in competition with the $^{13}C(\alpha,n)^{16}O$ reaction, a well-known neutron source for the s-process. Its cross section at low energy is dominated by the 511 keV resonance. Thus, we have measured the de-excitation of this level to the ground state and derived the S-factor. An *R*-matrix treatment has been applied to the data and the results are compared with previous works.

## 2. Experimental details

The experimental setup presenting the low background conditions has been detailed in a previous paper [1]. Briefly, it consists in a lead castle to reduce the natural radioactivity effects on the high efficiency (133%) germanium gamma-ray detector. An anticoincidence is applied between this detector and a plastic scintillator, placed above the castle, in order to decrease the impact of cosmic rays on the gamma-ray spectra. The setup allows a decrease of 70% of the background interference and provides a sensitivity gain of approximately 200 times between 600 and 3000 keV. A $^{13}C$ beam was produced by the LARN accelerator, which allows the utilization of nearly all elements of the periodic table, given that we have worked in reverse kinematics. Moreover, the target used is a hydrogen standard carried out by ion implantation in silicon [2]. It contains about $9 \times 10^{16}$ $^{1}H/cm^{2}$, as determined with the hydrogen profiles obtained by use of the $^{1}H(^{15}N,\alpha\gamma)^{12}C$ resonant reaction.

We measured the S-factor for thirty different energies around the 511 keV resonance up to 561 keV and down to 225 keV, corresponding to energies from 3200 to 7900 keV in the lab system. The effective energy is obtained with the mean energy approximation inside the hydrogen target. Only the gamma-rays from the full-energy (FE) peak were used, being detected at 8.062 MeV (if we neglect the Doppler effect). The only background that has to be subtracted is due to the cosmic-rays. Indeed, thanks to the work in reverse kinematics, no gamma-ray line has appeared above the FE peak. All points were obtained with more than 1000 counts in the ROI, except the two at lower energy (573 and 288 counts). The detection efficiency was obtained by means of MCNPX simulations, corrected with the data of a $^{152}Eu$ source. The errors take into account the errors on the integrated charge, the number of hydrogen atoms, the detection efficiency and the energy, in addition to the statistical error.

### 3. Results and discussions

As mentioned before, we focused on the transition from the 511 keV resonant level ($1^-$) to the ground state ($1^+$) of the $^{13}C(p,\gamma)^{14}N$ reaction. Though, we measured the branching ratios at the resonance energy for the different transitions from the resonant level. The transition to the ground state has a branching ratio of 0.814. In the following, we assume that this ratio is constant, when it is used. This is an approximation, given the fact that the branching ratio is energy-dependent, as demonstrated in the work of King et al. where they measured all transitions [3]. The energy dependence indicates a decreasing branching ratio, from the resonance towards the low energies, which presents variations between 0.69 and 0.81.

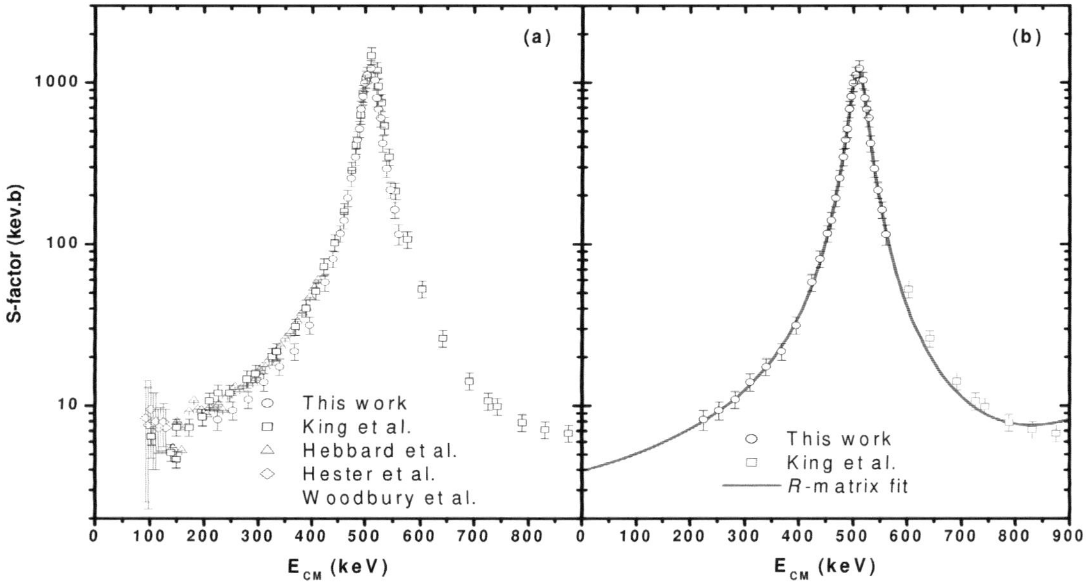


**Figure 1. (a)** S-factor for the $1^-$ to $1^+$ transition of the $^{13}C(p,\gamma)^{14}N$ reaction. The results of this work are compared with the data of King et al. [3], Hebbard et al. [4], Hester et al. [5] and Woodbury et al. [6].

**Figure 1. (b)** Extrapolation of the *R*-matrix fit (circles). Some points of King et al. [3] (squares) are also shown.

Figure 1 (a) presents the S-factor that we have obtained and we compare these results with previous measurements. Our data are in good agreement with those of King et al. [3] and of Hebbard et al. [4].

Nevertheless, it seems that they are a little bit lower in intensity, especially at low energy, but also above the resonance. This last point could be explained by a shift of the resonance towards the lower energies and a smaller width. Note that our error bars and those of King et al. (used in NACRE [7]) overlap for nearly all data points, while it is not the case with those of Hebbard et al. that are very small (statistical errors only).

We applied an *R*-matrix treatment to our data in order to derive the properties of the 511 keV resonance and an extrapolation of the S-factor at low energy. As reviewed by Lane and Tomas in 1958 [8], the philosophy of the *R*-matrix theory is to separate the space within two parts, defined by a sphere of radius $a$: an internal one where the nuclear interaction takes place and an external one where only the Coulomb interaction is taken into account. The radius $a$ is of the order of the radius of the colliding nuclei. The treatment consists in deriving an approximate solution of the Schrödinger equation by involving poles in the internal region. These poles are the bound states of the compound nucleus that we want to include in the description of the reaction. It is possible to use only the internal contribution if a strongly bound state is involved.

Figure 1 (b) presents the (extrapolated) fit obtained with $a=6.8\ fm$ and some points of King et al. [3] are added to show the evolution of the S-factor at higher energy but are not used in the fit. The treatment involved two poles, the 8.062 MeV level ($1^-$) and the 8.776 MeV level ($0^-$), with no external part. The parameters of the second level were used as fixed parameters, taken from Ajzenberg-Selove [9]. The fit to our experimental data is very good, providing a $\chi^2=0.14$. Moreover, the extrapolation at higher energy is pretty good, in comparison with the data of King et al, except above 800 keV where the contribution of the second resonance starts to dominate (the data in Figure 1 are only for the transition to the ground state).

| | $E_R$ (keV) | $\Gamma_R$ (keV) | $\omega\gamma$ (eV) |
|---|---|---|---|
| This work | 511.3±0.5 | 34.5±1.1 | 6.2±0.2 |
| Seagrave [10] | ~511 | 30.2±0.9 | ~8.6 |
| Hebbard [4] | 511±4 | 30±2 | \ |
| Milne [11] | ~511 | ~33 | \ |
| Ajzenberg-Selove [9] | 512±1 | 30±1 | 9.2±2.2 |
| King [3] | 517.8±0.5 | 40±1 | 8.8±1.1 |
| Galster [12] | 511.5±1.2 | 33.8±1.2 | 8.5±1.2 |

**Table 1.** Comparison of the results for the 511 keV resonance obtained with the *R*-matrix method ($a=6.8\ fm$) with those of the literature.

The properties of the 8.062 MeV level are compared to the literature in Table 1. In 1991, Galster et al. [10] published a review of existing values (Seagrave [11], Hebbard et al. [4], Milne [12] and Ajzenberg-Selove [9]) and provided some average values that, in addition, took into account measurements done in their work. The resonance energy is in very good agreement with the previous measurements, except with King et al., as already pointed out before. The resonance width is close to the one of Galster et al. The total resonance strength is too low. That one is evaluated from the resonance strength of the studied transition, taking into account the measured branching ratio. In fact, it is not clear if the resonance strength presented in the literature is the observed or the calculated one, as introduced by Lane and Thomas [8]. The one given in this work is the observed one, while the calculated one is about 20% higher, which could reduce the discrepancy.

Finally, we derived extrapolated S-factor values at low energy. For the transition to the ground state, we get S(0) = 3.94±0.59 keV.b which is lower than S(0) = 5.16±0.72 keV.b that Mukhamedzhanov et al. [13] obtained from a *R*-matrix fit to the data of King et al. [3]. However, they are consistent within the error bars. If we consider the total S-factor (with the approximation of the constant branching ratio) we have S(0) = 4.85±0.76 keV.b and S(25keV) = 5.12±0.80 keV.b which is still lower than S(0) = 7.60±1.10 keV.b and S(25keV) = 8.0±1.2 keV.b from Mukhamedzhanov et al. But our value is near the S(0) = 5.7±0.8 keV.b of Hebbard et al. [4].

## 4. Conclusions and perspectives

We measured the S-factor of the $^{13}C(p,\gamma)^{14}N$ reaction between 225 and 561 keV and fitted the data with the *R*-matrix technique. This measurement was made possible by a low background setup and the work in reverse kinematics. This technique provides very clean measurements and we suggest applying it in underground facilities, which could allow the study of more difficult reactions. In particular, the $^{15}N(p,\gamma)^{16}O$ reaction for which the recent S-factor measurement by Bemmerer et al. [14] presents relatively high error bars mainly because of the interaction between the proton beam and the boron contaminants.

## Acknowledgment

Gilles Genard and Pierre Descouvemont are supported as research fellow and research director, respectively, by the Belgian National Fund for Scientific Research (F.R.S.-FNRS). The targets used for the cross section measurement have been created within the SC-TEC-02 collaboration between the Wallonie-Bruxelles Community and the Québec government. In the context of this collaboration, we acknowledge Guy Ross and his team from INRS-EMT (Université du Québec).

## References

[1] Genard G, Nuttens V E, Bouchat V and Terwagne G 2009 *To be published in Nuclear Instruments and Methods in Physics Research B*
[2] Genard G, Yedji M, Ross G G and Terwagne G 2007 *Nucl. Instr. and Meth. B* **264** 156
[3] King J D, Azuma R E, Vise J B, Gorres J, Rolfs C, Trautvetter H P and Vlieks A E 1994 *Nucl. Phys. A* **567** 354
[4] Hebbard D F and Vogl J L 1960 *Nucl. Phys.* **21** 652
[5] Hester R E and Lamb W A S 1961 *Phys. Rev.* **121** 584
[6] Woodbury E J and Fowler W A 1952 *Phys. Rev.* **85** 51
[7] Angulo C et al. 1999 *Nucl. Phys. A* **656** 3
[8] Lane A M and Thomas R G 1958 *Rev. Mod. Phys.* **30** 257
[9] Ajzenberg-Selove F 1991 *Nucl. Phys. A* **523** 1
[10] Galster W et al. 1991 *Phys. Rev. C* **44** 2776
[11] Seagrave J D 1952 *Phys. Rev.* **85** 197
[12] Milne E A 1954 *Phys. Rev.* **93** 762
[13] Mukhamedzhanov A M, Azhari A, Burjan V, Gagliardi C A, Kroha V, Sattarov A, Tang X, Trache L and Tribble R E 2003 *Nucl. Phys. A* **725** 279
[14] Bemmerer D et al. 2009 *J. Phys. G* **36** 045202

# Stellar Helium Burning: Carbon and Oxygen Formation (Studied with Optical-TPC at HI$\gamma$S) [1]

**Moshe Gai**
**for the UConn-Yale-TUNL-Weizmann-PTB-UCL Collaboration**
LNS at Avery Point, 1084 Shennecossett Rd., University of Connecticut, Groton, CT 06340.
E-mail: <moshe.gai@yale.edu>

**Abstract.** An Optical Readout Time Projection Chamber (O-TPC) operating with the gas mixture of $CO_2(80\%)$ + $N_2(20\%)$ at 100 torr with gamma beams from the HI$\gamma$S facility of TUNL at Duke was used to study the formation of carbon and oxygen during helium burning. Measurements were carried out at beam energies: $E_\gamma$ = 9.51, 9.61, 9.72, 10.00, 10.54, 10.84 and 11.14 MeV. We have begun the process of extracting complete angular distributions for the $^{12}C(\alpha,\gamma)^{16}O$ reaction using energy bins of approximately 90 keV in order to determine the values of $S_{E1}$, $S_{E2}$ and the mixing phase $\phi_{12}$. The rate of carbon formation at high temperatures (T > 3 GK) may increase due to contributions from higher lying states. We searched for such contributions including a contribution from the predicted higher lying second $2^+$ state of $^{12}C$. No significant contributions from higher lying states including the known $1^-$ state at 10.84 MeV were found and specifically no evidence was found for the predicted second $2^+$ state of $^{12}C$.

## 1. Introduction

The outcome of helium burning is the formation of the two elements: carbon and oxygen [1]. The ratio of carbon to oxygen at the end of helium burning has been identified two decades ago as one of the key open questions in Nuclear Astrophysics [1] and it is still today an open question. To solve this problem one must extract the p-wave [$S_{E1}(300)$] and d-wave [$S_{E2}(300)$] cross section factors of the $^{12}C(\alpha,\gamma)^{16}O$ reaction at the Gamow peak (300 keV) with an accuracy of approximately 10% or better. Our current knowledge of these astrophysical cross section factors is not near the required accuracy.

Several new measurements of the $^{12}C(\alpha,\gamma)^{16}O$ reaction using gamma-ray detectors have been reported [2, 3] with energies in the vicinity of 1.0 MeV. However, the S-factors are measured with very low accuracy (±40-80%) and most importantly one cannot rule out a low value (close to 10 keV-b if not zero) of the E1 S-factor [4, 5]. These new experiments use some of the highest intensity alpha-particle beams (10 - 500 $\mu$Amp) with an impressive luminosity of $10^{33}$ [2] and $10^{31}$ $cm^{-2}sec^{-1}$ [3], and a $4\pi$ array of HPGe and BaF gamma-detectors, respectively, that provides large counting statistics. Yet the obtained accuracy of the S-factors is limited due to the limited accuracy of the measured angular distribution. A major disadvantage of measuring gamma rays is the large background. This large background is not expected in our proposed experiment using a Time Projection Chamber (TPC) and in addition we will measure complete

[1] Work Supported by the Yale-Weizmann Collaboration of the ACWIS, N.Y., and USDOE Grant No. DE-FG02-94ER40870, and DE-FG02-97ER41033.

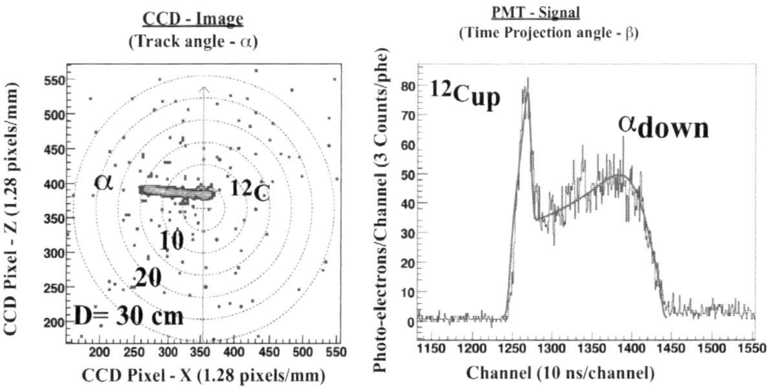


**Figure 1.** A typical $\alpha +^{12} C$ event registered by the CCD camera with the associated PMT signal and the predicted Time Projection line shape for the measured out of plane angle ($\beta$).

and detailed angular distributions at all angles (including the essential backward angles) and thus we anticipate a large sensitivity to the E2/E1 ratio and the mixing angle $\phi_{12}$.

## 2. Measurement of the $^{16}O(\gamma,\alpha)^{12}C$ Reaction With O-TPC.

We have used the upgraded optical readout system of the O-TPC, including the large area (142 mm diameter) design lens, to measure the $^{16}O(\gamma,\alpha)^{12}C$ reaction [6, 7, 8, 9]. Measurements were carried out at $E_\gamma$ = 9.512(10), 9.611(10) and 9.720(10) MeV with beam widths FWHM = 293, 285 and 294 keV, respectively. A typical $^{16}O$ dissociation event is shown in Fig. 1 and the measured spectra of the photodisintegration of oxygen nuclei are shown in Fig. 2. The detector resolution is considerably smaller than the beam width and we plan to bin the measured data in three bins of approximately 90 keV each. Preliminary angular distributions were measured and the data analysis is in progress to determine the values of: $S_{E1}$, $S_{E2}$ and the mixing phase $\phi_{12}$. We note that the rapid variation of $\phi_{12}$ across the $1^-$ resonance in $^{16}O$ which is required by unitarity, was not observed in [2, 3], suggesting a significant problem with gamma-ray measurements.

## 3. Measurement of the $^{12}C(\gamma,3\alpha)$ Reaction With O-TPC.

Carbon is formed during stellar helium burning in the "triple-alpha process", the $^8Be(\alpha,\gamma)^{12}C$ reaction, that is mostly governed by the $0^+$ state at 7.654 MeV. At high temperatures (T > 3 GK) higher lying states in $^{12}C$ may contribute. Indeed a broad ($\Gamma$ = 560 keV, $\Gamma_\gamma$ = 0.2 eV ) $2^+$ state at 9.11 MeV in $^{12}C$ was included in the NACRE compilation [10] following theoretical prediction for the $2^+$ member of the rotational band built on top of the $0^+$ (Hoyle) state at 7.654 MeV. It increases the production of carbon at temperatures in excess of 1 GK by up to a factor of 15. The non-existence of the $2^+$ member of the rotational band build on top of the Hoyle state would give credence to the conjectured alpha condensate and a spherical Hoyle state.

We used our Optical-Readout Time Projection Chamber (O-TPC) operating with $CO_2$ gas to search for such states via the identification of triple alpha events from the $^{12}C(\gamma,3\alpha)$ reaction as shown in Fig. 3. We have studied this reaction at E = 9.51, 9.61, 9.72, 10.00, 10.54, 10.84 and 11.14 MeV. The gamma width of the known $1^-$ state at 10.84 MeV was measured, as shown in Fig. 4, but no evidence was found for the predicted $2^+$ at 9.11 MeV, a shown in Fig. 4. Stringent upper limit on $\Gamma_\gamma$ of the predicted $2^+$ at 9.11 MeV as well as any other higher lying $2^+$ state in $^{12}C$ between 9.0 - 11.0 MeV were reported [11].

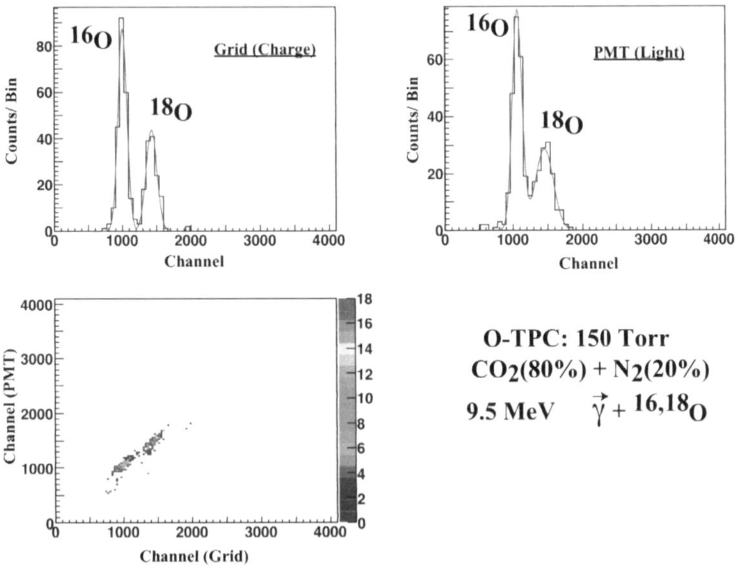


**Figure 2.** Typical oxygen spectra measured by the correlated grid (charge) and PMT (light) signals together with a two Gaussian fit.

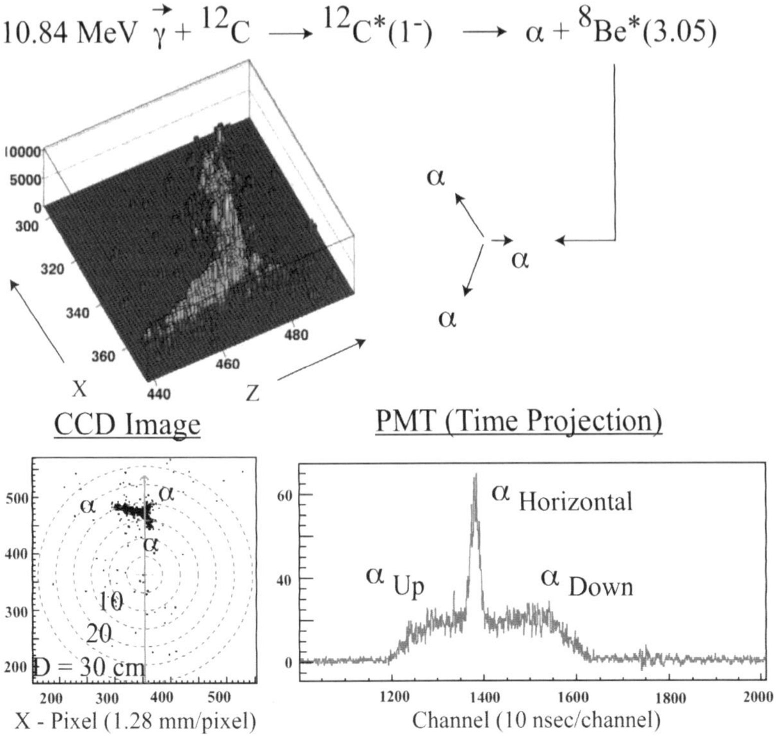


**Figure 3.** A typical three alpha event detected by the O-TPC.

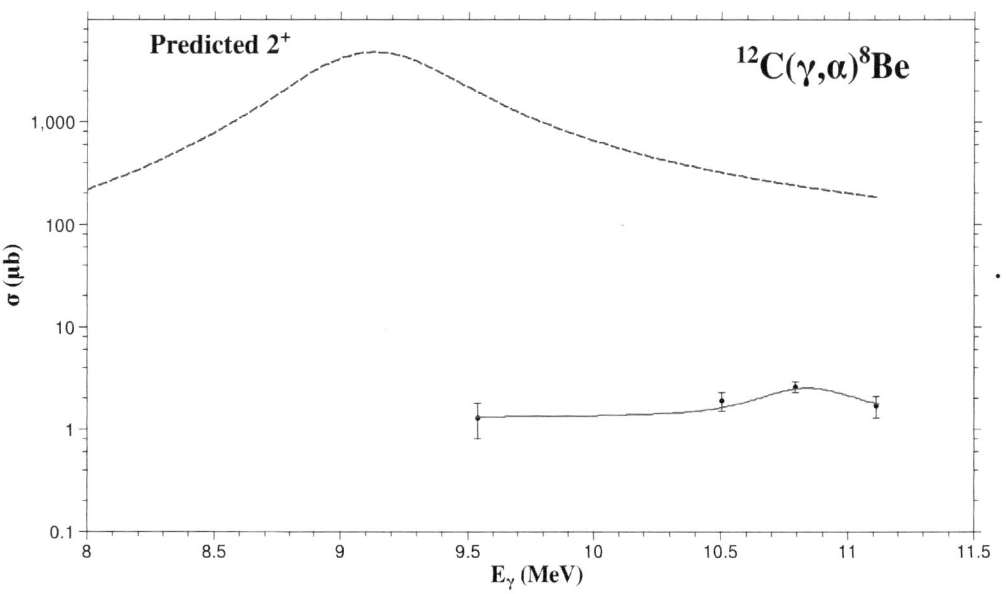


**Figure 4.** The measured cross section of the $^{12}C(\gamma, 3\alpha)$ reaction compared to the cross section for the predicted $2^+$ at 9.11 MeV. The line through the data point is the sum of the cross section due to the $1^-$ resonance at 10.84 MeV plus a constant background term.

## 4. References

[1] W.A. Fowler, Rev. Mod. Phys. 56(1984)149.
[2] M. Assuncao *et al.*; Phys. Rev. C**73**(2006)055801.
[3] R. Plag *et al.*; Nucl. Phys. **A758**(2005)415c.
[4] G.M. Hale; Nucl. Phys. **A621**(1997)177c.
[5] L. Gialanella *et al.*; Eur. Phys. J. **A11**(2001)357.
[6] M. Gai, A. Breskin, R. Chechik, V. Dangendorf and H.R. Weller; APH A, Heavy Ion Physics, **25**(2006)461; nucl-ex/0504003.
[7] L. Weissman, M. Gai, A. Breskin, R. Chechik, V. Dangendorf, K. Tittelmeier, H.R. Weller.; JINST **1**(2006)05002.
[8] H.R. Weller, M.W. Ahmed, H. Gao, W. Tornow, Y. Wu, M. Gai, R. Miskimen; Prog. Part. Nucl. Phys. **62**(2009)257.
[9] M. Gai, M.W. Ahmed, A. Breskin, B. Bromberger, R. Chechik, V. Dangendorf, Th. Delbar, R.H. France III, S.S. Henshaw, C.R. Howell, T.J. Kading, P.P. Martel, J.E.R. McDonald, P.-N. Seo, S.C. Stave, K. Tittelmeier, H.R. Weller, A.H. Young, and W.R. Zimmerman; to be submitted to JINST.
[10] C. Angulo *et al.*; Nucl. Phys. **A656**(1999)3.
[11] W.R. Zimmerman, P.-N. Seo, M. Gai, M.W. Ahmed, S.S. Hinshaw, C.R. Howell, S.C. Stave, H.R. Weller, P.P.Martel; Bull. Amer. Phys. Soc. **54#4**(2009)171.

Journal of Physics: Conference Series **202** (2010) 012016 doi:10.1088/1742-6596/202/1/012016

# Helium Burning: Carbon and Oxygen Formation Studied With an Optical Readout TPC (O-TPC) [1]

Moshe Gai; for the UConn-Yale-Duke-Weizmann-PTB-UCL collaboration.
LNS at Avery Point, University of Connecticut, Groton, CT 06340, USA;
<moshe.gai@yale.edu>

During stellar helium burning both oxygen and carbon are formed; the two elements essential for life as we know it. There is a renewed interest in carbon formation at high temperatures (above 1 GK) via the so called triple alpha process, the $^8Be(\alpha,\gamma)^{12}C$ reaction [1]. The formation of oxygen via the $^{12}C(\alpha,\gamma)^{16}O$ reaction that was declared by Willie Fowler twenty five years ago 'the holy grail' of Nuclear Astrophysics [2], is not yet well known, in spite of repeated strong statements in the literature. Estimates for $S_{E1}$ and $\frac{S_{E2}}{S_{E1}}$ vary in some cases by as much as a factor 8. Furthermore, the recent measurement of the beta-delayed alpha-particle emission of $^{16}N$ performed at Argonne [3] was found to be in disagreement with the TRIUMF data, but in agreement with the Yale-UConn data and the old Mainz data.

We have undertaken a completely new approach [4] to study the time reversed reactions with photon beams extracted from the HI$\gamma$S facility of TUNL at Duke University [5] with the target-detector being an optical readout Time Projection Chamber (O-TPC) operating with $CO_2(80\%)$ + $N_2(20\%)$ gas mixture [6]. Several in beam measurements with the O-TPC target/detector were already performed at $E_\gamma$ = 9.55, 10.54, 10.84, and 11.14 MeV to study the broad $1^-$ states in $^{16}O$ and $^{12}C$ at 9.58 and 10.84 MeV, respectively. The results of these measurements will be discussed.

1. H.O.U. Fynbo *et al.*, ISOLDE collaboration; Nature **433**(2005)136.
2. W.A. Fowler; Rev. Mod. Phys. **56**(1984)149.
3. X.D. Tang *et al.*; Phys. Rev. Lett. **99**(2007)052502.
4. M. Gai, A. Breskin, R. Chechik, V. Dangendorf and H.R. Weller; APH A, Heavy Ion Physics, **25**(2006)461; nucl-ex/0504003.
5. H.R. Weller, M.W. Ahmed, H. Gao, W. Tornow, Y.K. Wu, M. Gai, and R. Miskimen; Prog. Part. Nucl. Phys. (2008) in press, doi:10.1016/j.ppnp.2008.07.001.
6. L. Weissman, M. Gai, A. Breskin, R. Chechik, V. Dangendorf, K. Tittelmeier, H.R. Weller. JINST. **1**(2006)05002.

[1] Work Supported by the Yale-Weizmann Collaboration of the ACWIS, N.Y., and US-DOE Grant No. DE-FG02-94ER40870, and DE-FG02-97ER41033.

# Asymptotic Normalization Coefficient and Important Astrophysical Process $^{15}N(p,\gamma)^{16}O$

**A M Mukhamedzhanov[1], A Banu[1], P Bem[2], V Burjan[2], C A Gagliardi[1], V Z Goldberg[1], Z Hons[2], V Kroha[2], M La Cognata[3], Š Piskoř[2], R G Pizzone[3], S Romano[1], E Šimečková[2], C Spitaleri[3], L Trache[1] and R E Tribble[1]**

[1]Cyclotron Institute, Texas A&M University, College Station, Texas, 77843, USA
[2]Nuclear Physics Institute, Prague-Rez, Czech Republic
[3]INFN Laboratori Nazionali del Sud & DMFCI Universit'a di Catania, Catania, Italy

E-mail: akram@comp.tamu.edu

**Abstract.** In this work we report the application of the ANC method for the determination of the non-resonant radiative capture amplitude for the important astrophysical CNO cycle reaction $^{15}N(p,\gamma)^{16}O$, which provides a leak from the CN cycle into the CNO bi-cycle and CNO tri-cycle. It is contributed by the resonance capture to the ground state through two strong $1^-$ resonances and non-resonant capture to the ground state, which interferes with the resonant capture terms. To determine more accurately the contribution from the non-resonant capture we determined the proton ANCs for the ground and seven excited states of $^{16}O$ by measuring the angular distributions of the peripheral $^{15}N(^3He,d)^{16}O$ proton transfer reaction. Using these ANCs and proton and $\alpha$ resonance widths determined from an $R$-matrix fit to the data from the $^{15}N(p,\alpha)^{12}C$ reaction, we calculated the astrophysical $S$ factor for the $^{15}N(p,\gamma)^{16}O$ reaction. The results indicate that the direct capture contribution was previously overestimated. We find the astrophysical factor to be $S(0) = 36.0 \pm 6.0$ keVb, which is about a factor of two lower than the presently accepted value. We conclude that for every $2200 \pm 300$ cycles of the main CN cycle one CN catalyst is lost due to this reaction.

## 1. ANC in peripheral radiative capture and transfer reactions

It is well known that capture of a charged particle at stellar energies occurs at distances that are large compared to the nuclear radius. Direct radiative capture reaction rates depend on the normalization of the overlap function which is fixed by the appropriate ANCs. The connection between ANCs and direct proton capture rates at low energies is straightforward to obtain. The cross section for the direct capture reaction $A(p,\gamma)B$ can be written as

$$\sigma = \lambda \left| < I_{Ap}^{B}(r) \mid \hat{O}(r) \left| \psi_i^{(+)}(r) > \right|^2 , \right. \tag{1}$$

where $\lambda$ contains kinematical factors, $I_{Ap}^{B}(r)$ is the overlap function for $B \to A + p$, $\hat{O}$ is the electromagnetic transition operator and $\psi_i^{(+)}$ is the scattering wave in the incident channel. For peripheral reactions the overlap function may be replaced by

$$I_{Ap}^{B} \approx C_{Ap}^{B} \frac{W_{-\eta, l_B+1/2}(2\kappa r)}{r}, \tag{2}$$

where $C^B_{Ap}$, the ANC, defines the amplitude of the tail of the radial overlap function, $W_{-\eta,l+1/2}(2\kappa r)$ is the Whittaker function, $\eta$ the Coulomb parameter, $l_B$ orbital angular momentum for the bound state $B = (Ap)$ and $\kappa$ is the bound state wave number. Thus peripheral direct capture cross sections are directly proportional to the squares of the ANCs. In a similar way, ANCs can be used to determine reaction rates for subthreshold states [1, 2]. In addition, ANCs can be related to the external or channel part of the radiative width for resonant capture [3]. The internal part of the the radiative width, however, depends on the wave functions in the nuclear interior. If resonance parameters are known either from measurements or calculations and ANCs are known, the resonant and non-resonant components can be used together in an $R$-matrix calculation to obtain capture cross sections.

Peripheral transfer reactions provide an excellent way to determine ANCs. Consider the proton transfer reaction $A(a,b)B$ where $a = (bp)$ and $B = (Ap)$. As was previously shown [4] we can write the DWBA cross section in the form

$$\frac{d\sigma}{d\Omega} = \sum_{j_B\, j_a} \frac{(C^B_{Apl_Bj_B})^2 (C^a_{bpl_aj_a})^2}{b^2_{Apl_Bj_B}\, b^2_{bpl_aj_a}}\, \tilde{\sigma}^{DW}_{l_Bj_Bl_aj_a}, \tag{3}$$

where $\tilde{\sigma}^{DW}_{l_Bj_Bl_aj_a}$ is the reduced DWBA cross section and $j_i$, $l_i$ are the total and orbital angular momenta of the transferred proton in nucleus $i$. The factors $b_{bpl_aj_a}$ and $b_{Apl_Bj_B}$ are the single-particle ANCs of the bound state proton wave functions in nuclei a and B which are related to the corresponding ANC of the overlap function by

$$(C^a_{bpl_aj_a})^2 = S^a_{bpl_aj_a}\, b^2_{bpl_aj_a}, \tag{4}$$

where $S^a_{bpl_aj_a}$ is the spectroscopic factor. If the reaction is peripheral, the ratio

$$R_{l_Bj_Bl_aj_a} = \frac{\tilde{\sigma}^{DW}_{l_Bj_Bl_aj_a}}{b^2_{Apl_Bj_B}\, b^2_{bpl_aj_a}} \tag{5}$$

is independent of the single particle ANCs $b_{bpl_aj_a}$ and $b_{Apl_Bj_B}$. Thus for surface reactions where Eq. (5) holds, the DWBA cross section is best parametrized in terms of the product of the square of the ANCs of the initial and final nuclei $(C^B_{Apl_Bj_B})^2 (C^a_{bpl_aj_a})^2$. If the ANC $C^a_{bpl_aj_a}$ is known from the independent measurements the second ANC $C^B_{Apl_Bj_B}$ can be determined from the peripheral transfer reaction $A(a,b)B$.

In this paper we apply the ANC method to determine the astrophysical factor for the important astrophysical $^{15}\mathrm{N}(p,\gamma)^{16}\mathrm{O}$ reaction, which provides a path from the CN cycle to the CNO bi-cycle and CNO tri-cycle. The measured astrophysical factor for this reaction is dominated by resonant capture through two strong $J^\pi = 1^-$ resonances at $E_R = 312$ and 962 keV and direct capture to the ground state [5, 6]. This reaction can be analyzed using the $R$-matrix method in which the reaction amplitude is written as the sum of the resonant and non-resonant ones. The contribution from the nuclear interior in the $R$-matrix approach is entirely included into the resonant amplitudes, so that the non-resonant amplitude takes into account only the external contribution. That is why the normalization of the non-resonant amplitude in the $R$-matrix approach for the reaction under consideration is determined by the ANC for $^{15}\mathrm{N} + p \to {}^{16}\mathrm{O}$. In [6] the normalization of the non-resonant amplitude was varied to fit the experimental data. In order to better understand the non-resonant capture part of the low-energy astrophysical factor for $^{15}\mathrm{N}(p,\gamma)^{16}\mathrm{O}$, we have measured the ANCs for proton removal from the ground state and 7 excited states of $^{16}\mathrm{O}$ using the $^{15}\mathrm{N}(^3\mathrm{He,d})^{16}\mathrm{O}$ proton transfer reaction at incident energies of around 10 MeV/A. Here we report a measurement of the ANCs using this reaction. The ANCs are used to determine, within the $R$-matrix approach, the direct capture astrophysical $S$ factors to the corresponding eight bound states and calculate the total astrophysical factor.

## 2. ANC for $^{15}\mathrm{N}+p\rightarrow{}^{16}\mathrm{O}$ from the $^{15}\mathrm{N}(^{3}\mathrm{He},\mathrm{d})^{16}\mathrm{O}$ reaction

The experiment was performed using a momentum analyzed 25.74 MeV $^{3}$He beam from the U-120M cyclotron at the Nuclear Physics Institute of Prague-Rez, CAS, incident on a nitrogen gas target. The target gas chamber contained nitrogen gas enriched to 99.99% $^{15}$N. Angular distributions were obtained for the $^{3}$He elastic scattering and fourteen deuteron groups from the $^{15}\mathrm{N}(^{3}\mathrm{He},\mathrm{d})^{16}\mathrm{O}$ reaction populating states below and above the particle emission threshold of $^{16}$O, but only eight deuteron groups corresponding to proton transfer to the bound states in $^{16}$O were further analyzed. From the optical model analysis of the measured elastic scattering in the entrance channel the optical potential describing the $^{3}\mathrm{He}+{}^{15}\mathrm{N}$ scattering was determined. Optical potentials for the exit channel were taken from Fulbright [7] or calculated from global formulas derived by Daehnick et al. [8].

The analysis of the transfer reaction has been done within the DWBA. By normalization of the calculated DWBA cross section to the experimental one the proton ANCs for the eight bound states of $^{16}$O were determined using the ANC for $^{3}\mathrm{He}\rightarrow d+p$ from [9]. The determined square of the ANC for the ground state is $192\pm26.0\ \mathrm{fm}^{-1}$. The transitions to the excited bound states are not important for astrophysical reactions and here we don't present the measured ANCs for the excited states.

## 3. Astrophysical factor for the $^{15}\mathrm{N}(p,\gamma)^{16}\mathrm{O}$ radiative capture process

We have calculated the $S$ factor for this reaction using the two-channel, two-level $R$- matrix method. The contribution from the $\alpha-{}^{12}\mathrm{C}$ channel is also taken into account. In the case under consideration, only two levels with $J^{\pi}=1^{-}$ produce strong energy dependence corresponding to two resonances at 312 and 962 keV. If the channel radius is large enough, then the levels calculated for the Woods-Saxon potential will be close to the single-particle shell model ones. The closest state to the subthreshold $1^{-}$ level is located at $E_x=9.585$ MeV, but its ANC (or reduced width) is so small that we were not able to observe it. Hence we neglect this level. The other level at $E_x=7.1169$ MeV is 5.06 MeV away from the threshold and also does not affect the behavior of the S factor at low energies. The levels located at higher energies can be included into the background, which has a very smooth energy dependence. The total reaction amplitude in the $R$-matrix approach for the capture to the ground state of $^{16}$O (in the case of the interfering resonant and non-resonant terms) is

$$M(E)=i\,\Omega_p(E)\sum_{\nu,\lambda}A_{\nu\,\lambda}(E)\,[\Gamma_{\nu\gamma}(E)]^{1/2}\,\tilde{\Gamma}_{\lambda p}^{1/2}(E)+M^{NR}(E). \tag{6}$$

Here, $\Omega_p(E)$ is the phase factor in the initial channel of the reaction, $\mathbf{A}(E)$ is the level matrix, $[\Gamma_{\nu\gamma}(E)]^{1/2}$ is the amplitude of the radiative width of the level $\nu$ decaying to the bound state, $\tilde{\Gamma}_{\lambda p}(E)=2\,P_l(E)\,\gamma_{\lambda p}$ is the formal proton width of the level $\lambda$, $\gamma_{\lambda p}$ its reduced width and $P_l(E)$ is the barrier penetrability for protons; $E$ is the relative energy in the channel $p-{}^{15}\mathrm{N}$ and $M^{NR}$ is the amplitude of the non-resonant radiative capture occuring in the external region. In the two-level approximation $\lambda,\nu=1,2$, the resonant amplitude in Eq. (6) contains four terms rather than two terms used in Ref. [6]. It constitutes one of the differences between our fit and fit in [6]. We note that in the $R$-matrix method the non-resonant capture matrix element in the internal region does not explicitly appear. This non-resonant part is given by the radial integral taken from the channel radius $r_0$ to $\infty$ and its absolute normalization is entirely determined by the ANC of the bound state. Note that the ANC and the reduced width used in the $R$-matrix approach are related by Eq. (60) of Ref. [1].

The resonant parameters in Eq. (6) and the channel radius are fitting parameters to reproduce the experimental data. We choose the channel radius in the proton channel to be $r_0=5.0$ fm. The proton and alpha reduced widths can be expressed in terms of the observable widths. These widths are available in compilations [10]. But they are determined from different reactions with different uncertainties. First we have determined the proton and alpha reduced widths by fitting the measured astrophysical factor for the $^{15}N(p,\alpha)^{12}C$ reaction in the $R$-matrix approach. In this fit the channel radii for the proton and alpha channels were taken to be 5.0 and 7.0, correspondingly. The search region for proton and alpha widths was originally taken from [10] and then extended for the alpha widths for the second resonance. Since the cross section for this reaction is significantly larger than for the $^{15}N(p,\gamma)^{16}O$ reaction, it has been measured with significantly higher accuracy. This allows us to determine the proton and alpha partial widths with higher accuracy. We used two different boundary conditions. First, we fixed the second level energy at $E_2 = E_{R_2} = 962$ keV, where $E_{R_2}$ is the energy of the second resonance, and determined from the fit to the experimental data that the first level energy is $E_1 = 152$ keV. It means that in the R matrix approach the two levels are separated by $\approx 800$ keV. The $\chi^2$ fit per degree of freedom is $\chi^2/N = 1.27$. If we adopt the first level energy to be $E_1 = E_{R_1} = 312$ keV, then the best fit is achieved for $E_2 = 1070$ keV with $\chi^2/N = 1.51$. The results of the fit are shown in Fig 1. After determining the particle reduced widths, we made the $R$-matrix fit

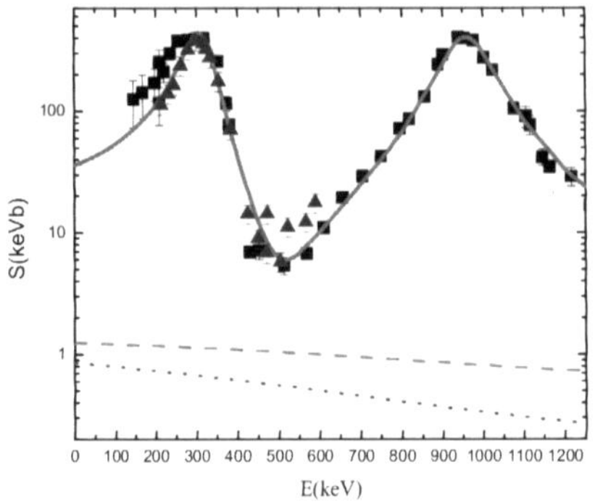


**Figure 1.** (Color online) The astrophysical $S(E)$ factor for the $^{15}N(p,\alpha)^{12}C$ reaction. The black squares and blue triangles are experimental data from [6] and [5], correspondingly. The red solid line is our $R$ matrix fit with the boundary condition $E_2 = 962$ keV, which takes into account two interfering $1^-$ resonances at $E_{R_1} = 312$ and $E_{R_2} = 962$ keV and non-resonant capture. The magenta dashed line is the total non-resonant capture $S$ factor given by the sum of the nonresonant capture $S$ factors to all eight bound states observed in the present $(^3He, d)$ experiment, and the violet dotted line is the non-resonant capture $S$ factor to the ground state.

of the $^{15}N(p,\gamma)^{16}O$ data. To achieve a better fit for this reaction, we slightly readjusted the reduced widths. Both boundary conditions gave slightly different reduced widths and similar fits to the data. The main fitting parameters in this case are the complex radiative width amplitudes, $[\Gamma_{1\gamma}]^{1/2}$ and $[\Gamma_{2\gamma}]^{1/2}$. The radiative width amplitude consists of the internal and external (channel) parts. The internal radiative width amplitude is real while the external one is complex since the resonant scattering wave function in the external region is described by the outgoing wave. We find that the imaginary parts of the radiative width amplitudes are important for the fit of the experimental data. The best fit is achieved if we assume that the sign of the real parts of both radiative width amplitudes is the same and opposite to the sign of their imaginary parts. At fixed particle reduced widths the radiative width amplitudes determine the absolute value of the astrophysical $S$ factor at the resonant peaks. According to [6] the $S$ factors at the resonance peaks are 400 keVb. The absolute cross section was determined in [6] by normalization of the $^{15}N(p,\alpha_1\gamma_1)^{12}C$ excitation function to the known cross section of

$\sigma = 250 \pm 35$ mb at $E_R = 1134.40$ keV. Then from the observed relative intensity of $\gamma$-rays from the $^{15}N(p, \alpha_1\gamma_1)^{12}C$ and $^{15}N(p, \gamma_0)^{16}O$ reactions, the capture cross section for the latter reaction was determined. To determine accurately the uncertainty of the S factors at the resonance peaks one needs to know information, such as the errors of the measured $\gamma$-ray intensities and the uncertainty in the detector efficiency, which are not given in Ref. [6]. Thus we believe that a 14% uncertainty for the S factors at the resonance peaks is too optimistic. We also assigned a 16% uncertainty to the S factors at the resonances from the $\approx 16\%$ uncertainties for the radiative widths for the first and second resonances given in Ref. [6]. We assigned a 16% uncertainty to the radiative widths of both resonances in our fit. The $\chi^2$ fit for the radiative capture is worse than for the $(p, \alpha)$ reaction because we are not able to reproduce the low-energy data from [6]. For the boundary condition $E_2 = E_{R_2} = 962$ keV we get $\chi^2/N = 2.8$ and for $E_1 = E_{R_1} = 312$ keV we get $\chi^2/N = 3.0$. We find that one of the alpha reduced widths (we cannot determine which one) has an opposite sign relative to three other particle reduced widths. In our fit we assumed that $\gamma_{1\alpha} < 0$. The level energies used in [5] are not given, which is why we cannot compare directly our parameters and the set from [5]. The observable widths of each resonance can be expressed in terms of the reduced widths when the corresponding level energy coincides with the resonance energy. From the R matrix fit with the first level energy $E_1 = 312$ keV we get the observable partial widths for the first resonance, $\Gamma_{1p} = 1.1$ keV and $\Gamma_{1\alpha} = 90.5$ keV. From the R matrix fit with the second level energy $E_2 = 962$ keV, we get the observable partial widths of the second resonance, $\Gamma_{2p} = 99.7$ keVb and $\Gamma_{2\alpha} = 50.4$ keV. The low energy tail of our calculated $S$ factor fitting the data from [5] goes along the lower limit of the data from [6]. Our astrophysical factor $S(0) = 36.0 \pm 6.0$ keVb obtained for the proton channel radius $r_0 = 5.0$ fm is lower then $S(0) = 64.0 \pm 6.0$ keVb [6] but agrees with $S(0) = 29.8 \pm 5.4$ keVb [5] within uncertainties. We note that the $R$-matrix calculations show very little sensitivity to the variation of the channel radius $r_0$. Decreasing (increasing) $r_0$ increases (decreases) the integration region of the radial matrix element for the non-resonant capture, i.e. increases (decreases) the direct capture amplitude and decreases (increases) the resonant part, so that the total sum remains nearly constant. A variation of the channel radius by 33% changes the $S(0)$ factor by 5%. The 17% uncertainty of our $S(0)$ astrophysical factor comes from the (assumed) 16% uncertainty of the experimental data, the 10% uncertainty of the ANCs, which results in about a 2% uncertainty in the $S(0)$ factor, and a 5% uncertainty due to the dependence of the $S$ factor on the channel radius.

The most important difference between our fit and the result in [6] is in the contribution of the non-resonant capture to the ground state. The absolute normalization of the non-resonant capture terms is entirely determined by the corresponding ANC in the $R$-matrix approach. Since we have measured the ANCs, we can quite accurately determine the contribution from the non-resonant capture terms. At zero energy it contributes about 3.0% to the total astrophysical factor. But the non-resonant capture to the ground state, which contributes about 69% to the total non-resonant $S$ factor, is much more important when calculating the total astrophysical factor due to its interference with the resonant capture terms. We find that our calculated non-resonant astrophysical factor for the capture to the ground state is $S(0) = 0.86$ keVb, i.e. about 9 times smaller than the $S(0)$ obtained in [6]. It is the main reason why our low-energy tail of the $S$ factor goes lower than the data in [6]. Since normalization of the non-resonant capture amplitude is determined by the ANC, which has been determined with 10% uncertainty, we do not agree with the result obtained in [6] within the hard sphere approach. Unfortunately, there is no explanation of the method used to calculate the non-resonant capture terms in [6]. We note that we also tried to fit the $^{15}N(p, \gamma)^{16}O$ data using the expression given in [6] and failed to reproduce the low-energy tail of the S factor for the same reason. Recent low-energy direct measurements by LUNA [12] are in a very good agreement with our $S$ factor prediction confirming the power of the indirect ANC method.

## 4. Summary and conclusion

To determine more accurately the contribution from the non-resonant capture to the astrophysical factor for the $^{15}N(p,\gamma)^{16}O$ CNO cycle process we determined the proton ANCs for the ground and seven excited states of $^{16}O$ by measuring the angular distributions of the peripheral $^{15}N(^{3}He,d)^{16}O$ proton transfer reaction. To fit the available experimental astrophysical factors from [6, 5] we have determined the resonant proton and $\alpha$ partial widths by fitting the available experimental data for the stronger and better measured reaction $^{15}N(p,\alpha)^{12}C$. Our astrophysical factor in the energy interval 150 − 300 keV agrees well with the data from [5] but goes slightly lower than the low limit of data reported in [6]. This new astrophysical factor, $S(0) = 36.0 \pm 6.0$ keVb, obtained for $^{15}N(p,\gamma_0)^{16}O$ allows us to reevaluate the rate of leak from the CN cycle due to this reaction. In Ref. [6] it has been estimated as the ratio of the $S(0)$ factors for $^{15}N(p,\alpha)^{12}C$ and $^{15}N(p,\gamma_0)^{16}O$. The $S$ factor $S(0) = 57$ MeVb was used for $^{15}N(p,\alpha)^{12}C$. However, the later measurements [11] gave the higher value of the astrophysical factor $S(0) = 65.0 \pm 4.0$ MeVb. Our new fit for the $^{15}N(p,\alpha)^{12}C$ data gives $S(0) = 73.0 \pm 5.0$ MeVb [13].

This result overlaps, within the experimental uncertainties, with the $S$ factor recently measured via the indirect Trojan Horse method [13]. Using the data from [11] we can re-evaluate the loss of catalyst in the CN cycle at 25 keV proton energy, which corresponds to the relative $p - {}^{15}N$ energy $E = 23.44$ keV. We find, using our new astrophysical factor $S(E = 23.44\text{keV}) = 38.8 \pm 6.6$ keVb for the $^{15}N(p,\gamma_0)^{16}O$ reaction and $84.1 \pm 5.9$ MeVb for the $^{15}N(p,\alpha)^{12}C$ reaction , that for every $2200 \pm 300$ cycles of the main CN cycle one CN catalyst is lost due to the $^{15}N(p,\gamma_0)^{16}O$ reaction, rather than $1200 \pm 100$ cycles determined from data of [6] ($S(E = 23.44\text{keV}) = 70.0 \pm 11.0$ keVb). Our result agrees with the leak rate $2600 \pm 400$ cycles obtained with $S(E = 23.44\text{keV}) = 32.0 \pm 5.8$ keVb for the $^{15}N(p,\gamma)^{16}O$ reaction in [5]. Our calculated reaction rates for $^{15}N(p,\gamma)^{16}O$ for temperatures $T_9 < 0.15$ are lower by approximately a factor of 2 than the adopted reaction rates given in the NACRE compilation [14], which were calculated using data from [6]. Extension of the LUNA measurements to the region covering the first resonance peak will allow us to update our $S$ factor fit.

## 5. Acknowledgments

This work was supported by the U. S. Department of Energy under Grant No. DE-FG02-93ER40773 and DE-FG52-06NA26207, the Robert A. Welch Foundation under Grant No. A-1082, NSF under Grant No. PHY-0852653.

## References

[1] Mukhamedzhanov A M and Tribble R E (1999) *Phys. Rev.* **C 59** 3418
[2] A. M. Mukhamedzhanov *et al.* 2003 *Phys. Rev.* **C 67** 065804.
[3] Banu A *et al.* 2009 *Phys. Rev.* **C 79** 025805.
[4] Mukhamedzhanov A M 1997 *Phys. Rev.* **C 56** 1302.
[5] Hebbard D F 1960 *Nucl. Phys.* **15** 289.
[6] Rolfs C and Rodney D F 1974 *Nucl. Phys.* **A235** 450.
[7] Fulbright H W 1969 *Phys. Rev.* **184** 1068.
[8] Daehnick W W 1980 *Phys. Rev.* **C 21** 2253.
[9] Mukhamedzhanov A M, Tribble R E and Timofeyuk N K (1995) *Phys. Rev.* **C 51** 3472.
[10] Tilley D R, Weller H R and Cheves C M 1993 *Nucl. Phys.* **A565** 1.
[11] Redder A *et al.* 1982 *Z. Phys.* **A305** 325.
[12] D Bemmerer *et al.* 2009 *J. Phys. G: Nucl. Part. Phys* **36** 045202
[13] La Cognata M *et al.* 2009 *Phys. Rev.* **C 80** 012801(R).
[14] Angulo C *et al.* 1999 *Nucl. Phys.* **A656** 3.

Journal of Physics: Conference Series **202** (2010) 012018 doi:10.1088/1742-6596/202/1/012018

# Trojan Horse Method: a tool to explore electron screening effect

**R.G. Pizzone, C. Spitaleri, S. Cherubini, M. La Cognata, L. Lamia, S. Romano, M.L. Sergi**

Laboratori Nazionali del Sud-INFN, Catania, Italy & Dipartimento di Metodologie Fisiche e Chimiche per l'Ingegneria, Università di Catania, Catania, Italy

**C. Rolfs, F. Strieder**

Ruhr Universität Bochum, Germany

**V. Burjan, V. Kroha, J. Mrazek**

Cyclotron Institute, Academy of Science, Rez, Czech Rep.

**C. Li, Q. Wen, S. Zhou**

CIAE, Beijing, China

**A. Tumino**

Universitá Kore, Enna, Italy

E-mail: rgpizzone@lns.infn.it

**Abstract.** Owing the presence of the Coulomb barrier at astrophysically relevant energies, it is very difficult, or sometimes impossible to measure reaction rates for charged particle induced reactions. Moreover due to the presence of the electron screening effect in direct measurements, the relevant nuclear input for astrophysics, i.e. the bare nucleus S(E)-factor, can hardly be extracted. This is why different indirect techniques are being used along with direct measurements. The THM is an unique indirect technique which allows one to measure reactions cross sections of astrophysical interest down the thermal energies typical of the different scenarios. The basic principle and a review of the main applications of the Trojan Horse Method are given. The applications aiming at the extraction of the bare $S_b(E)$ astrophysical factor and electron screening potentials $U_e$ for several two body processes are discussed.

## 1. Introduction

Nuclear fusion reactions, that take place in the hot interiors of remote and long-vanished stars over billions of years, are the origin in the universe of nearly all the chemical elements and their isotopes [7, 11, 21]. The detailed understanding of the origin of the chemical elements and their isotopes has combined astrophysics and nuclear physics, and forms what is called nuclear astrophysics. In turn, nuclear reactions are the heart of nuclear astrophysics: they influence sensitively the nucleosynthesis of the elements in the earliest stages of the universe and in all the objects formed thereafter, and control the associated energy generation (by processes called nuclear fusion or nuclear burning), neutrino luminosity, and evolution of stars. A good knowledge of the rates of these fusion reactions is essential for understanding this broad picture [29]. Moreover only understanding how the electron screening goes in the laboratory will provide knowledge for astrophysical application as well as fusion reactor physics.

## 2. Limit of the two-body measurements

In a stellar plasma the constituent nuclei are usually in thermal equilibrium at some local temperature T. Occasionally they collide with other nuclei, whereby two different nuclei can emerge from collision A+x→ c+C. The cross section σ(E) of nuclear fusion reaction A(x,c)C is of course governed by the laws of quantum mechanics where, in most cases, the Coulomb and centrifugal barriers arising from nuclear charges and angular momenta in the entrance channel of the reaction strongly inhibit the penetration of one nucleus into another. This barrier penetration leads a steep energy dependence of the cross section. It is the challenge for the experimentalist to make precise σ(E) measurements over a wide range of energies, as our fragmented knowledge of nuclear physics prevents us from predicting σ(E) on purely theoretical grounds. For these reasons bare nucleus cross section measurements $\sigma_{pl}(E)$(for stellar plasma) of the $(p,\alpha)$ reaction at the Gamow energy $(E_G)$ should be known with an accuracy better than 10% [21, 22] because of their crucial role in understanding the first phases of the Universe history and the subsequent stellar evolution. Unfortunately the presence of Coulomb barrier, in the reactions with charge particles, is a limit, often insuperable, to perform measurements of the cross sections at ultralow energies. Indeed the Coulomb barrier of height $E_C$ in charged-particle induced reactions cause an exponential decrease of the cross section $\sigma_b(E)$ at $E < E_C$, $\sigma_b(E) \sim exp(-2\pi\eta)$, leading to a low-energy limit of direct $\sigma_b(E)$ measurements. Owing to the strong Coulomb suppression, the behavior of the cross section at $E_G$ is usually extrapolated from the higher energies by using the definition of the smoother astrophysical factor S(E):

$$S_b(E) = E\sigma_b(E)exp(2\pi\eta) \tag{1}$$

where $exp(2\pi\eta)$ is the inverse of the Gamow factor, which removes the dominant energy dependence of $\sigma(E)$ due to the barrier penetrability.
Although the $S_b$(E)-factor allows for an easier extrapolation, large uncertainties to $\sigma_b(E_G)$ may be introduced due to for instance the presence of unexpected resonances, or high energy tails of sub-threshold resonances. In order to avoid the extrapolation procedure, a number of experimental solutions were proposed in direct measurements for enhancing the signal-to-noise ratio at $E_G$.
In recent years the availability of hight-current low-energy accelerators, such as that at the underground Laboratories together with improved target and detection techniques have allowed us to perform $\sigma_b(E)$ measurements in some cases down to $E_G$ or at least close to $E_G$ [6]. Then in principle no $\sigma_b(E)$ extrapolation would be needed anymore for these reactions. However, the measurements in laboratory at ultralow energies suffer from the complication due to the effects of electron screening [1, 29]. This leads to an exponential increase of the laboratory measured cross section $\sigma_s(E)$ [or equivalently of the astrophysical factor $S_s(E)$] with decreasing energy relative to the case of bare nuclei. This can be described by an enhancement factor defined by the relation

$$f_{lab}(E) = \sigma_s(E)/\sigma_b(E) \approx exp(\pi\eta U_e/E) \tag{2}$$

In this equation $U_e$ is the electron screening potential in the laboratory which is different from the $U_{pl}$ present in the stellar environment. Clearly, a good understanding of $U_e$ is needed in order to calculate $\sigma_b$ from the experimental data $\sigma_s$ using equation (2). The effective cross section $\sigma_{pl}(E)$ in the stellar plasma, is connected to the bare nucleus cross section $\sigma_b(E)$ and to the the stellar electron screening enhancement factor $f_{pl}$ by the relation

$$\sigma_{pl}(E) = \sigma_b(E) f_{pl}(E) \approx \sigma_b(E) \cdot exp(\pi\eta U_{pl}/E) \tag{3}$$

with $U_{pl}$ is the plasma potential energy, $\eta$ the Sommerfeld parameter.
If $\sigma_b(E)$ is measured at the ultralow energies $E_G$ and $U_{pl}$ is estimated within the framework of the Debye-Hückel theory, it is possible estimate from eq.(3) the effective cross section $\sigma_{pl}(E)$ in the stellar plasma. In turn, the understanding of $U_e$ may help to better understand $U_{pl}$, needed to calculate $\sigma_{pl}$.
Then, although it is possible to measure cross sections in the Gamow energy range, the bare nucleus cross section $\sigma_b$ is extracted by extrapolating the direct data behavior at higher energies where negligible electron screening contribution is expected. In order to decrease uncertainties in the case of charged particle induced reactions a rather striking conclusion could be achieved: to avoid extrapolations, experimental techniques were improved. After improving measurements (at very low energies), electron screening effects were discovered. Finally to extract from direct (shielded) measurements the bare astrophysical $S_b$(E)-factor, extrapolation were performed at higher energy. In any case the extrapolation procedure is necessary and in consequence we find again the uncertainties problem in direct measurements.

## 3. The Trojan Horse Method

Alternative methods for determining bare nucleus cross sections of astrophysical interest are needed. In this context a number of indirect methods, e.g. the Coulomb dissociation (CD) [4, 5], the Asymptotic Normalization Coefficient method (ANC)[30, 2, 17, 15, 16] and the Trojan-horse method (THM) were developed.
In particular, the THM is a powerful tool which selects the quasi-free (QF) contribution of an appropriate three-body reaction performed at energies well above the Coulomb barrier to extract a charged particle two-body cross section. In the framework of the extrapolation problems linked to the presence of electron screening effects a number of experimental measurements were carried out in order to measure the bare nucleus cross section in reactions of astrophysical interest.

The idea of the THM [3] is to extract the cross section of an astrophysically relevant two-body reaction

$$A + x \rightarrow c + C \tag{4}$$

at low energies from a suitable chosen three-body quasifree reaction

$$A + a \rightarrow c + C + S \tag{5}$$

This is done with the help of direct theory assuming that the nucleus $a$ has a strong $x \oplus S$ cluster structure. In many applications [27, 28, 14], this assumption is trivially fulfilled e.g. $a$ = deuteron, $x$ = proton, $S$= neutron. This three-body reaction can be described by a Pseudo-Feynmam diagram, where only the first term of the Feynman series is retained The upper pole describes the virtual break-up of the target nucleus $a$ into the clusters $x$ and $S$; $S$ is then considered to be spectator to the $A + x \rightarrow c + C$ reaction which takes place in the lower pole.
We will refer to other papers for the THM theory [27, 28, 14]. In Plane Wave Impulse Approximation (PWIA) the cross section of the three body reaction can be factorized into two terms corresponding to the two poles and it is given by [18]:

$$\frac{d^3\sigma}{dE_c d\Omega_c d\Omega_C} \propto KF \left(\frac{d\sigma}{d\Omega_{cm}}\right)^{off} \cdot |\Phi(\vec{p}_s)|^2 \tag{6}$$

where:

- $[(d\sigma/d\Omega)_{cm}]^{off}$ is the half-off-energy-shell differential cross section for the two body A(x,c)C reaction at the center of mass energy $E_{cm}$ given in post collision prescription by:

$$E_{cm} = E_{c-C} - Q_{2b} \tag{7}$$

where $Q_{2b}$ is the two body Q-value of the $A + x \rightarrow c + C$ reaction and $E_{c-C}$ is the relative energy between the outgoing particles $c$ and $C$;

- KF is a kinematical factor containing the final state phase-space factor and it is a function of the masses, momenta and angles of the outgoing particles:

$$KF = \frac{\mu_{Aa} m_c}{(2\pi)^5 \hbar^7} \frac{p_C p_c^3}{p_{Aa}} \left[ \left( \frac{\vec{p}_{Bx}}{\mu_{Bx}} - \frac{\vec{p}_{Cc}}{m_c} \right) \cdot \frac{\vec{p}_c}{p_c} \right]^{-1} \tag{8}$$

- $\Phi(\vec{p}_s)$ is the Fourier transform of the radial wave function $\chi(\vec{r})$ for the x-S inter-cluster motion, usually described in terms of Hänkel, Eckart and Hulthén functions depending on the x-S system properties.

We stress that with THM, one cannot extract the absolute value of the two-body cross section. However, the absolute value can be extracted through normalization to the direct data available at energies above and/or below the Coulomb barrier. Thanks to this, since we select the region of low momentum $p_s$ for the spectator ($p_s \leq 40$ MeV/c) the PWIA approach can be used for further analysis of the experimental results. If $|\Phi(\vec{p}_s)|^2$ is known and $KF$ is calculated, it is possible to derive $[(d\sigma/d\Omega)_{cm}]^{exp}$ from a measurement of $d^3\sigma/dE_c d\Omega_c d\Omega_C$ by using Eq. 6.

$$\left(\frac{d\sigma}{d\Omega}\right) \propto \left[\frac{d^3\sigma}{dE_c d\Omega_c d\Omega_C}\right] \cdot [KF|\Phi(\vec{p}_s)|^2]^{-1} \tag{9}$$

If the bombarding energy $E_A$ is chosen high enough to overcome the Coulomb barrier in the entrance channel of the three-body reaction, both Coulomb barrier and electron screening effects are negligible. In this way it is possible to extract the two-body cross section from Eq.(9) after inserting the appropriate penetration function $G_l$ in order to account for the penetrability effects affecting the direct data below the Coulomb barrier [9, 25]. The complete formula is given by:

$$\left(\frac{d\sigma}{d\Omega}\right) \propto \left[\frac{d^3\sigma}{dE_c d\Omega_c d\Omega_C}\right] \cdot [KF|\Phi(\vec{p}_s)|^2]^{-1} \cdot G_l \tag{10}$$

As shown above, since in the experimental works the IA validity conditions are fulfilled, the PWIA was applied for the extractions of the two-body cross-section. In this approximation the differential two-body cross-section of Eq.9 is expressed by:

$$\left(\frac{d\sigma}{d\Omega}\right) = \sum_l C_l G_l \left(\frac{d\sigma_l}{d\Omega}\right) \tag{11}$$

As already mentioned, the THM data are not affected by electron screening effects. Therefore, once the behavior of the absolute bare $S_b(E)$ factor from the two-body cross-section is extracted, a model-independent estimate of the screening potential $U_e$ can be obtained from comparison with the direct screened S(E)-factor.
Then, after normalization of the indirect data, the comparison between the two data sets can be performed down to the low energy region.

**Table 1.** Two-body reactions results in recent experiments as regards the astrophysical S(E)factor and the electron screening potential

| | Reaction | $U_e^{Adiab.}$ (eV) | $U_e^{Dir}$ (eV) | $U_e^{THM}$ (eV) | $S(0)^{THM}$ (MeVb) |
|---|---|---|---|---|---|
| [1] | $^7$Li + p $\to \alpha + \alpha$ | 186 | 300±160 | 330± 40 | 0.055 |
| [2] | $^6$Li + d $\to \alpha + \alpha$ | 186 | 330±120 | 340±50 | 16.9 |
| [3] | $^6$Li + p $\to^3 He + \alpha$ | 186 | 440±150 | 450±100 | 3.0 |
| [4] | $^9$Be + p $\to^6 Li + \alpha$ | 240 | 830±130 | 676±86 | 21.0 |
| [6] | d + $^3He \to p + \alpha$ | 115 | 219±7 | 180±40 | 6.08±1.42 |

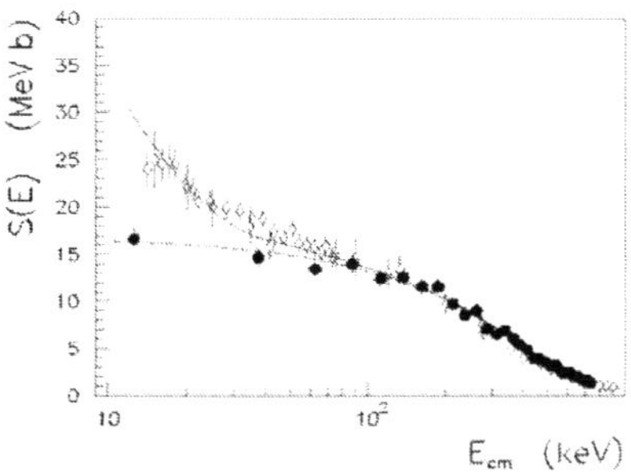


**Figure 1.** Bare nucleus astrophysical S(E) factor for $^6$Li + d $\to \alpha + \alpha$

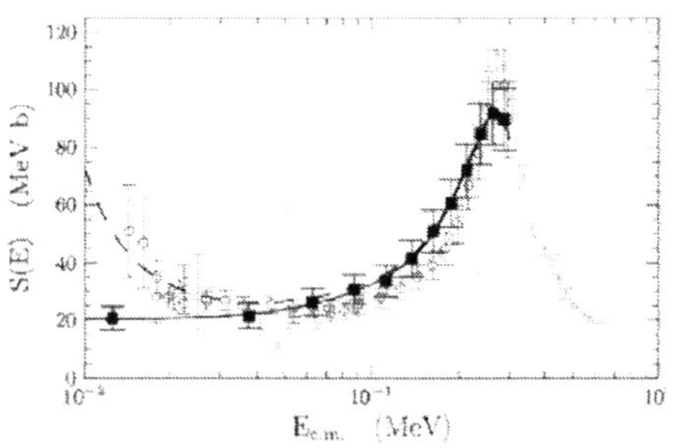


**Figure 2.** Bare nucleus astrophysical S(E) factor for $^9Be + p \to^6 Li + \alpha$

## 4. Conclusions

An experimental program has already been undertaken to study p-capture reactions on $^{6,7}Li$, main responsible for their destruction [20, 19]. The extracted S(E) factors as well as electron screening potentials, extracted as stated above are shown in table I. Recently, $^3He(d,p)^4He$ [12] and $^9Be(p,\alpha)^6Li$ [23] were also investigated. Their importance is indeed strongly related to the cosmology as well as to stellar structure and evolution. The bare nucleus S(E)-factor for the $^3He(d,p)^4He$ and $^9Be(p,\alpha)^6Li$ are reported in table 1 as well.

In picture 1 and 2 we present the results for the $^9Be(p,\alpha)^6Li$ and $^6Li(d,\alpha)^4He$. The bare nucleus S(E)-factor is reported as black dots together with the solid line that represents a fit. We can notice that at higher energies the two data sets agree pretty well while at lower ones they do not, due to the electron screening effect. The present paper presents the basic features of the THM and a review of recent applications to several reactions of importance in astrophysics.In particular these results show the possibility of extracting the bare nucleus two-body cross section via THM.However, a lot remains to do in the future to achieve reliable information for many key reactions and processes. New theoretical developments are strongly needed especially for the study of electron screening effects in fusion reactions in order to meet progress in the application field (fusion reactors). This information and the evaluation of fusion reaction cross sections will help to determine the basic properties of future reactors as well as plasma confinement devices.

## References

[1] H.J Assenbaum, K.Langanke, C.Rolfs, Z.Phys.,**327** 461 (1987)

[2] A.Azhari, V.Burjan, F.Carstoiu, C.A.Gagliardi, V.Kroha, A.M.Mukhamedzhanov, F.M.Nunes, X.Tang, L.Trache, R.E.Tribble, Phys.Rev., **C63** , 055803 (2001)
[3] G. Baur, Phys. Lett. B **178**, 135 (1986)
[4] G. Baur and H. Rebel, J. Phys. G **20**, 1 (1994) and references therein
[5] G. Baur and H. Rebel, Annu. Rev. Nucl. Part. Sci. **46**, 321 (1996)
[6] R.Bonetti et al., Phys. Rev. Lett., **82**, 5205 (1999)
[7] E.M.Burbige,G.R.Burbige,W.A.Fowler,F.Hoyle, Rev.Mod.Phys.**29**, 547 (1957)
[8] G. Calvi, M. Lattuada, D. Miljanić, F. Riggi, C. Spitaleri, M. Zadro,Phys. Rev. C **41**, 1848 (1990)
[9] S. Cherubini, V. N. Kondratyev, M. Lattuada, C. Spitaleri, Dj. Miljanić, M. Zadro, G. Baur, Ap. J. **457**, 855 (1996).
[10] G.F. Chew, Phys.Rev. **80**, 196(1950)
[11] W.A.Fowler, Rev. Mod. Phys.**56**, 149 (1984)
[12] M.La Cognata, C.Spitaleri, A.Tumino, S.Typel, S.Cherubini, L.Lamia, A.Musumarra, R:G.Pizzone, A.Rinollo, C.Rolfs, S.Romano, D.Schiurmann, F.Streider, Phys. Rev. C **72** 065802(2005)
[13] L.Lamia, S.Romano, N.Carlin, S.Cherubini, V.Crucillá M.De Moura, M.Gamiero Munhoz, M.Gulino, R.Liguori Neto, M.La Cognata, F.Mudó,R.G,Pizzone, S.M.R.Puglia.A.Suaide, E.Szanto, A.Szanto de Toledo, M.L.Sergi, C.Spitaleri,S.Tudisco, A.Tumino, Nucl. Phys. A.**787** 309 (2007)
[14] M. Lattuada, R.G. Pizzone, S. Typel, P. Figuera, Dj. Miljanić, A. Musumarra, M.G. Pellegriti, C. Rolfs, C. Spitaleri, H.H. Wolter, Ap.J. **562**, 1076 (2001)
[15] A.M.Mukhamedzhanov, A.M.Clark, C.A.Gagliardi,Y. W. Lui, L.Trache,R.E.Tribble,H.M.Xu,V.Cejpek ,V.Burjan, V.Kroha, F.Castiou, Phys. Rev. C **56**, 1302 (1997)
[16] A.M.Mukhamedzhanov, R.E.Tribble Phys. Rev. C**59**, 3418(1999)
[17] A.M.Mukhamedzhanov, C.A.Gagliardi, R.E.Tribble Phys. Rev. C**63**, art.n.024612 (2001)
[18] U.G.Neudatchin, Y.F.Smirnov, At. Energy Rev., **3** , 157 (1965)
[19] R.G. Pizzone, C. Spitaleri, M. Lattuada, S. Cherubini, A. Musumarra, M.G. Pellegriti, S. Romano, A. Tumino, V. Castellani, S. Degl'Innocenti, A. Imperio, Astron. & Astroph. 398, (2003) 423-427;
[20] R.G. Pizzone, A. Tumino, S. Degl'Innocenti, C. Spitaleri, S. Cherubini, A. Musumarra, S. Romano, Astronomy & Astrophysics 438, 779784 (2005)
Rinollo et al., 2005] A.Rinollo, S.Romano, C.Spitaleri, C.Conomo, S.Cherubini,A Del Zoppo, P.Figuera, M.La Cognata, L.Lamia, A.Musumarra,M.G.Pellegriti, R.G.Pizzone, c.Rolfs, D.Schurmann, F.Streider,S.Tudisco, A.Tumino, Nucl. Phys. A **758**146c (2005)
[21] C.Rolfs, W.S. Rodney, Cauldrons in the cosmos, University of Chicago Press, Chicago (1988)
[22] C.Rolfs, Prog. Part. Nucl. Phys., **153** , 23 (2001)
[23] Qun-Gang Wen, Cheng-Bo Li, Shu-Hua Zhou, Qiu-Ying Meng, Jing Zhou, Xiao-Mei Li, Shou-Yang Hu, and Yuan-Yong Fu C. Spitaleri, A. Tumino, R. G. Pizzone, and G. G. Rapisarda, PHYS. REV. C 78, 035805 (2008)
[24] C. Spitaleri, Problems of Fundamental Modern Physics II, World Scientific,p. 21- 35, 1990 and references therein
[25] C. Spitaleri, M. Aliotta, S. Cherubini, M. Lattuada, Dj. Miljanić, S. Romano, N. Soić, M. Zadro, R. A. Zappalà, Phys. Rev. C **60**, 055802 (1999)
[26] C. Spitaleri, M. Aliotta, P. Figuera, M.Lattuada,R. G. Pizzone, S. Romano, A. Tumino, C. Rolfs, L. Gialanella, F. Strieder, S. Cherubini, A. Musumarra, D. Miljanić, S. Typel and H. H. Wolter, Eur. Phys. J **A 7**, 181 (2000)
[27] C. Spitaleri, S. Typel, R.G. Pizzone, M.Aliotta, S. Blagus, M. Bogovac, S. Cherubini, P. Figuera, M. Lattuada, M. Milin, Dj. Miljanić, A. Musumarra, M.G. Pellegriti, D. Rendić, C. Rolfs, S. Romano, N. Soić, A. Tumino, H.H. Wolter, M. Zadro, Phys. Rev. C **63**, 005801 (2001)
[28] Spitaleri, A. Di Pietro, P. Figuera, M. Lattuada, A.Musumarra, M.G. Pellegriti, R.G. Pizzone S. Romano, C. Rolfs, S. Tudisco, S. Typel, Phys. Rev.C . **69** 055806 (2004)
[29] F. Streider, C.Rolfs, C.Spitaleri,P.Corvisiero, Naturwissenschaften **88**, 461, (2001)
[30] X.D.Tang, A.Azhari, C.A.Gagliardi, A.M.Mukhamedzhanov, F.Pirlepeson, L.Trache, R.E.Tribble, V.Burjan, V.Kroha, F.Castiou, Phys. Rev. C**67**, art.n.015804 (2003)

# First measurement of the $^{18}O(p,\alpha)^{15}N$ cross section at astrophysical energies

**M La Cognata[1,2,3], C Spitaleri[1,2], A Mukhamedzhanov[4], R E Tribble[4], T Al-Abdullah[4], A Banu[4], S Cherubini[1,2], A Coc[5], V Crucillà[1,2], V Goldberg[4], M Gulino[1,2], B Irgaziev[6], G G Kiss[7], L Lamia[1,2], J Mrazek[8], R G Pizzone[1,2], S M R Puglia[1,2], G G Rapisarda[1,2], S Romano[1,2], M L Sergi[1,2,3], G Tabacaru[4], L Trache[4], W Trzaska[9], S Tudisco[1,2] and A Tumino[1,10]**

[1]INFN Laboratori Nazionali del Sud, Catania, Italy
[2]Dipartimento di Metodologie Chimiche e Fisiche per l'Ingegneria, Università di Catania, Italy
[3]CSFNSM Centro Siciliano di Fisica Nucleare e Struttura della Materia, Catania, Italy
[4]Cyclotron Institute, Texas A&M University, College Station, TX, USA
[5]CSNSM, CNRS/IN2P3, Universitè Paris Sud, Orsay, France
[6]GIK Institute of Engineering Sciences and Technology, Topi, District Swabi, N. W. F. P., Pakistan
[7]ATOMKI, Debrecen, Hungary
[8]Nuclear Physics Institute of ASCR, Rez near Prague, Czech Republic
[9]Physics Department, University of Jyvaskyla, Finland
[10]Università Kore, Enna, Italy

E-mail: LaCognata@lns.infn.it

**Abstract.** The $^{18}O(p,\alpha)^{15}N$ reaction rate has been deduced by means of the Trojan horse method. For the first time the contribution of the 20 keV resonance has been directly evaluated, giving a value about 35% larger than the one in the literature. Moreover, the present approach has allowed to improve the accuracy by a factor 8.5, as it is based on the measured strength instead of spectroscopic measurements. The contribution of the 90 keV resonance has been also determined, which turned out to be of negligible importance to astrophysics.

## 1. Introduction

Fluorine is one of the few elements whose nucleosynthesis is still debated as three possible astrophysical sites for fluorine production have been identified, namely Type II Supernovae (SNe II), Wolf-Rayet (WR) stars, and asymptotic giant branch (AGB) stars [1]. In particular, in AGB stars fluorine abundance is enhanced with respect to the solar one by up to a factor 30 [2]. In such stars, $^{19}F$ nucleosynthesis takes place at the same evolutionary stage and in the same region as the s-process nucleosynthesis. For these reasons, AGB stars play an extremely important role in astrophysics and the understanding of fluorine production, allowing to constrain the existing models [3], would make predictions on AGB star nucleosynthesis and s-process element yields more accurate. This is because $^{19}F$ abundance is very sensitive to the temperatures and the mixing processes taking place inside AGB stars. Anyway, if standard theoretical abundances are compared to the observed ones [2], a remarkable discrepancy shows up because the largest

$^{19}$F abundances cannot be matched for the typical $^{12}$C/$^{16}$O ratios [3]. It has been shown that extra-mixing phenomena, such as the cool bottom process [4], could help to pin down the origin of this discrepancy [3]. A complementary way to explain $^{19}$F abundance can be provided by nuclear physics, in particular by an improved measurement of the $^{18}$O$(p,\alpha)^{15}$N reaction rate. In fact, this reaction represents the main $^{15}$N production channel, which is burnt to $^{19}$F via the $^{15}$N$(\alpha,\gamma)^{19}$F reaction during thermal pulses, at temperatures of the order of $10^8$ K. Thus a larger $^{18}$O$(p,\alpha)^{15}$N reaction rate would lead to an increase of the $^{19}$F supply, while the $^{12}$C/$^{16}$O ratio would not change.

## 2. The measurement

In order to reduce the nuclear uncertainties affecting its reaction rate we have performed an experimental study of the $^{18}$O$(p,\alpha)^{15}$N reaction by means of the Trojan horse method (THM), which is an indirect technique to measure the relative energy-dependence of a charged-particle reaction cross section at energies well below the Coulomb barrier ([5, 6, 7] and references therein). The cross section of the $^{18}$O$(p,\alpha)^{15}$N reaction is deduced from the $^2$H($^{18}$O$,\alpha^{15}$N$)n$ three-body process, performed in quasi-free (QF) kinematics. The beam energy is chosen larger than the Coulomb barrier for the interacting nuclei, so the break-up of the deuteron (acting as the Trojan-horse nucleus) takes place inside the $^{18}$O nuclear field. Therefore, the cross section of the $^{18}$O$(p,\alpha)^{15}$N reaction is not suppressed by the Coulomb interaction of the target-projectile system, while no electron screening enhancement is spoiling the nuclear information because the reaction is performed at high energies (several tens of MeV). The THM cross section for the $^2$H($^{18}$O$,\alpha^{15}$N$)n$ reaction proceeding through a resonance in the subsystem $^{19}$F = $^{18}$O+$p$ = $^{15}$N+$\alpha$ can be obtained if the process is described as a transfer to the continuum, where the emitted neutron keeps the same momentum as the one it has inside deuteron (QF condition). If such a hypothesis is satisfied, the cross section for the QF $^2$H($^{18}$O$,\alpha^{15}$N$)n$ three-body reaction is [5, 8]:

$$\frac{d^2\sigma}{dE_{\alpha^{15}\mathrm{N}}\,d\Omega_n} \propto \frac{\Gamma_{(\alpha^{15}\mathrm{N})_i}(E)\,|M_i(E)|^2}{(E-E_{R_i})^2+\Gamma_i^2(E)/4} \tag{1}$$

Here, $M_i(E)$ is the direct transfer reaction amplitude for the binary reaction $^{18}\mathrm{O}+d \to {}^{19}\mathrm{F}+n$ leading to the population of the $i-th$ resonant state of $^{19}$F with resonance energy $E_{R_i}$, $E$ is the $^{18}$O-$p$ relative kinetic energy related to $E_{\alpha^{15}\mathrm{N}}$ by the energy conservation law, $\Gamma_{(\alpha^{15}\mathrm{N})_i}(E)$ is the partial resonance width for the decay $^{19}\mathrm{F} \to \alpha + {}^{15}\mathrm{N}$ and $\Gamma_i$ is the total resonance width of the $i-th$ resonance. The appearance of the transfer reaction amplitude $M_i(E)$ instead of the entry channel partial resonance width $\Gamma_{(p^{18}\mathrm{O})_i}(E)$ is the main difference between the THM cross section and the cross section for the resonant binary sub-reaction $^{18}$O$(p,\alpha)^{15}$N [5, 8, 7]. Therefore the cross section of the three-body process can be easily connected to the one for the two-body reaction of interest by evaluating the transfer amplitude $M_i(E)$. The experiment was performed at Laboratori Nazionali del Sud, Catania (Italy). The SMP Tandem Van de Graaff accelerator provided the 54 MeV $^{18}$O beam which was accurately collimated to achieve the best angular resolution. The intensity was 5 enA on the average and the relative beam energy spread was about $10^{-4}$. Thin self-supported deuterated polyethylene (CD2) targets, about 100 $\mu$g/cm$^2$ thick, were adopted in order to minimize angular straggling. The detection setup consisted of a telescope (A), to single out Z=7 particles, made up of an ionization chamber and a silicon position sensitive detector (PSD A). Negligible angular straggling was introduced on the $^{15}$N detection by the ionization chamber. Three additional silicon PSD's (B, C and D) were placed on the opposite side, with the aim of detecting alpha particles from the $^2$H($^{18}$O$,\alpha^{15}$N$)n$ QF three-body process. No $\Delta$E detectors were put in front of PSD's B, C and D to decrease detection thresholds and to achieve the best energy and angular resolution.

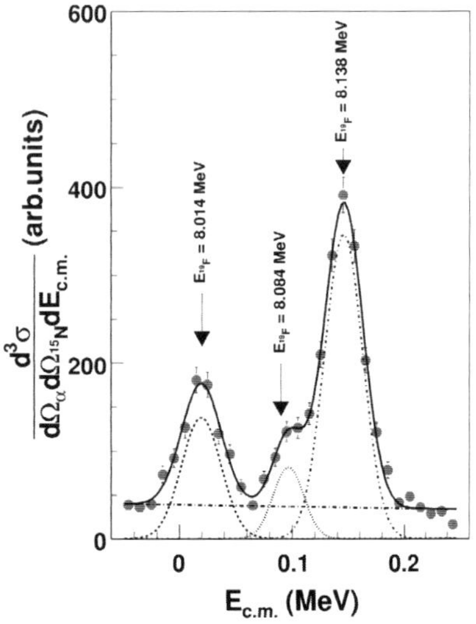


**Figure 1.** Cross section of the $^2H(^{18}O, \alpha^{15}N)n$. The arrows mark the corresponding $^{19}F$ excited states.

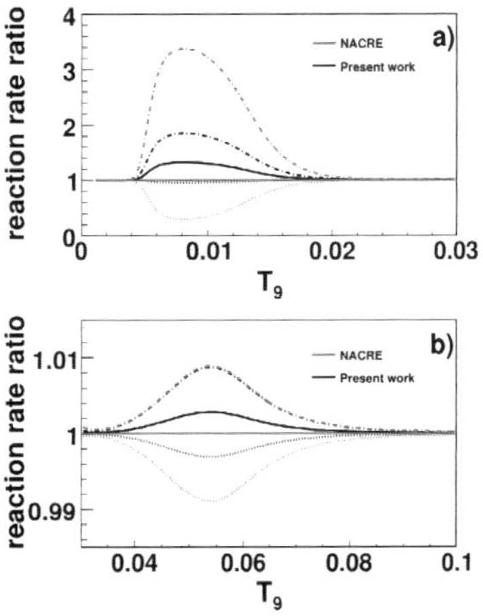


**Figure 2.** Comparison of the THM reaction rate (black lines) of the $^{18}O(p, \alpha)^{15}N$ reaction with the NACRE one [9] (red lines). The full lines are the ratio of the recommended rate to the NACRE one. The dot-dashed and dotted black lines represent the upper and lower limits respectively, allowed by the experimental uncertainties. $T_9$ is the temperature in billion kelvin.

### 3. Extraction of the cross section

A description of the data analysis is reported in [8], here we show the main results. The extracted three-body cross section has been integrated in the whole angular range. The resulting $^2H(^{18}O, \alpha^{15}N)n$ reaction cross section is shown in Fig. 1 (full circles). The experimental energy resolution turned out to be about 40 keV (FWHM). Horizontal error bars represent the integration bin while the vertical ones arise from statistical uncertainty and angular distribution integration. The solid line in the figure is the sum of three Gaussian functions to fit the resonant behaviour and a straight line to account for the non-resonant contribution. The resonance energies were then deduced: $E_{R1} = 19.5 \pm 1.1$ keV, $E_{R2} = 96.6 \pm 2.2$ keV and $E_{R3} = 145.5 \pm 0.6$ keV (in fair agreement with the ones reported in the literature, see [9]) as well as the peak values of each resonance in arbitrary units: $N_1 = 138 \pm 8$, $N_2 = 82 \pm 9$ and $N_3 = 347 \pm 8$. The peak values were used to derive the resonance strengths:

$$(\omega\gamma)_i = \frac{2J_{^{19}F_i} + 1}{(2J_{^{18}O} + 1)(2J_p + 1)} \frac{\Gamma_{(p^{18}O)_i} \Gamma_{(\alpha^{15}N)_i}}{\Gamma_i}. \tag{2}$$

In this work we did not extract the absolute value of the cross section. Anyway, the proton and alpha partial widths for the third resonance are well known [9], thus we can determine the strength for the 20 keV and 90 keV resonances from the ratio of the peak values of the THM cross sections, as discussed by [8]. The electron screening gives a negligible contribution around 144 keV (4% maximum [10]), thus no systematic uncertainty is introduced by normalizing to the highest energy resonance. If $\omega\gamma_3$ is taken from [11], one gets $\omega\gamma_1 = 8.3^{+3.8}_{-2.6}\, 10^{-19}$ eV, which is well within the confidence range established by NACRE, $6^{+17}_{-5}\, 10^{-19}$ eV [9]. The largest contribution to the error is due to the uncertainty on the resonance energy, while statistical and normalization errors sum up to about 9.5%. With the same procedure, we got $\omega\gamma_2 = 1.76 \pm 0.33\,10^{-7}$ eV for the 90 keV resonance, in good agreement with the result in NACRE [9]. The significant improvement of the accuracy of the deduced resonance strength is mainly due to the fact that NACRE recommended value is based on spectroscopic data while the present result is obtained from experimental ones. We underscore that the improvement comes from a reduced systematic uncertainty strictly connected with the spectroscopic approach. In fact, the resonance strength deduced by means of spectroscopic measurements is affected by large and not-well-defined uncertainties, because they are strongly model dependent. They rely on the optical model potentials adopted in the data analysis, and different set of potentials or of parameters, though giving a reasonable account of the experimental data, lead to the extraction of spectroscopic factors that can differ even by one order of magnitude.

## 4. Reaction rate

The reaction rate for the $^{18}O(p,\alpha)^{15}N$ reaction has been deduced by using the narrow resonance approximation [9], which is fulfilled for the resonances under investigation. The resulting rate is given in Fig. 2 as a function of the temperature. In order to compare with the one reported in NACRE [9], the ratio of the THM reaction rate to the NACRE one is deduced and shown as a full black line in Fig. 2. In this representation, the NACRE rate is given by a full red line, that is by 1 in the whole examined range. The dot-dashed and dotted black lines represent the upper and lower limits respectively, allowed by the experimental uncertainties. As before, black and red lines mark THM and NACRE data. In the low temperature region (below T= $3\,10^7$ K, Fig. 2a) the reaction rate can be about 35% larger than the one given by NACRE, while the indetermination is greatly reduced with respect to the NACRE one, by a factor of 8.5, in the case the error on the NACRE rate is supposed to come entirely from the uncertainty on the 20 keV resonance strength, to make the comparison homogeneous. Those temperatures are typical of the bottom of the convective envelope, thus an increase of this reaction rate might have important consequences on the cool bottom process [4] and, in turn, on the surface abundances and isotopic ratios in AGB stars. The 8.084 MeV excited state of $^{19}F$ (corresponding to the 90 keV resonance) provides a negligible contribution to the reaction rate in agreement with the previous estimate by [12]. This is clearly displayed by Fig. 2b), where an increase of less than 1% is obtained due to the THM measurement of the 90 keV level resonance strength.

As a next step, the astrophysical consequences of the present work are to be evaluated, both onto the scenarios sketched in the introduction and on alternative environments. In addition, at higher temperatures, higher energy resonances in the $^{18}O(p,\alpha)^{15}N$ reaction can play a role. These studies will be the subject of forthcoming works.

## References

[1] Renda A, et al. 2004 *MNRAS* **354** 575
[2] Jorissen A, et al. 1992 *ApJ* **261** 164
[3] Lugaro M, et al. 2004 *ApJ* **615** 934
[4] Nollett K M, et al. 2003 *ApJ* **582** 1036
[5] La Cognata M, et al. 2007 *Phys. Rev. C* **76** 065804
[6] Spitaleri C, et al. 1999 *Phys. Rev. C* **90** 055802
[7] Mukhamedzhanov A, et al. 2008 *J. Phys. G: Nucl. Part. Phys.* **35** 014016
[8] La Cognata M, et al. 2008 *Phys. Rev. Lett.* **101** 152501
[9] Angulo C, et al. 1999 *Nucl. Phys. A* **656** 3
[10] Assenbaum H J, et al. 1987 *Z. Phys. A* **327** 461
[11] Becker H W, et al. 1995 *Z. Phys. A* **351** 453
[12] Champagne A E, et al. 1986 *Nucl. Phys. A* **457** 367

Journal of Physics: Conference Series **202** (2010) 012020 doi:10.1088/1742-6596/202/1/012020

# Precise Measurement of the Neutron Capture Reaction $^{54}$Fe(n,γ)$^{55}$Fe via AMS

**A Wallner[1], K Buczak[1], T Belgya[2], M Bichler[3], L Coquard[4], I Dillmann[4,5], O Forstner[1], R Golser[1], F Käppeler[4], W Kutschera[1], C Lederer[1], A Mengoni[6], A Priller[1], R Reifarth[7,8], P Steier[1], L Szentmiklosi[2]**

[1] VERA Laboratory, Faculty of Physics, University of Vienna, Austria
[2] Dep. of Nuclear Research, Institute of Isotopes, Hungarian Academy of Sciences
[3] Atominstitut, Vienna University of Technology, Austria
[4] Institut für Kernphysik, Forschungszentrum Karlsruhe, Germany
[5] Physik Department E12 and Excellence Cluster Universe, Technische Universität München, Germany
[6] Nuclear Data Section, IAEA, PO Box 100, Vienna, Austria
[7] GSI Helmholtzzentrum für Schwerionenforschung GmbH, Darmstadt, Germany
[8] J.W. Goethe Universität, Frankfurt a.M., Germany

E-mail: anton.wallner@univie.ac.at

**Abstract.** The measurement of cross sections relevant to nuclear astrophysics has become one main research topic at the VERA (Vienna Environmental Research Accelerator) facility. The technique applied, accelerator mass spectrometry (AMS), offers excellent sensitivity for the detection of long-lived radionuclides through ultra-low isotope ratio measurements. We discuss the potential and preliminary results of ongoing precision measurements of neutron-capture cross sections of $^{54}$Fe. Such measurements might help to clarify the recently found discrepancy of s-process nucleosynthesis at lower-mass nuclei (A<120), which might be attributed to a systematic offset in previous experimental data. Samples were irradiated with neutrons from thermal to MeV energies. After the irradiations, the amount of produced long-lived $^{55}$Fe ($t_{1/2}$ = 2.72 yr) was analyzed using AMS. At VERA, detection of $^{55}$Fe was developed with a reproducibility of about 1%, which makes the $^{54}$Fe(n,γ)$^{55}$Fe reaction a precise and unique laboratory measurement, which can serve as a reference to complementary techniques. In this regard a new $^{55}$Fe standard for AMS measurements was produced. The final cross-section data are expected to be accurate to better than 3%. We report a preliminary, however, already significantly improved thermal neutron cross section value of (2.32 ± 0.10) barn, and a value of (6.3 ± 0.6) mbarn for $E_n$ = (520 ± 50) keV.

## 1. Introduction

Recent accurate abundance observations in very rare, ultra metal-poor (UMP) stars in the galactic halo showed abundance patterns that scale exactly with the solar r component [1]. This holds for elements heavier than barium, however, for lighter elements significant discrepancies of the order of 20% were found from this observation. A similar discrepancy is found for s-only isotopes lighter than barium (see [2]). These facts hint to some deficiency in the standard description of s- or r-process nucleosynthesis or there is a systematic offset in the experimental data. The latter can be tested by precision measurements of neutron-capture cross sections, e.g. the $^{54}Fe(n,\gamma)^{55}Fe$ reaction: Via the weak s-process massive stars contribute dominantly to the abundance distribution between Fe and Zr. For those nuclides, however, neutron capture cross sections are poorly known compared to the required accuracy of 3-5% [3].

For specific reactions accelerator mass spectrometry (AMS) offers a powerful tool to measure cross sections independent of the half-lives of the reaction products. Activation with subsequent AMS measurements have been applied for isotopes where off-line decay counting is difficult or impossible due to long half-lives or due to the absence of suitable $\gamma$-ray transitions (see e.g. [4,5], for a detailed list see [6].) In particular, measurements on $^{55}Fe$ allow producing highly precise data. AMS will therefore allow one to identify possible hidden systematic uncertainties associated with different detection methods and, in special cases, to reduce the total uncertainty of cross section measurements to a few %. A summary of AMS measurements at the VERA facility (Vienna Environmental Research Accelerator) studying neutron-induced cross-sections of relevance in nuclear astrophysics is given in Tab. 1.

**Table 1.** Summary of AMS measurements at VERA related to neutron capture studies of interest to nuclear astrophysics. FZK, ATI and IKI denote neutron irradiations performed at Forschungszentrum Karlsruhe ("keV neutrons"), at the Atominstitut, Vienna ("thermal neutrons"), and at the Budapest Research Reactor ("cold neutrons"), respectively.

| Reaction | Irradiation facility | neutron energy |
|---|---|---|
| $^{9}Be(n,\gamma)^{10}Be$ | FZK, ATI | thermal, 25 keV, 520 keV |
| $^{13}C(n,\gamma)^{14}C$ | FZK, ATI | thermal, 25 keV, 128 keV, 167 keV |
| $^{14}N(n,p)^{14}C$ | FZK, ATI | thermal, 25 keV, 123 keV, 178 keV, 14 MeV |
| $^{40}Ca(n,\gamma)^{41}Ca$ | FZK | 25 keV |
| $^{54}Fe(n,\gamma)^{55}Fe$ | FZK, ATI, IKI | cold, thermal, 25 keV, 520 keV, 14 MeV |
| $^{209}Bi(n,\gamma)^{210m}Bi$ | FZK, ATI | thermal, 25 keV |

## 2. Comparison of AMS measurements to other techniques

Cross section measurements can be classified into two complementary techniques: direct and indirect methods. The direct method makes use of the detection of the prompt and characteristic radiation associated with the production of a specific nuclide, or selectively detects the reaction product itself. A second and independent method makes use of the activation technique, with sample irradiation and subsequent measurement of the reaction product. After the irradiation the number of produced radioactive nuclei can be quantified either by decay counting or by mass spectrometric methods. This method is mostly restricted to radioactive products; however, it represents a very sensitive technique due to potential long irradiation periods. With the dedicated activation setup at Karlsruhe, stellar conditions can directly be simulated by generating a quasi-stellar spectrum (see e.g. [7,8]), and consequently, that reaction rate directly scales with stellar rates.

The determination of the stellar cross sections via the combination of the activation technique and AMS is complementary to direct measurements, since this independent approach implies different systematic uncertainties. In this regard, precision measurements of neutron-capture cross sections of $^{54}Fe$ are in progress to clarify the recently found discrepancy of s-process nucleosynthesis at lower-mass nuclei ($A<120$; see above). The potential and power of AMS for detecting the product nuclei $^{55}Fe$ has been demonstrated recently in a first irradiation campaign [9,10].

## 3. Activation and AMS measurement

Several neutron activations in combination with subsequent AMS measurements at VERA were performed recently for studying neutron capture on $^{54}$Fe: Thermal neutrons and cold neutrons were produced at the Atominstitut of the Vienna University of Technology, and the Budapest Research Reactor (IKI, Institute of Isotopes, Hungarian Academy of Sciences) [11]. "keV neutrons" were produced at Forschungszentrum Karlsruhe (FZK) [7] via the $^{7}$Li(p,n)$^{7}$Be reaction: By choosing a proper irradiation geometry, a quasi-stellar neutron spectrum was generated (by simulating a Maxwellian-Boltzmann distribution) of kT = 25 keV; and by higher-energetic protons neutron energies of 520 keV were produced as well. In addition, neutron activations in the fast neutron energy range (14-MeV) were performed in cooperation with TU Dresden. Depending on the cross-sections, isotope ratios $^{55}$Fe/$^{54}$Fe between $10^{-11}$ and $10^{-12}$ were produced in such activations. After the neutron irradiations, the subsequent AMS measurements were performed at the VERA facility in Vienna.

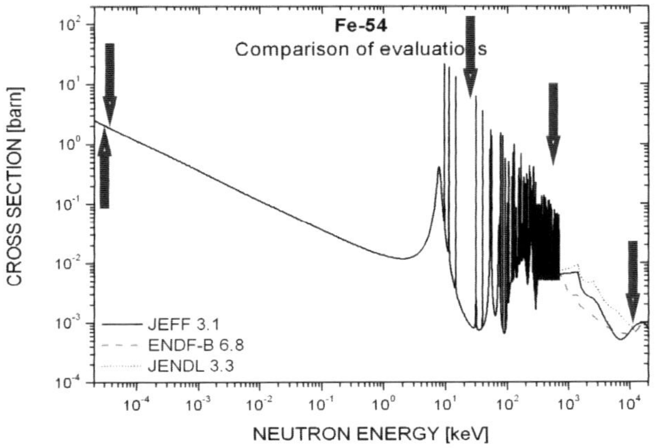


**Figure 1.** Neutron capture cross section for $^{54}$Fe(n,γ)$^{55}$Fe. Plotted are evaluations [12]; the energies of new activation/AMS data are indicated by arrows.

In Fig. 1 these neutron energies are schematically indicated by arrows. Also plotted are results from semi-empirical evaluations [12], which indicate discrepancies in the energy region above a few hundred keV. With these "activation"-anchors, distributed over a wide energy range, AMS data can be compared to continuous data obtained with time-of-flight techniques.

## *4. AMS measurement of $^{55}$Fe*

AMS represents a mass spectrometric method based on a tandem accelerator. Negatively charged ions are produced in an ion source requiring mg mass of solid sample. Extraction of Fe$^{-}$ completely suppressed the stable isobar of $^{55}$Fe, because $^{55}$Mn does not form stable negative ions [13]. At VERA $^{55}$Fe-measurements on the level of 1% reproducibility with a background $^{55}$Fe/$^{56}$Fe < $2\times10^{-15}$ can be performed (for more details see [10]). In this regard, an important aspect is the availability of accurate $^{55}$Fe reference materials for AMS. In general, a typical iron sample with an isotopic ratio $^{55}$Fe/$^{56}$Fe of $10^{-12}$ will result in a $^{55}$Fe count rate of typically one event per 10 second. Accordingly, a background ratio $^{55}$Fe/$^{56}$Fe of $1\times10^{-15}$, corresponds to a rate of ≈ 1 count per 3 hours measuring time.

Important ingredients for AMS measurements are reference samples to which the measured results of unknowns are normalized. AMS does not allow deducing accurate data directly from the measured quantities, particle count rate ($^{55}$Fe) and ion currents ($^{54,56}$Fe), because mass fractionation, machine instabilities etc. are difficult to quantify at levels better than 5 – 10%. For $^{55}$Fe measurements no well-known reference material was available. Therefore, we produced our own in-house standard. Best results were obtained from a $^{55}$Fe activity solution, certified by PTB Braunschweig, Germany [14]. This solution was stepwise diluted with Fe of natural isotopic composition. From the certified activity solution we produced $^{55}$Fe AMS reference materials with an isotope ratio known to ±1.5%, and basically limited by the uncertainty of the certified value.

## 5. Results

The neutron capture cross section of the $^{54}$Fe(n,γ)$^{55}$Fe reaction can directly be calculated from the neutron fluence and the isotope ratio measured with AMS. Interestingly, the recommended value for the thermal capture cross section of (2.25±0.18) barn [15] is known with an accuracy of only 8%. Our preliminary result for the thermal cross section is based on samples activated at the NIPS station [11] of the Budapest Research Reactor and at the TRIGA Mk II research reactor at the Atominstitut, Vienna. Fe metal pellets of natural isotopic composition, and metallic Fe foils, consisting of highly enriched $^{54}$Fe, were exposed to a neutron fluence of $10^{13}$ – $10^{14}$ n cm$^{-2}$. From the AMS measurements, we obtained a thermal value of (2.33 ± 0.10) barn, in good agreement with the recommended value for $^{54}$Fe(n,γ)$^{55}$Fe of (2.25 ± 0.18) barn [15]. Full data analysis will result in a final uncertainty of ≈ 2 %.
At Forschungszentrum Karlsruhe Fe pellets were activated with neutrons of kT ≈ 25 keV; and with neutrons of 520 keV energy. In a previous publication [9] we presented first preliminary results for 25 keV. Here, we add new data from AMS measurements performed relative to the new $^{55}$Fe reference material. Our new results for 520 keV indicate a cross section value of (6.3 ± 0.6) mbarn. These preliminary results confirm previous ones, however, already with improved accuracy. The full set of acquired data is being analyzed now, with the goal to reduce uncertainties to 2-3%.

## 6. Conclusion

The neutron capture reaction $^{54}$Fe(n,γ)$^{55}$Fe represents an excellent candidate for comparing different and independent methods for cross section measurements. AMS detection of $^{55}$Fe combined with neutron activations has been performed at the VERA laboratory, where $^{55}$Fe detection was demonstrated to be precise at a level of 1%. The preparation of a well-known reference material allows now to deduce accurate cross-section data over a wide neutron energy range. While activation allows one only to gain information on cross section values for selected neutron energies, time-of-flight based techniques will provide continuous data over a wide energy range. However, this more complicated technique can be checked and normalized with AMS data. In particular, such comparisons are important to clarify the recently debated discrepancy of measured solar-system abundances and observed abundances in ultra-metal poor stars.

**Acknowledgement**: Part of this work was funded by the Austrian Science Foundation (FWF), No. P20434.

## References

[1] C Sneden et al, The Astrophys. Journal **591** (2003) 936
[2] JL Tain, C Domingo et al., n_TOF proposal for campaign 2006-2010 (http://pceet075.cern.ch)
[3] F Käppeler, Nucl. Instrum. Meth. B**259** (2007) 663
[4] H Nassar, M Paul et al., Phys. Rev. Lett. **96** (2006) 041102
[5] A Arazi et al., Phys. Rev. C**74** 025802 (2006)
[6] A Wallner, Nucl. Instrum. Meth. B (in press, NIMB 57047), (2009)
[7] W Ratynski, F. Käppeler, Phys. Rev. C**37** (1988) 595
[8] I Dillmann et al., Phys. Rev. C**79** (2009) 065805
[9] L Coquard et al., Proc. 9th Conference "Nuclei in the Cosmos", Geneva (2006), PoS(NIC-IX)274.
[10] A Wallner et al., Nucl. Instr. and Meth. B**259** (2007) 677
[11] T Belgya, Z Revay, L Szentmiklosi, 12th Int. Symp. On Capture Gamma-Ray Spectroscopy and Related Topics, ed. by A Woehr and A Aprahamian (AIP, New York, 2006) AIP Conf. Proc. **819**, 300 & Z. Révay, T. Belgya, Z. Kasztovszky, J.L. Weil, G.L. Molnár, Nucl. Instr. Meth. B 213 (2004) 385
[12] see http://www-nds.iaea.org/exfor/exfor.htm
[13] G Korschinek et al., Nucl. Instr. Meth. B**52** (1990) 498
[14] see: http://www.ptb.de/en/org/6/61/611/_index.htm
[15] H Pomerance, Phys. Rev. **88** (1952) 412

Nuclear Physics in Astrophysics IV IOP Publishing
Journal of Physics: Conference Series **202** (2010) 012021 doi:10.1088/1742-6596/202/1/012021

# Indirect measurement of $^{17}$O(p,$\alpha$)$^{14}$N cross section at ultra-low energies

**M L Sergi[1,2,3], C Spitaleri[1,2], A Coc[4], A Mukhamedzhanov[5], V Burjan[6], S Cherubini[1,2], V Crucillá[1,2], M Gulino[1,2], F Hammache[7], V Z Hons[6], B Irgaziev[8], G Kiss[9], V Kroha[6], M La Cognata[1,2,3], L Lamia[1,2], R G Pizzone[1,2], S M R Puglia[1,2], G G Rapisarda[1,2], S Romano[1,2], N de Séréville[7], E Somorjai[9], S Tudisco[1,2] and A Tumino[1,2,10]**

[1] INFN Laboratori Nazionali del Sud, Catania, Italy
[2] Dipartimento di Metodologie Fisiche e Chimiche per l'Ingegneria- Universit di Catania, Italy
[3] CSFNSM, Catania, Italy
[4] CSNSM, CNRS/IN2P3/UPS, Bâtiment 104, 91405 Orsay Campus, France
[5] Texas A&M College Station,USA
[6] Nuclear Physics Institute of ASCR, Rez near Prague, Czech Republic
[7] IPN, IN2P3-CNRS et Université de Paris-Sud, 91406 Orsay Cedex, France
[8] GIK Institute of Engineering Sciences and Technology Topi District Swabi NWFP, Pakistan
[9] Institute of Nuclear Research of Hungarian Academy of Sciences (ATOMKI), Debrecen
[10] Università degli Studi di Enna "Kore", Enna, Italy

E-mail: `sergi@lns.infn.it`

**Abstract.** The indirect measurement of $^{17}$O(p,$\alpha$)$^{14}$N cross section was performed by means of the Trojan Horse Method. This approach allowed to investigate the ultra-low energy range ($E_{c.m.}$= 0-300 keV) relevant for several astrophysics environments, where two resonant levels of $^{18}$F at $E^R_{c.m.}$= 65 keV and $E^R_{c.m.}$= 183 keV play a significant role in the reaction rate determination.

## 1. Introduction

The $^{17}$O+$p$ reactions are of paramount importance for the nucleosynthesis in a number of stellar sites, including red giants (RG), asymptotic giant branch (AGB) stars, massive stars and classical novae [1, 2]. In particular the $^{17}$O(p,$\alpha$)$^{14}$N and $^{17}$O(p,$\gamma$)$^{18}$F reaction govern the destruction of $^{17}$O and the formation of the short-live radio-isotope $^{18}$F which is of special interest for gamma ray astronomy [1, 2].

The importance of the $^{17}$O(p,$\alpha$)$^{14}$N reaction in stellar nucleosynthesis theory comes from its occurrence in the carbon-nitrogen-oxygen cycle (CNO cycle) in which it acts as a feedback into the main part of the cycle (see Fig.1). Knowledge of the ratio of the amount of $^{16}$O to the amount of $^{17}$O formed in CNO cycle then depends on a knowledge of the cross section for $^{16}$O(p,$\gamma$), which forms the $^{17}$O, and for $^{17}$O(p,$\alpha$) which destroys it. Actually, while the nuclear reaction rate $^{16}$O(p,$\gamma$)$^{17}$F($\beta^+$)$^{17}$O is rather well known [3], the experimental status for its destruction dominant channel, $^{17}$O(p,$\alpha$)$^{14}$N reaction, is much less satisfactory [2, 3].
Stellar temperatures of primary importance for $^{17}$O nucleosynthesis are typically in the ranges

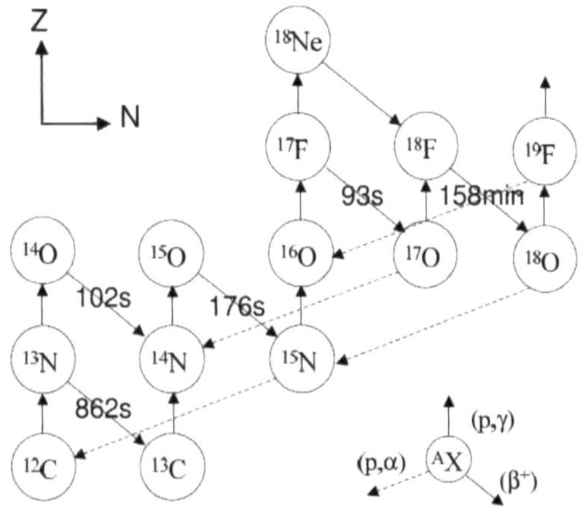


**Figure 1.** Scheme of the carbon-nitrogen-oxygen (CNO) cycle of hydrogen burning which operates in nova explosions. The lifetime of the $\beta^+$unstable nuclei are displayed

T=0.01-0.1 GK for RG, AGB, and massive stars, and T=0.1-0.4 GK for classical nova explosion [2]. Thus, the $^{17}$O(p,$\alpha$)$^{14}$N and $^{17}$O(p,$\gamma$)$^{18}$F reaction cross sections have to be precisely known in the energy range $E_{c.m.}$=0.017-0.37 MeV. In this energy range, the $^{17}$O(p,$\alpha$)$^{14}$N reaction cross section is dominated by two resonances: one at $E^R_{c.m.}$=65 keV above the $^{18}$F proton threshold, corresponding to the 5.673 MeV $^{18}$F level and the other one at $E^R_{c.m.}$=183 keV ($E_X$=5.786 MeV). While, in the last years, several measurements ([2] and references therein) of the $E^R_{c.m.}$=183 keV resonance for both (p,$\alpha$) and (p,$\gamma$) channels have drastically reduced the uncertainties on both $^{17}$O(p,$\alpha$)$^{14}$N and $^{17}$O(p,$\gamma$)$^{18}$F rates in the context of explosive H-burning, only one direct measurement for the $E^R_{c.m.}$=65 keV resonance was performed [4]. In fact, direct measurements of low-lying resonance strengths such that of the 65 keV resonant level, are very difficult because the Coulomb barrier and large uncertainties are still present on the available direct data. In addition, some sub-threshold levels could contribute to the total reaction rate and then a further study of this reaction in the energy region relevant for astrophysics is necessary. In the present work we report on the indirect measurement of $^{17}$O(p,$\alpha$)$^{14}$N reaction at energies below 300 keV by using the Trojan Horse Method [5, 6].

## 2. The experiment: measurement and data analysis

In order to reduce the uncertainties affecting the direct measurements, in the last twenty years many indirect methods have been developed. In particular the Trojan Horse Method (THM) [5, 6] is a powerful tool which selects, under appropriate kinematical conditions, the quasi-free (QF) contribution of a suitable three-body reaction performed at energies well above the Coulomb barrier to extract a charged particle two-body cross section at astrophysical energies, free of Coulomb suppression and electron screening effects.
The present study of the $^{17}$O(p,$\alpha$)$^{14}$N reaction in the energy window relevant for astrophysics was performed by selecting the QF-contribution of the $^2$H($^{17}$O,$^{14}$N$\alpha$)n reaction [7]. The deuteron was used as 'trojan horse nucleus' because of its p-n cluster structure: the proton is brought in the nuclear field of $^{17}$O while the neutron acts as a spectator to the reaction [7]. The experiment was performed at the Laboratori Nazionali del Sud in Catania. The SMP Tandem Van de Graaf accelerator provided a 41 MeV $^{17}$O beam with a spot size on target of about 1.5 mm ad intensities up to 2-3 nA. Deuterated polyethylene targets (CD2) of about 150 $\mu$g/cm$^2$ were placed at 90°

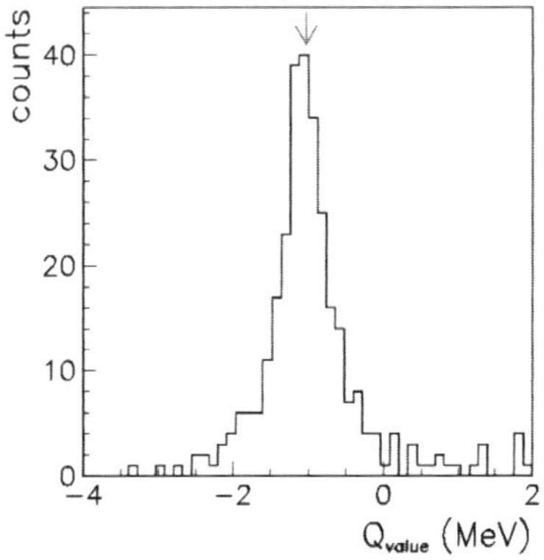


**Figure 2.** Experimental Q-value for the three body reaction $^{2}H(^{17}O,^{14}N)n$. The peak at about -1.0 MeV must be compared with the -1.033 MeV theoretical Q-value (red arrow).

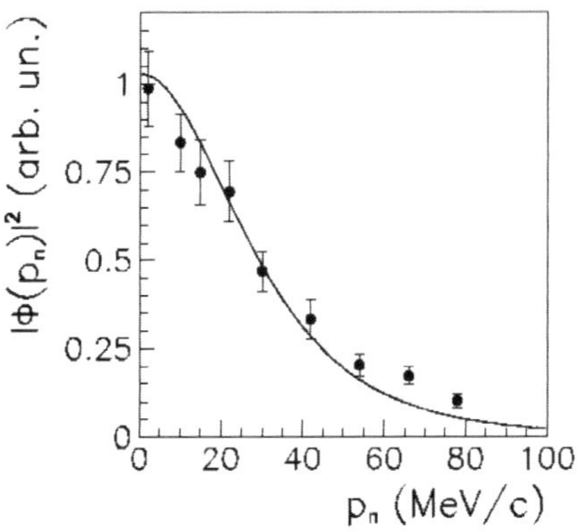


**Figure 3.** Experimental neutron momentum distribution. The full line represents the shape of the theoretical Hulthén function in momentum space.

with respect to the beam axis. The detection setup consisted of six Position Sensitive Detector (PDS) and two ionization chambers filled with 60 mbar of isobuthane gas as $\Delta$E detector. The angular ranges were chosen in order to cover momentum values $p_s$ of the undetected neutron ranging from -100 MeV/c to 100 MeV/c. This allows us to select the kinematical region where a strong contribution of the Quasi Free ( QF) mechanism is expected.

The first step in a THM analysis is the selection of the three-body reaction channel. After the energy and position calibration of the detectors, nitrogen particles were selected with the standard $\Delta$E-E technique and for these selected events the experimental Q-value spectrum was reconstructed and compared with the theoretical value (Fig. 2). A sharp peak centered at about -1 MeV shows up, which corresponds to our three-body reaction, according to the expected theoretical value ($Q_{theor}$ = -1.033 MeV). The agreement between the two values is a test of the goodness of the adopted calibration.

The selection of the QF-condition is related to the behavior of the spectator momentum values. In fact, the QF-processes are characterized by the presence in the exit channel of a particle that acts as a spectator; this means that in the exit channel the spectator particle, the neutron in this case, must have the same momentum distribution that it had before the interaction of oxygen beam with the deuteron target. By following the Plain Wave Impulse Approximation (PWIA) [8] the experimental momentum distribution was deduced as reported in [9] and the result is shown in Fig. 3. The full line superimposed onto the data represents the shape of the theoretical Hulthén function which is normalized to the experimental maximum. A quite good agreement shows up, making us confident that in the experimentally selected kinematical region the QF mechanism gives the main contribution to the $^{2}H+^{17}O$ reaction. The further analysis was performed by considering coincidence events corresponding to neutron momentum values ranging between -30 and 30 MeV/c.

## 3. Results and Conlusion

In the framework of the PWIA, the three-body cross section can be factorized into different terms [10]:

$$\frac{d^3\sigma}{dE_{cm} d\Omega_{^{14}N} d\Omega_\alpha} \propto KF \cdot |\Phi(\vec{p}_s)|^2 \cdot \frac{d\sigma^N}{d\Omega} \tag{1}$$

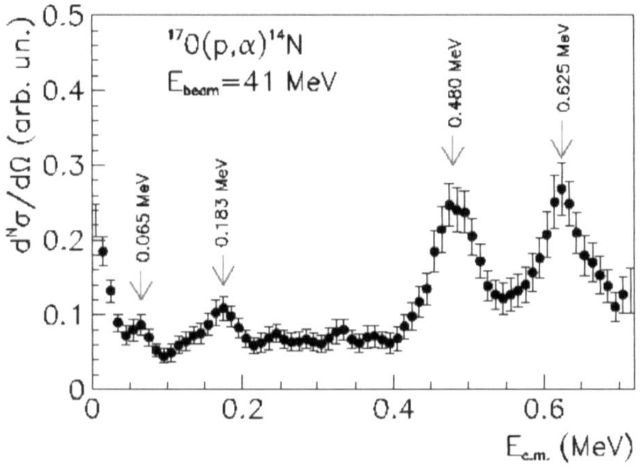


**Figure 4.** Nuclear excitation function $d\sigma^N/d\Omega$ of the $^{17}O(p,\alpha)^{14}N$ reaction by means of the THM in the PWIA approach. The red arrows are placed in correspondence of the resonance peaks in the excitation function.

where $KF$ is a kinematical factor containing the final state phase-space factor, $\Phi(\vec{p}_s)$ is the Fourier transform of the radial wave function $\chi(\vec{r})$ for the p-n intercluster relative motion. By applying the previous formula, the behavior of the nuclear cross section for the reaction studied here is obtained and the result is shown Fig.4 where a clear evidence of both levels at $E_{c.m.}$=65 and 183 keV is given. For this reason this figure represents the first experimental measurement of the $^{17}O(p,\alpha)^{14}N$ two-body reaction at very low energy, well below the Coulomb barrier (~2.1 MeV) where several excited states of $^{18}F$ have been populated. In particular, all populated states have a great importance on reaction rate calculation being in the energetic region of Gamow peak. Moreover, the presence of two sub-threshold states at energy $E_X$=5.605 MeV and $E_X$=5.603 MeV should be taken into account because their contribution and because their high energy tails could be interference with the resonant level at 65 keV. The error bars represent the statistical uncertainties. The 'nuclear' two-body cross section is expressed in arbitrary units since the current TH investigation doesn't allow to extract the results in absolute units.
However, besides the encouraging results obtained through the indirect investigation of such reaction, our results are affected by a statistical error of ~25%. In order to increase the statistics and to improve the resolution, a further experiment was performed at Nuclear Structure Lab of the University of Notre Dame (Indiana, USA) in November 2008. The data analysis is still in progress.

## References

[1] Coc A, et al. 2000 *Astronomy & Astrophysics* **357** 561
[2] Chafa A, et al. 2007 *Phys. Rev.* C **75** 035810
[3] Angulo C, et al. 1999 *Nucl. Phys.* A **656** 3
[4] Blackmon J C, et al. 1995 *Phys. Rev. Lett* **74** 2642
[5] Spitaleri C, et al. 1999 *Phys. Rev.* C **60** 55802
[6] Spitaleri C, et al. 2004 *Phys. Rev.* C **69** 55806
[7] Zadro M, et al. 1989 *Phys. Rev.* C **C40** 181
[8] Chew G F and Wick G C 1952 *Phys. Rev.* **85** 636
[9] Sergi M L, et al. 2008 *AIP Conf. Proc.* **1016** 433
[10] Jain M, et al. 1970 *Nuclear Physics* A **153** 49

Journal of Physics: Conference Series **202** (2010) 012022 doi:10.1088/1742-6596/202/1/012022

# The s-process - overview and selected developments

**René Reifarth**

Gesellschaft für Schwerionenforschung mbH, Darmstadt, D-64291, Germany
J.W. Goethe Universität, Frankfurt a.M, D-60438, Germany

E-mail: r.reifarth@gsi.de

**Abstract.** Almost all of the heavy elements are produced via neutron capture reactions in a multitude of stellar production sites. The predictive power of the underlying stellar models is currently limited because they contain poorly constrained physics components such as convection, rotation or magnetic fields.

Neutron captures measurements on heavy radioactive isotopes provide a unique opportunity to largely improve these physics components, and thereby address important questions of nuclear astrophysics. Such species are branch-points in the otherwise uniquely defined path of subsequent neutron captures along the s-process path in the valley of stability. These branch points reveal themselves through unmistakable signatures recovered from pre-solar meteoritic grains that originate in individual element producing stars.

Measurements on radioactive isotopes for neutron energies in the keV region represent a stringent challenge for further improvements of experimental techniques. This holds true for the neutron sources, the detection systems and the technology to handle radioactive material.

## 1. Introduction

About 50% of the element abundances beyond iron are produced via slow neutron capture nucleosynthesis (*s* process). Starting at iron-peak seed, the *s*-process mass flow follows the neutron rich side of the valley of stability via a sequence of neutron captures and $\beta^-$ decays (see Figure 1) synthesizing the elements between iron and bismuth.

If different reaction rates are comparable, the *s*-process path branches and the branching ratio reflects the physical conditions in the interior of the star. Such nuclei are most interesting, because they provide the tools to effectively constrain modern models of the stars where the nucleosynthesis occurs. As soon as the $\beta^-$ decay is significantly faster than the typically competing neutron capture, no branching will take place. Therefore experimental neutron capture data for the *s* process are only needed, if the respective neutron capture time under stellar conditions is similar or smaller than the $\beta^-$ decay time, which includes all stable isotopes. Depending on the actual neutron density during the *s* process, the "line of interest" is closer to or farther away from the valley of stability.

The modern picture of the main *s*-process component refers to the He shell burning phase in AGB stars [1]. Nuclei with masses between 90 and 209 are mainly produced during the main component. The highest neutron densities in this model occur during the $^{22}$Ne($\alpha$,n) phase and are up to $10^{12}$ cm$^{-3}$ with temperatures around $kT$ = 30 keV. The other extreme can be found during the $^{13}$C($\alpha$,n) phase where neutron densities as low as $10^7$ cm$^{-3}$ and temperatures around $kT$ = 5 keV are possible. Similarly to the main component, also the weak component referring to different evolutionary stages in massive stars has to phases [2, 3]. Mainly nuclei with masses

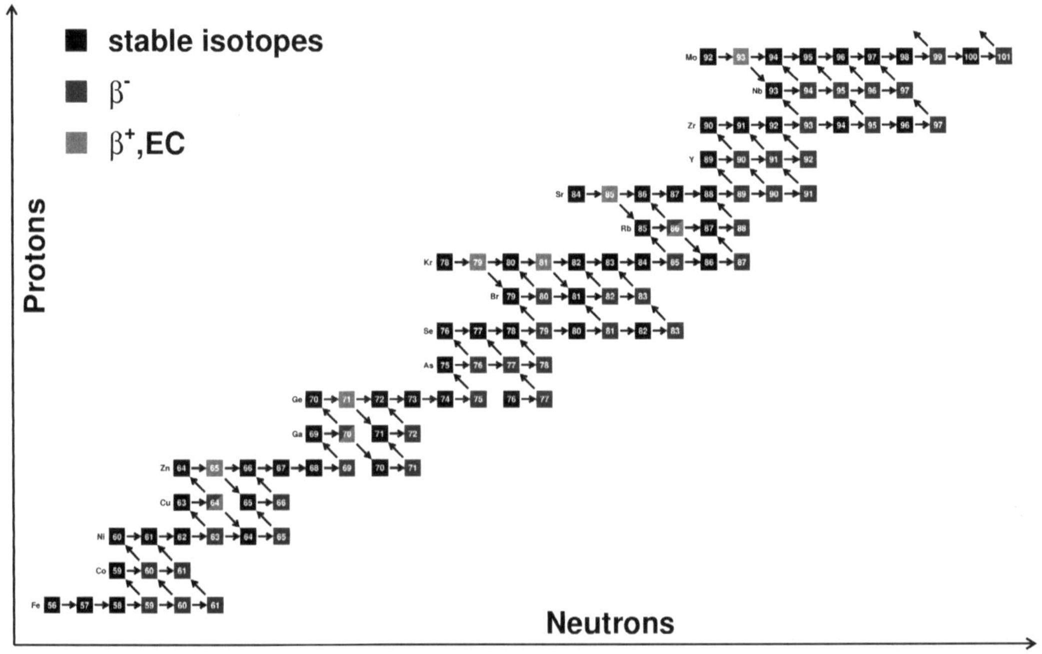


**Figure 1.** The *s*-process path between Fe and Mo, the domain of the weak component.

between 56 and 90 are mainly produced during the weak component. The first phase occurs during the helium core burning with neutron densities down to $10^6$ cm$^{-3}$ and temperatures around $kT = 25$ keV. The second phase happens during the carbon shell burning with neutron densities up to $10^{12}$ cm$^{-3}$ at temperatures around $kT = 90$ keV.

The left part of Figure 2 shows a summary of the neutron capture and $\beta^-$ decay times for radioactive isotopes on the neutron rich side of the valley of stability, under the condition that the neutron capture during the $^{13}$C($\alpha$,n) occurs faster than than the $\beta^-$ decay at stellar temperature. Obviously the vast majority of isotopes, where an experimental neutron capture cross section is desirable, have terrestrial $\beta^-$ half-lives of at least thousands of days. The right part Figure 2 shows the same as the left part, but for the higher neutron density and temperature. Now isotopes with half-lives down to a few days can be of interest for the *s*-process reaction network.

Improved experimental techniques, especially as far as the neutron source and sample preparation are concerned, are necessary to perform direct neutron capture measurements on such isotopes [6]. Though the activation method or accelerator mass spectroscopy of the reaction products could be applied in a limited number of cases, experimental facilities like DANCE at LANL (USA) [7], n-TOF at CERN (Switzerland) [8] and the upcoming projects like SARAF (Israel) [9] and FRANZ at the Goethe University in Frankfurt (Germany) [10] are addressing the need for such measurements on the basis of the more universal method of detecting the prompt capture $\gamma$-rays, which is required for the application of the neutron time-of-flight (TOF) technique. In particular the FRANZ facility, which will be located close to the new FAIR facility might allow the investigation of radioactive isotopes with half-lives down to tens of days, while present facilities require half-lives of a few hundred days.

In section 2 the time-of-flight method is explained at the example of the DANCE facility,

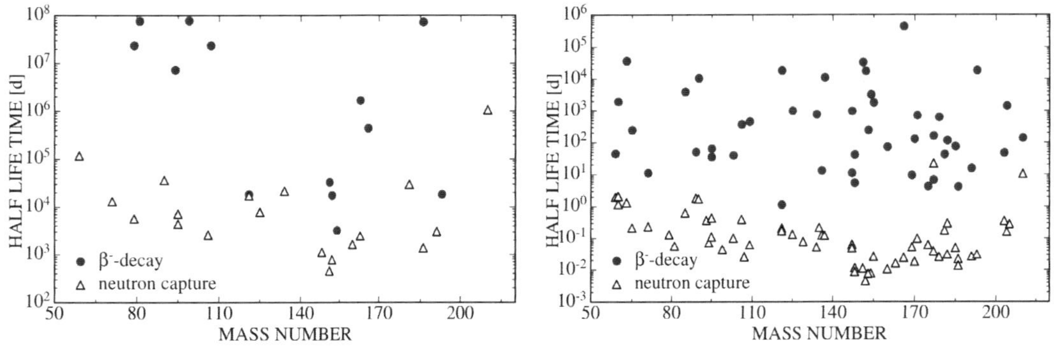


**Figure 2.** Left: Half life times with respect to neutron capture (open triangles) for a neutron of density $10^7$ cm$^{-3}$ and terrestrial $\beta^-$ half life times (filled circles) for unstable isotopes on the s-process path as a function of mass number. Shown are only isotopes where the neutron capture is faster than the $\beta^-$-decay at $kT = 5$ keV. The neutron capture cross sections are taken from [4] and the half lives under stellar conditions from [5]. Right: Same as left, but for a neutron of density $10^{12}$ cm$^{-3}$ and $\beta^-$-decays at $kT = 30$ keV

the currently most sensitive facility for (n,$\gamma$) experiments and a new approach (FRANZ) currently realized at the Goethe University Frankfurt, Germany is described. The FRANZ facility will allow energy-dependent neutron capture cross section and activation experiments in the astrophysically interesting energy region with significantly higher neutron fluxes then currently available elsewhere [11]. In section 3 alternative approaches like the activation or indirect methods are described.

## 2. The time-of-flight method

### *2.1. DANCE*

The Detector for Advanced Neutron Capture Experiments (DANCE) is designed as a high efficiency, highly segmented $4\pi$ $BaF_2$ detector for calorimetrically detecting $\gamma$-rays following a neutron capture. DANCE is located on the 20 m neutron flight path 14 (FP14) at the Manuel Lujan Jr. Neutron Scattering Center at the Los Alamos Neutron Science Center (LANSCE). The design of the detector is such that a full $4\pi$ array would consist of 162 crystals of four different shapes, each shape covering the same solid angle [12]. Two of the 162 crystals are left out in order to leave space for the neutron beam pipe, see Figure 3.

Depending on the experiment, one crystal can be replaced by a sample changer mechanism, which makes it possible to exchange up to 3 samples without closing the beam shutter and breaking the vacuum of the beam pipe. Thus the full array is designed to host 159 or 160 out of 162 possible $BaF_2$ crystals. The dimensions of the bare crystals are designed to form a $BaF_2$ shell with an inner radius of 17 cm and a thickness of 15 cm. Thanks to the fairly low repetition rate of 20 Hz, measurements can be carried out over the whole energy range from 10 meV to 500 keV. This combination of a strong neutron source and a high efficiency $\gamma$-ray detector allows to measure (n,$\gamma$) cross sections of radioactive isotopes down to a few hundred days half-life. Further details on the overall performance of the array can be found in [7]. For details on the analysis and the neutron flux see [13]).

One of the first reactions investigated at DANCE was the neutron capture on $^{62}$Ni, which is related to the weak *s*-process component during helium burning in massive stars of 10 to 25 $M_\odot$ [14]. The production of $^{62}$Ni in a 25 $M_\odot$ stellar model [15] motivated a closer investigation of

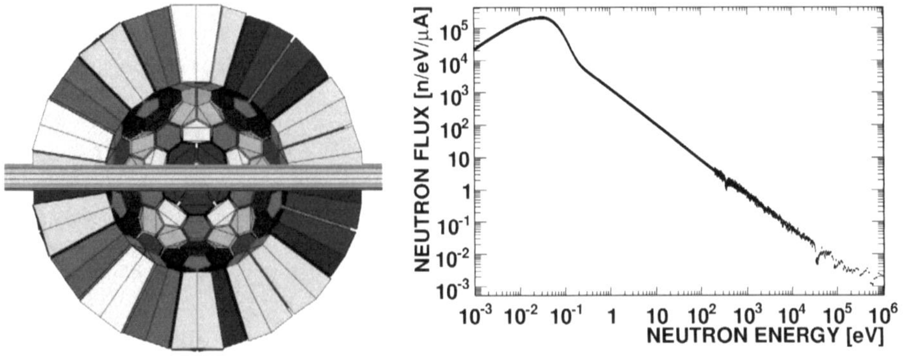


**Figure 3.** Left: Schematic drawing of one half of in total 160 crystals of the DANCE array including the neutron beam pipe. Right: The neutron flux given per $\mu$A at DANCE. Typically 100 $\mu$A of beam are available during experiments.

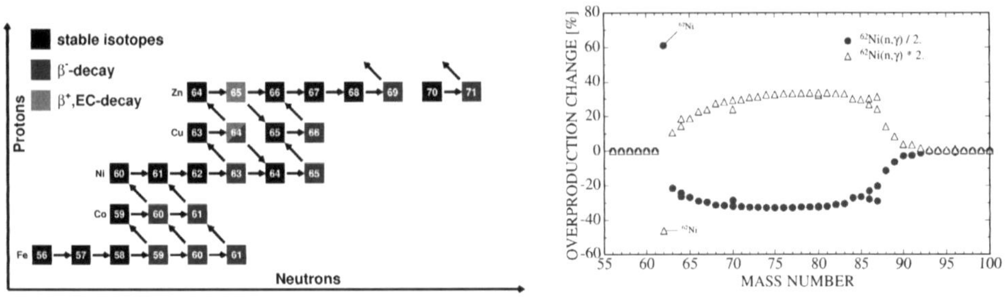


**Figure 4.** Left: $s$-process nucleosynthesis in the region between iron and tin with the important branching at $^{63}$Ni. Right: Example of the effect of changing one cross section in the weak s-process path. The standard case was calculated for a $25M_{\odot}$ star using the cross section data from [16]. The two curves show the relative change in overproduction after changing this cross section by a factor of two down (blue, filled circles) or up (open, black triangles). Changing one cross section in the weak s-process path affects the production of the respective isotope and all following ones.

the neutron capture rates involved (see Figure 4). Details can be found in Ref. [16]. The derived Maxwellian averaged cross section at $kT = 25$ keV ($31.5 \pm 2.5_{stat} \pm 2.2_{sys}$ mb) is in agreement with recently performed activation measurements (see section 3), but seems to disagree with a recent TOF measurement [17]. One reason for the disagreement could be the determination of the efficiency for detecting the $\gamma$-cascades after the neutron capture on $^{62}$Ni in the recent TOF measurment [17]. A new measurement in this interesting mass region around nickel is scheduled for 2009. The DANCE array will be used to investigate the radioactive branch point nucleus $^{63}$Ni, which has an half life of 100 years.

### *2.2. FRANZ*

The Stern-Gerlach-Zentrum SGZ, recently founded at the University of Frankfurt, allows to build and operate larger experiments now in accelerator physics, astrophysics and material science research. It was decided to develop an intense neutron generator, the "Frankfurter

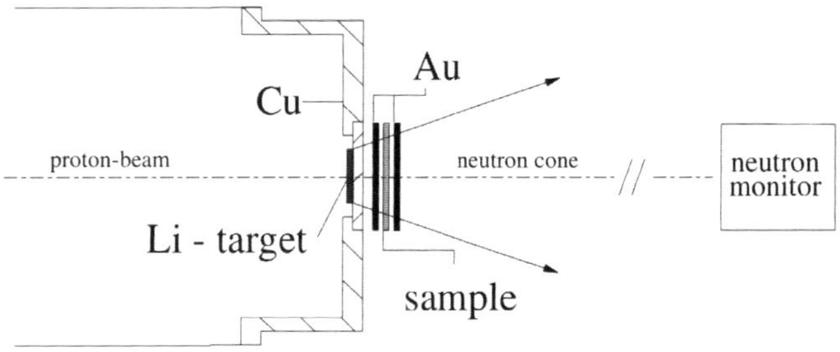


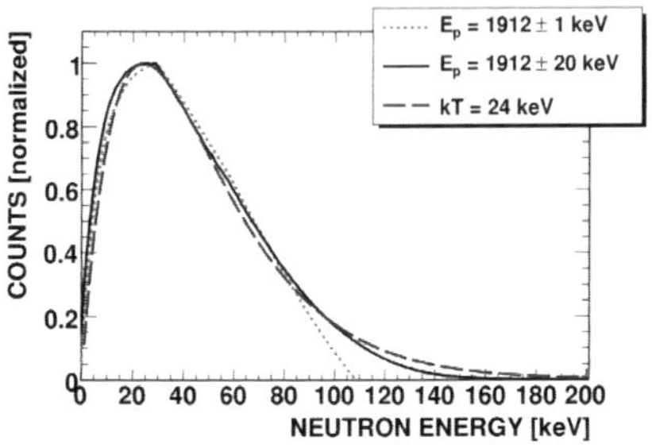


**Figure 5.** Left: Typical activation setup at the Forschungszentrum Karlsruhe. Neutrons are produced via the $^{7}$Li(p,n) reaction just above the production threshold. The emitted neutrons are then kinematically focussed into a cone with an opening angle of 120°. The sample is usually sandwiched by two gold foils in order to determine the neutron flux just before and behind the sample. Right: Comparison of simulated activation spectra including two gaussian proton energy profiles with a theoretical spectrum corresponding to neutron fluxes at $kT = 24$ keV [25].

Neutronenquelle am Stern-Gerlach-Zentrum" (FRANZ). The planned proton driver LINAC consists of a high voltage terminal already under construction to provide primary proton beam energies of up to 150 keV. A volume type ion source will deliver a DC beam current of 100-250 mA at a proton fraction of 90%. A low energy beam transport using four solenoids will inject the proton beam into an RFQ, while a chopper at the entrance of the RFQ will create pulse lengths in the range of 100 ns at a repetition rate of up to 250 kHz. A drift tube cavity, which delivers variable end energies between 1.9 and 2.1 MeV will be installed downstream of the RFQ. Finally a bunch compressor of the Mobley type forms a proton pulse length of 1 ns at the Li target, where neutrons will be produced via the $^{7}$Li(p,n) reaction. The maximum energies of the neutrons are then be adjustable between ≈30 keV and ≈500 keV by varying the primary proton beam energy (for details see also [18]).

## 3. Alternative methods

### *3.1. The activation method*

If the product of the neutron capture is radioactive, the freshly produced nuclei can be detected after an extended period of irradiation either via their unique decay signatures (if the half life time is rather short) or by applying accelerator mass spectroscopy (AMS).

While other neutron-energy distributions were used on occasion [19, 20], the quasi-stellar neutron spectrum, which can be obtained by bombarding a thick metallic Li target with protons of 1912 keV, slightly above the reaction threshold at 1881 keV, was the working horse at the Forschungszentrum Karlsruhe [4]. Under such conditions, the $^{7}$Li(p,n)$^{7}$Be reaction yields a continuous energy distribution with a high-energy cutoff at $E_n = 106$ keV. The produced neutrons are emitted in a forward cone of 120° opening angle. The angle-integrated spectrum closely resembles a distribution necessary to determine the Maxwellian averaged cross section (MACS) for a thermal energy of kT = 25 keV [21]. The samples are typically sandwiched between gold foils and placed directly on the backing of the lithium target. A typical setup is sketched in Fig. 5.

Triggered by the significant uncertainty for the (n,$\gamma$) cross section of $^{62}$Ni [4, 22], two activation experiments have been carried out detecting the number of produced $^{63}$Ni nuclei via AMS [23, 24]. The activation measurements agree with the TOF measurement carried out at DANCE within the quoted uncertainties.

### 3.2. Indirect methods

Indirect methods play an important role in the determination of nuclear reaction cross sections that are hard to measure directly. There are two main approaches in this context. The first is the determination of the desired A(n,$\gamma$)B via the inverse reaction B($\gamma$,n)A. If the product of the neutron capture B is itself radioactive, the Coulomb dissociation (CD) method can be applied at radioactive beam facilities [26, 27]. If the product is stable or very long-lived real photons can be used is addition to the CD method [28]. Another method that is used successfully for neutron-induced fission cross sections is the surrogate method. So far, no successful proof of principle could be performed for neutron capture cross sections, but it might be an interesting additional approach [29].

## 4. Summary

Nowadays, the s-process as a nucleosynthesis process is well understood and established. Current research uses the s-process as a link between abundance observations and stellar models. Data on radioactive branch point nuclei are urgently needed to enhance the reliability of that approach. Current facilities can measure some, upcoming facilities will investigate a suite of radioactive isotopes. There will always be the need for other than TOF methods, especially for isotopes with short half lives.

## Acknowledgments

R.R. is supported by the HGF Young Investigators Project VH-NG-327.

## References

[1] Lugaro M, Herwig F, Lattanzio J C, Gallino R and Straniero O 2003 *Ap.J.* **586** 1305
[2] Travaglio C, Gallino R, Arnone E, Cowan J, Jordan F and Sneden C 2004 *Ap. J.* **601** 864
[3] Heil M, Käppeler F, Uberseder E, Gallino R and Pignatari M 2007 *Prog. Nucl. Part. Phys.* **59** 174
[4] Bao Z Y, Beer H, Käppeler F, Voss F, Wisshak K and Rauscher T 2000 *At. Data Nucl. Data Tables* **76** 70
[5] Takahashi K and Yokoi K 1987 *Atomic Data Nucl. Data Tables* **36** 375
[6] Reifarth R, Haight R C, Heil M, Käppeler F and Vieira D J 2004 *Nucl. Instr. Meth. A* **524** 215
[7] Reifarth R, Bredeweg T A, Alpizar-Vicente A, Browne J C, Esch E I *et al.* 2004 *NIM A* **531** 528
[8] Abbondanno U, Aerts G *et al.* (n-TOF Collaboration) 2004 *PRL* **93** 161103
[9] Nagler A, Mardor I, Berkovits D, Piel C, vom Stein P and Vogel H 2006 *LINAC 2006* (Knoxville, TN: Joint Accelerator Conferences Website) p MOP054
[10] Meusel O, Chau L P, Mueller I, Ratzinger U, Schempp A, Volk K and Zhang C 2006 *LINAC 2006* (Knoxville, TN: Joint Accelerator Conferences Website) p MOP051
[11] Koehler P 2001 *Nucl. Instr. Meth. A* **460** 352 – 361
[12] Heil M, Reifarth R, Fowler M M, Haight R C, Käppeler F *et al.* 2001 *Nucl. Instr. Meth. A* **459** 229
[13] Esch E I, Reifarth R, Bond E M, Bredeweg T A, Couture A *et al.* 2008 *Phys. Rev. C* **77** 034309
[14] Raiteri C, Gallino R, Busso M, Neuberger D and Käppeler F 1993 *Ap. J.* **419** 207
[15] Rauscher T, Heger A, Hoffman R and Woosley S 2002 *Ap. J.* **576** 323
[16] Alpizar-Vicente A M, Bredeweg T A, Esch E I, Greife U, Haight R C *et al.* 2008 *Phys. Rev. C* **77** 015806
[17] Tomyo A, Temma Y, Segawa M, Nagai Y, Makii H *et al.* 2005 *Ap. J.* **623** L153
[18] Reifarth R, Chau L P, Heil M, Käppeler F, Meusel O *et al.* 2009 *PASA* **26(3), in press**
[19] Reifarth R, Schwarz K and Käppeler F 2000 *Ap. J.* **528** 573 – 581
[20] Reifarth R, Heil M, Forssén C, Besserer U, Couture A *et al.* 2008 *Phys. Rev. C* **77** 015804
[21] Ratynski W and Käppeler F 1988 *Phys. Rev. C* **37** 595 – 604
[22] Rauscher T and Guber K H 2002 *Phys. Rev. C* **66** 028802
[23] Nassar H, Paul M, Ahmad I, Berkovits D, Bettan M *et al.* 2005 *Phys. Rev. Let.* **94** 092504
[24] Domingo-Pardo C, Dillmann I, Faestermann T, Giesen U, Görres J *et al.* 2009 *American Institute of Physics Conference Series* vol 1090 ed Jolie J, Zilges A, Warr N and Blazhev A p 230
[25] Reifarth R, Heil M, Käppeler F and Plag R 2009 *Nucl. Inst. Meth. A* **608** 139
[26] Nakamura T, Fukuda N, Aoi N, Imai N, Ishihara M *et al.* 2009 *Phys. Rev. C* **79** 035805
[27] Datta Pramanik U, Aumann T, Boretzky K, Carlson B, Cortina D *et al.* 2003 *Phys. Lett. B* **551** 63
[28] Sonnabend K, Mohr P, Vogt K, Zilges A, Mengoni A *et al.* 2003 *Ap. J.* **583** 506
[29] Forssén C, Dietrich F S, Escher J, Hoffman R D and Kelley K 2007 *Physical Review C* **75** 055807

# The effect of $^{12}$C + $^{12}$C rate uncertainties on s-process yields

**M E Bennett[1], R Hirschi[1,2], M Pignatari[7,3,4], S Diehl[5], C Fryer[6], F Herwig[7], A Hungerford[6], G Magkotsios[3,8], G Rockefeller[6], F Timmes[8], M Wiescher[3] and P Young[8].**

[1] Astrophysics Group, Keele University, ST5 5BG, UK
[2] IPMU, University of Tokyo, Kashiwa, Chiba 277-8582, Japan
[3] Joint Institute for Nuclear Astrophysics, University of Notre Dame, IN, 46556, USA
[4] TRIUMF, 4004 Wesbrook Mall, Vancouver, BC, Canada, V6T 2A3
[5] Theoretical Astrophysics Group (T-6), LANL, Los Alamos, NM, 87544, USA
[6] Computation, Physics and Methods (CCS2), LANL, Los Alamos, NM, 87544, USA
[7] Dept. of Physics & Astronomy, Victoria, BC, V8W 3P6,Canada
[8] School of Earth and Space Exploration, University of Arizona, Tempe, AZ 85287, USA

E-mail: meb@astro.keele.ac.uk

**Abstract.** The slow neutron capture process in massive stars (the weak s-process) produces most of the s-only isotopes in the mass region $60 < A < 90$. The nuclear reaction rates used in simulations of this process have a profound effect on the final s-process yields. We generated 1D stellar models of a $25M_{\odot}$ star varying the $^{12}$C + $^{12}$C rate by a factor of 10 and calculated full nucleosynthesis using the post-processing code PPN. Increasing or decreasing the rate by a factor of 10 affects the convective history and nucleosynthesis, and consequently the final yields.

## 1. Introduction

Elements in the solar system are formed from a variety of nucleosynthesis processes in stars, such as the slow and rapid neutron-capture processes (the s-process and r-process respectively). The s-process signature of the solar system abundances can be split into three components; the weak-component, which involves nuclei with $60 < A < 90$, the main-component, which involves nuclei having atomic mass between $90 < A < 208$, and the strong component, which accounts for the production of the solar $^{208}$Pb [1]. The s-process site is attributed to massive stars for the weak component, and to AGB stars having initial mass between 1 and $3M_{\odot}$ at solar-like metallicity for the main component and at low metallicity for the strong component [2]. Stellar evolution models of massive stars can be used to determine information on the conditions within stellar interiors and calculate the s-process yields relevant for the weak-component. Any changes to the input physics, such as improved laboratory nuclear reaction rates, can affect the evolution of the simulated star and consequently affect the s-process yields. Therefore, different nuclear reaction rates can be tested for their astrophysical impact. In this work, we will show how variations in the rate of the $^{12}$C + $^{12}$C reaction affect the evolution of a $25M_{\odot}$ star and the consequential s-process yields of the star. The motivation for this work originates from nuclear physics experiments and theory (see for instance Spillane et al. 2007 [3] and Gasques et al.

2007 [4]), which are probing low enough energies to investigate the reaction within the Gamow window for $^{12}C+^{12}C$ fusion.

## 2. Carbon burning in massive stars and the s-process

Carbon-core burning occurs at $T = 0.6 - 0.8$ GK. The dominating reactions are $^{12}C(^{12}C,\alpha)^{20}Ne$ and $^{12}C(^{12}C,p)^{23}Na$ which, combined with the efficient $^{23}Na(p,\alpha)^{20}Ne$ reaction, leaves $^{16}O$ and $^{20}Ne$ as the dominant isotopes in the core (the $^{16}O$ being a remnant from the previous helium burning stage). In addition to core burning, there are typically multiple stages of shell burning later in the evolution of the star, where the number of shells differs depending on the initial mass of the star. The main neutron source for the s-process in massive stars is $^{22}Ne$ by the reaction $^{22}Ne(\alpha,n)^{25}Mg$. $^{22}Ne$ is produced in helium burning from the $^{14}N(\alpha,\gamma)^{18}F(\beta^+)^{18}O(\alpha,\gamma)^{22}Ne$ reaction chain during the supergiant phase. $^{22}Ne$ is partially burnt at the end of helium burning with a neutron density, $n_n \sim 10^6$ cm$^{-3}$ (see, for example, Raiteri et al. 1991 [5]). The $^{22}Ne$ left in the CO-core is then burnt during carbon-shell burning with higher neutron densities ($n_n \sim 10^{11} - 10^{12}$cm$^{-3}$), with the $^{12}C(^{12}C,\alpha)^{20}Ne$ reaction providing the $\alpha$ particles [6]. The carbon-shell burning is sensitive to the profile of $^{12}C$ after core-carbon burning, which is in turn sensitive to the still uncertain $^{12}C(\alpha,\gamma)^{16}O$ rate and the choice of convection physics in the models [6][7][8]. In addition, the yields are sensitive to changes in reactions involving $^{22}Ne$ or its formation [6][9]. Most of the s-process-rich material ejected by the supernova event of a $25M_\odot$ star is formed by carbon-shell burning, affecting the weak s-component [10][11].

Stellar models of a $25M_\odot$ star with metallicity $Z = 0.01$ were generated using the Geneva Stellar Evolution Code [12], with the nuclear network post-processed using the NuGrid PPN tool [13]. The post-processing was computed using the KHAOS cluster at Keele University. Variations in the carbon burning rate were chosen in relation to the 'standard' Caughlan and Fowler rate (CF88 from now on). Models were generated with the CF88 rate (referred to as model C12s in this work), the CF88 rate multiplied by 10 (C12t10) and divided by 10 (C12d10). The ratio of the p- and $\alpha$-channels was chosen to be 35%:65% [14]. The additional uncertainty associated with this choice of ratio will be investigated in a forth-coming paper.

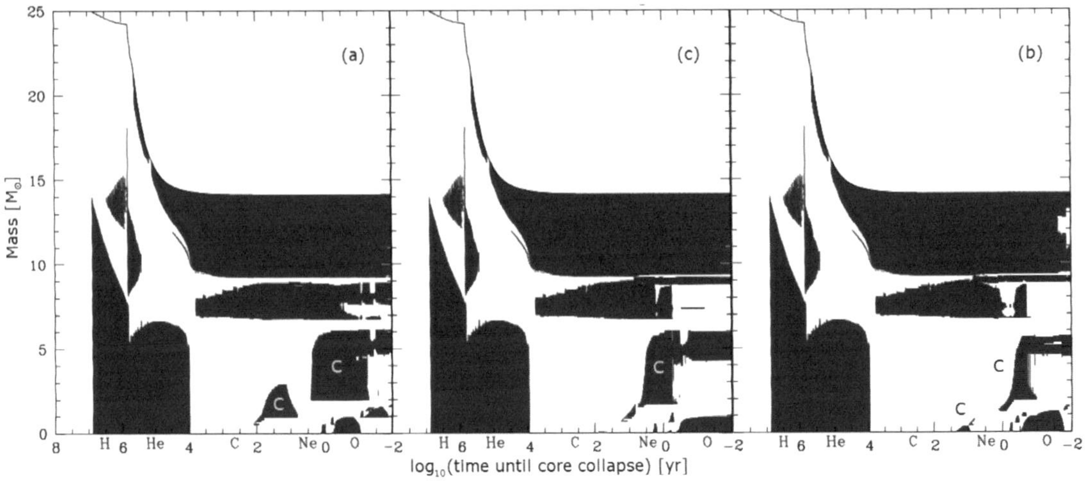


**Figure 1.** Structure evolution diagrams of $25M_\odot$ models at solar metallicity using different $^{12}C$ burning rates. Here, (a) is model C12t10, (b) is model C12s and (c) is model C12d10. The shaded regions correspond to convection zones present in the star, with the type of burning indicated by labels.

Differences in the convective history can be seen in each of the different models. Fig. 1 shows structure evolution diagrams for the models C12t10, C12s and C12d10. In each case the convection zones for carbon shell burning at $\log_{10}$(time in years until core collapse) $\sim$ 1 to -2 differ in size and duration. For model C12t10, the duration of the second carbon-shell burning process is much longer and the size of the convection zones is larger. A large fraction of overlap between the first shell and the second shell is observed and isotopes produced in the first shell by the s-process will be mixed into the second. The presence of overlapping carbon shells was previously noted by Chieffi et. al. [15]. For model C12d10, the second carbon-shell burning episode occurs later in time than for the other two models, which is due to the star contracting further before it can reach a temperature high enough so that carbon burning is activated in the shell.

Fig. 2 shows the relative abundances of stable isotopes in the second carbon shell of the C12d10 model with respect to the C12s model. Fig. 3 shows the equivalent plot for the C12t10 model with respect to the C12s model. In fig. 2, a clear signature of a higher neutron density is shown, e.g. higher production of r-only species $^{70}$Zn, $^{76}$Ge and lower production of $^{80}$Sr due to the $^{79}$Se branching. The reason is that the star in model C12d10 contracts further, resulting in carbon shells that burn at a higher temperature than the C12s model. Therefore the $^{22}$Ne neutron source is burned at a higher temperature, increasing the neutron density. The s-process efficiency is quite similar in model C12s and C12d10 (see, for example, the similar production of neutron magic isotope $^{88}$Sr). Concerning model C12t10, in fig. 3, a general increase in the s-process efficiency for isotopes with $60 < A < 90$ is shown, compared to model C12s. The overlap between the first and second convective shells causes the initial distribution of isotopes at the start of the second carbon-shell burning to be affected by the products of the first shell. In the first shell, neutrons are also efficiently produced by an additional neutron source, $^{17}$O($\alpha$,n)$^{21}$Ne, where $^{17}$O is mostly produced by $^{16}$O(p,$\gamma$)$^{17}$F($\beta^+$)$^{17}$O.

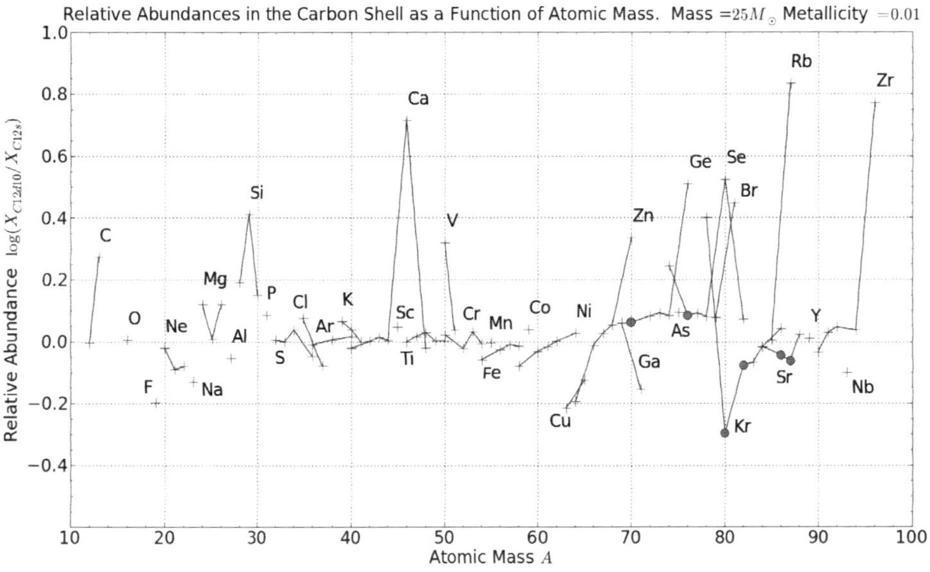


**Figure 2.** The relative abundances of stable isotopes up to Niobium between models C12d10 and C12s. S-only isotopes are indicated with a red circle.

In this paper, it has been shown that changes to the carbon burning rate by a factor of 10 in

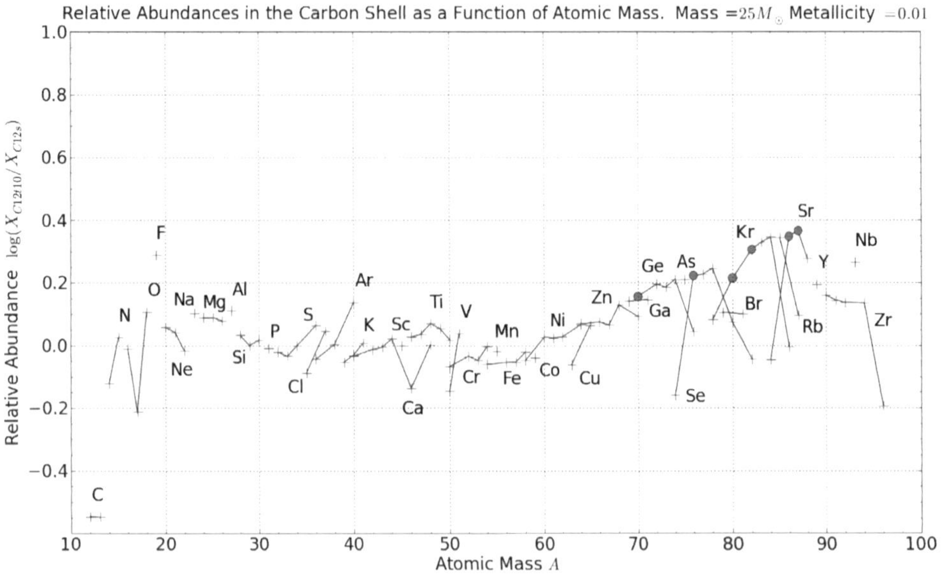


**Figure 3.** The relative abundances of stable isotopes up to Niobium between models C12t10 and C12s. S-only isotopes are indicated with a red circle.

stellar models of a $25M_\odot$ star significantly affect the s-process yields. The overlap of convective carbon shells active at different burning temperatures also has important implications for the s-process yields. Further analysis will be conducted in a forth-coming paper.

## References

[1] Kappeler F, Beer H and Wisshak K 1989 *Reports on Progress in Physics* **52** 945–1013
[2] Gallino R, Arlandini C, Busso M, Lugaro M, Travaglio C, Straniero O, Chieffi A and Limongi M 1998 *Astrophysical Journal* **497** 388
[3] Spillane T, Raiola F, Rolfs C, Schürmann D, Strieder F, Zeng S, Becker H W, Bordeanu C, Gialanella L, Romano M and Schweitzer J 2007 *Physical Review Letters* **98** 122501
[4] Gasques L R, Brown E F, Chieffi A, Jiang C L, Limongi M, Rolfs C, Wiescher M and Yakovlev D G 2007 *Physical Review C* **76** 035802
[5] Raiteri C M, Busso M, Picchio G, Gallino R and Pulone L 1991 *Astrophysical Journal* **367** 228–238
[6] The L S, El Eid M F and Meyer B S 2007 *Astrophysical Journal* **655** 1058–1078
[7] Imbriani G, Limongi M, Gialanella L, Terrasi F, Straniero O and Chieffi A 2001 *Astrophysical Journal* **558** 903–915
[8] El Eid M F, Meyer B S and The L S 2004 *Astrophysical Journal* **611** 452–465
[9] Kaeppeler F, Wiescher M, Giesen U, Goerres J, Baraffe I, El Eid M, Raiteri C M, Busso M, Gallino R, Limongi M and Chieffi A 1994 *Astrophysical Journal* **437** 396–409
[10] Raiteri C M, Busso M, Picchio G and Gallino R 1991 *Astrophysical Journal* **371** 665–672
[11] Hoffman R D, Woosley S E and Weaver T A 2001 *Astrophysical Journal* **549** 1085–1092
[12] Eggenberger P, Meynet G, Maeder A, Hirschi R, Charbonnel C, Talon S and Ekström S 2008 *Astrophysics and Space Science* **316** 43–54
[13] Herwig F, Diehl S, Fryer C L, Hirschi R, Hungerford A, Magkotsios G, Pignatari M, Rockefeller G, Timmes F X, Young P and Bennett M E 2008 *Proceedings of the 10th Symposium on Nuclei in the Cosmos (NIC X). July 27 - August 1, 2008 Mackinac Island, Michigan, USA. Available online at http://pos.sissa.it/cgi-bin/reader/conf.cgi?confid=53*
[14] Aguilera E F, Rosales P, Martinez-Quiroz E, Murillo G, Fernández M, Berdejo H, Lizcano D, Gómez-Camacho A, Policroniades R, Varela A, Moreno E, Chávez E, Ortíz M E, Huerta A, Belyaeva T and Wiescher M 2006 *Physical Review C* **73** 064601
[15] Chieffi A, Limongi M and Straniero O 1998 *Astrophysical Journal* **502** 737

Nuclear Physics in Astrophysics IV IOP Publishing
Journal of Physics: Conference Series **202** (2010) 012024 doi:10.1088/1742-6596/202/1/012024

# A unified model for nucleosynthesis of heavy elements in stars

**Miklós Kiss[1] and Zoltán Trócsányi[2]**
[1]Berze N.J. Gimnázium, Kossuth 33, H-3200 Gyöngyös, Hungary
[2]University of Debrecen and Institute of Nuclear Research of the Hungarian Academy of Sciences, H-4001 Debrecen P.O.Box 51, Hungary
E-mail: kiss-m@chello.hu, z.trocsanyi@atomki.hu

**Abstract.** We prospose a unified model for the nucleosynthesis of heavy ($A > 57$) elements in stars. The neutron flux can be set to describe neutron capture in arbitrary neutron flux. Our approach solves the coupled differential equations, that describe the neutron capture and decays of 2696 nuclei, numerically without truncating those to include only either capture or decay as traditionally assumed in weak neutron flux ($s$ process). As a result the synthesis of heavy nuclei always evolves along a wide band in the valley of stable nuclei. The observed abundances in the Solar system are reproduced reasonably already in the simplest version of the model. The model predicts that the nucleosynthesis in weak or modest neutron flux produces elements that are traditionally assumed to result in the high neutron flux of supernovae explosions ($r$ process).

## 1. Introduction

The traditional approach for nucleosynthesis in stars that maintain weak neutron flux is the slow neutron capture process ($s$ process) that occurs along a path in the stability valley of nuclei. After the pioneering works of Burbidge et al. [1] and of Cameron [2], this process has been studied extensively in the literature [3]. The $s$-process approach starts with the coupled differential equations that describe the change (increase and decrease) of the abundance $N_A$ of a certain nucleus of atomic number $A$ at time $t$ due to neutron capture and the decrease of the abundance due to $\beta$ decay,

$$\frac{\mathrm{d}N_A}{\mathrm{d}t}(t) = N_n(t)\langle\sigma v\rangle_{A-1}N_{A-1}(t) - N_n(t)\langle\sigma v\rangle_A N_A(t) - \Gamma_A^{(\beta)}N_A(t)\,, \qquad (1)$$

where $N_n(t)$ is the neutron density at time $t$, $\langle\sigma v\rangle_A$ is the neutron-capture reaction rate per neutron on a nucleus with mass number $A$ and $\Gamma^{(\beta)} = \ln 2/T_{1/2}^{(\beta)}$ denotes the $\beta$-decay width of the isotope ($T_{1/2}^{(\beta)}$ is the half life). For the large number of possible nuclei involved in this process, the standard way of finding an approximate solution is to assume that either the capture is much faster than the decay, or vice versa. With this assumption the nucleosynthesis of heavy elements evolves along a line in the valley of stable nuclei, called the $s$-process path.

The reaction rate depends on time through the variation of temperature. It is usually assumed that the temperature is constant during neutron irradiation, the typical value being $T = 3.5 \cdot 10^8\,\mathrm{K}$, corresponding to $kT = 30\,\mathrm{keV}$. Then one may use $\langle\sigma v\rangle_A = \langle\sigma_T\rangle_A v_T$ where $\langle\sigma_T\rangle_A$ is the Maxwellian-averaged cross section of neutron capture, that has the typical value of $100\,\mathrm{mb}$, and $v_T = \sqrt{2kT/m_n} \simeq 2.4 \cdot 10^8\,\mathrm{cm/s}$.

The classical $s$-process model can describe the observed abundances of $s$-process heavy elements surprisingly well provided one assumes two components of exponentially decaying neutron irradiation, the larger exposure *main* component and the smaller exposure *weak* component. A usual experimental confirmation of the $s$-process abundance is the comparison of the model prediction for the neutron capture cross section times abundance as a function of the mass number (see e.g. Fig. 19. in Ref. [3].) We note that this comparison involves only those nuclei which belong to the $s$-process path.

## 2. The model

While the classical $s$-process model gives a simple and fairly accurate description of the observed abundances of heavy elements, it certainly has some simplifying features that are worth a closer look. One point is that no matter how low the probability of a certain capture or decay process, if the sample of nuclei is sufficiently large, such events will occur in reality. This means that the evolution of nucleosynthesis along the $s$-process path can only be a simple approximation and one has to discuss carefully the conclusions drawn from the model. With present day computing capacity it is actually not necessary to make the simplifying assumptions that lead to the $s$-process path of evolution. Instead, we choose to solve the full system of differential equations numerically with some different kind of simplifying assumptions.

If we choose to keep all terms in Eq. (1), then we as well can include more terms that may have some relevance. If we do not force the evolution onto a path, then we can find increment of a given abundance $N_A$ by $\alpha$ decay from an already existing nucleus with two more proton and neutron number as well as decrement by possible $\alpha$ decay. This suggests that actually, the more correct abundance to follow is $N_{Z,N}(t)$, the abundance of the nuclei of a certain element with atomic number $Z$. Therefore, instead of Eq. (1), we choose to solve the coupled differential equations with one element as

$$\begin{aligned}\frac{\mathrm{d}N_{Z,N}}{\mathrm{d}t} &= N_n(t)\langle\sigma v\rangle_{A-1}N_{Z,N-1}(t) + \Gamma^{(\beta)}_{Z-1,N+1}N_{Z-1,N+1}(t) + \Gamma^{(\alpha)}_{Z+2,N+2}N_{Z+2,N+2}(t) \\ &- N_n(t)\langle\sigma v\rangle_{A}N_{Z,N}(t) - \Gamma^{(\beta)}_{Z,N}N_{Z,N}(t) - \Gamma^{(\alpha)}_{Z,N}N_{Z,N}(t)\,,\end{aligned} \tag{2}$$

where $\Gamma^{(\alpha)} = \ln 2/T^{(\alpha)}_{1/2}$ denotes the $\alpha$-decay width of the nuclei.This equation reflects the fact that the spectral analysis for observing a particular nucleus applies to fixed atomic numbers, not mass numbers.

In order to solve the coupled system of equations numerically, we introduce discrete time steps $\tau$. If the rate of a certain process is sufficiently large compared to the time step, then the differential equations can be linearized

$$\begin{aligned}N_{Z,N}(t+\tau) = N_{Z,N}(t) &+ p^{(n)}_{Z,N-1}N_{Z,N-1}(t) + p^{(\beta)}_{Z-1,N+1}N_{Z-1,N+1}(t) + p^{(\alpha)}_{Z+2,N+2}N_{Z+2,N+2}(t) \\ &- p^{(n)}_{Z,N}N_{Z,N}(t) - p^{(\beta)}_{Z,N}N_{Z,N}(t) - p^{(\alpha)}_{Z,N}N_{Z,N}(t)\,,\end{aligned} \tag{3}$$

where $p^{(x)}_{Z,N} = \tau\lambda^{(x)}_{Z,N}$, with $\lambda^{(x)}$ being the capture or decay rate for process $x$. This equation could be used only if the time step is (much) smaller than any of the capture or decay inverse rates, so that all probabilities $p^{(x)}_{Z,N}$ are (much) smaller than one. This condition would make the computer code uselessly slow because of the often very small decay life times. In order to be able to choose a sufficiently large time step, we perform the evolution in two steps, a neutron capture period followed by a decay period in an alternating manner.

Given a distribution of abundances at a given moment $t$, $N_{Z,N}(t)$, in the first step we allow only neutron capture with (possibly time dependent) probability $p^{(n)}_{Z,N}$, which increases each

$N_{Z,N}(t)$ with $p^{(n)}_{Z,N-1}N_{Z,N-1}(t)$ and decreases them with $p^{(n)}_{Z,N}N_{Z,N}(t)$, so the net change in the capture period is

$$\Delta N^{\text{capt.}}_{Z,N} = p^{(n)}_{Z,N-1}N_{Z,N-1} - p^{(n)}_{Z,N}N_{Z,N}\,, \tag{4}$$

and we update each value of the abundances accordingly.

In the next time step the nuclei present at the end of the previous capture period are allowed to decay only. This decay will be described by the usual exponential decay, and the number of decayed nuclei of a given isotope at the end of the decay period will be

$$\Delta N^{\text{dec.}}_{Z,N}(t+\tau) = N_{Z,N}(t)[1 - \exp(-\Gamma_{Z,N}\,\tau)]\,. \tag{5}$$

In Eq. (5) $\Gamma_{Z,N}$ is the total decay width of an unstable isotope with atomic number $Z$ and neutron number $N$, the sum of the partial widths for the various possible decays, $\Gamma_{Z,N} = \sum_i \Gamma^{(i)}_{Z,N}$. If the half life of the isotope is long as compared to the time step, we can use the approximation

$$\Delta N^{\text{dec.}}_{Z,N}(t+\tau) \simeq N_{Z,N}(t)\,\Gamma_{Z,N}\,\tau\,, \tag{6}$$

and the combined effect of the two time steps (Eqs. (4) and (5)) is equivalent to Eq. (3). If however, $\Gamma_{Z,N}\tau$ is not small the expansion in Eq. (6) cannot be used. We use Eq. (6) if $\Gamma_{Z,N}\tau < -\ln 0.99$, while Eq. (5) if $-\ln 0.99 < \Gamma_{Z,N}\tau < -\ln 0.01$ and $\Delta N^{\text{dec.}}_{Z,N}(t+\tau) = N_{Z,N}(t)$ if $\Gamma_{Z,N}\,\tau > -\ln 0.01$ (i.e. all nuclei decay immediately). The total number of nuclei does not change, only redistributed in this step because the nuclei disappearing from a certain type will turn into other types of nuclei in proportion to the partial widths of the possible decay modes.

We took the relevant neutron capture cross sections for given $Z$ and $N$ from Ref. [5], the nuclear decay rates from Ref. [6] and in the region around bismuth we used the nuclear decay data from Ref. [7]. We included the relevant nuclear data of 2696 nuclei of which 178 are stable.

## 3. Predictions

The neutron-capture rate of a nucleus is proportional to its neutron-capture cross section $\lambda = \Phi\langle\sigma_T\rangle$, where $\Phi = N_n v_T$ is the neutron flux. Using the value of the neutron density typically assumed in the $s$ process (the case of weak neutron flux), $N_n = 10^8\,\text{n}\,\text{cm}^{-3}$ and average speed of neutrons, $v_T = 2.4\cdot 10^8\text{cm}\,\text{s}^{-1}$, the flux is $\Phi = 2.4\cdot 10^{-11}\text{mb}^{-2}\text{s}^{-1}$. The evolution band obtained with this parameter is shown on Fig. 1 after five cycles of 2.4 million steps with successively decreasing time steps $\tau_i = 10^{5-i}\,\text{s}$ $(i = 1,\ldots,5)$. Thus the total evoulution time is about 845 y. The smaller time steps, the wider band. Thus large time steps approximate the $s$ process. The evolution starts with $3\cdot 10^{45}$ $^{56}\text{Fe}$ nuclei and ends up producing 1090 nuclei. The evolution always progresses along a band and increasing the neutron flux, the band gets wider.

Fig. 2 shows the predicted abundances of elements summed over isobars after two steps: (i) an initial evolution of 845 y with $\Phi = 2.4\cdot 10^{-8}\,\text{mb}^{-2}\text{s}^{-1}$ followed by (ii) a decay of 4.5 billion years. We compare our prediction to abundances resulting from the classical $s$ process (obtained with the same nuclear input). We see that our model gives identical results to the classical $s$ process.

The nucleosynthesis in our model evolves along a band in the stability valley until it seemingly stops at the $\alpha$-decaying isotopes of polonium, the $Z = 84$ line. However, with existing neutron flux the number of heavy nuclei constantly increases. The decay process is probabilistic, therefore, not all nuclei decay in a given period. If the neutron flux is maintained for sufficiently long time, the band widens below the $Z = 84$ line until it reaches $^{218}\text{Bi}$. The time needed for the appearance of $^{218}\text{Bi}$ depends on the neutron flux. For instance, choosing $\Phi = 2.4\cdot 10^{-8}\,\text{mb}^{-2}\text{s}^{-1}$ (corresponding to $N_n = 10^{11}\,\text{n}\,\text{cm}^{-3}$), it takes about one year. The larger the flux, the shorter the required time. Once $^{218}\text{Bi}$ appears, heavier nuclei can easily be produced and appear quickly up to fermium ($^{260}\text{Fm}$).

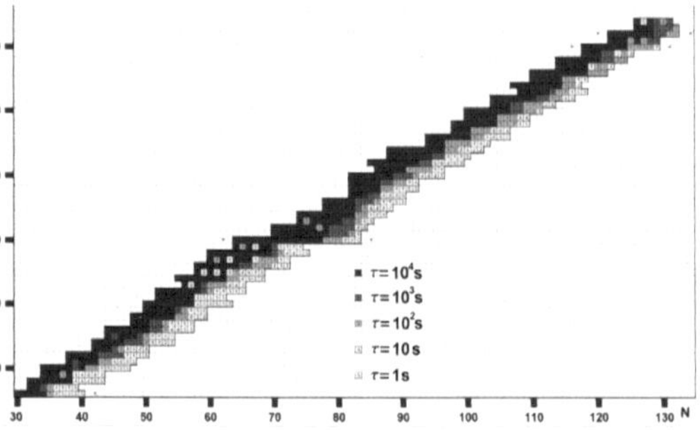


**Figure 1.** The evolution band obtained with $\Phi = 2.4 \cdot 10^{-11}\,\mathrm{mb}^{-2}\mathrm{s}^{-1}$. The numbers in the boxes denote the cycle in which the given nucleus appeared.

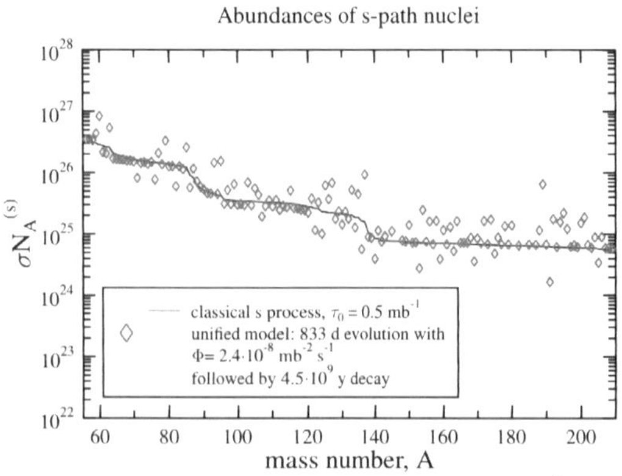


**Figure 2.** The abundances of elements obtained with $\Phi = 2.4 \cdot 10^{-8}\,\mathrm{mb}^{-2}\mathrm{s}^{-1}$ in 845 y of evolution and 4.5 billion years of decay compared to the abundances obtained in the classical $s$ process.

The rate of the evolution of trans-bismuth elements also correlates with the neutron flux: the larger $\Phi$, the faster evolution. After the neutron irradiance ceases, the unstable heavy nuclei decay quickly. However, some nuclei with long life time, such as $^{235}U$ and $^{238}U$, remain for a long time. The abundances of such nuclei depend strongly on the neutron density. For the relatively modest value of $N_n = 10^{11}\,\mathrm{n\,cm}^{-3}$ the abundances of uranium isotopes is negligible as compared to the observable abundances of elements [8]. The observed abundances can be obtained with the model with the parameter value $\Phi \simeq 3 \cdot 10^{-5}\,\mathrm{mb}^{-2}\mathrm{s}^{-1}$ corresponding to $N_n \simeq 10^{14}\,\mathrm{n\,cm}^{-3}$ that is still six orders of magnitude smaller than the typical neutron density in supernovae [4].

Another interesting observation is that some elements, that do not belong to the $s$-process path (traditionally called $r$-only nuclei), also appear in the band with noticible abundances. This means that the band-like evolution reaches certain $r$-only nuclei even in the case of weak neutron flux, corresponding to $\Phi = 2.4 \cdot 10^{-11}\,\mathrm{mb}^{-2}\mathrm{s}^{-1}$, although for such value the abundances of $r$-only nuclei are negligible. However, the observed abundances can be obtained with the still small value of $\Phi \simeq 10^{-9}\,\mathrm{mb}^{-2}\mathrm{s}^{-1}$.

We proposed a model of nucleosythesis of heavy elements in stars. The main features of the model are the following: (i) all known neutron-capture and decay data of 2696 nuclei are taken into account, (ii) the coupled system of differential equations that describe the change of abundances of these nuclei is solved numerically. The model is very simple and does not take into account many effects that are considered standard in current analyses. Most importantly, the implementation of time-varying neutron flux is in progress.

This research was supported by the Hungarian Scientific Research Fund OTKA K-60432.

## References

[1] M. E. Burbidge, G. R. Burbidge, W. A. Fowler and F. Hoyle, "Synthesis of the elements in stars," Rev. Mod. Phys. **29**, 547 (1957).
[2] A. G. W. Cameron, "Nuclear astrophysics," Ann. Rev. Nucl. Part. Sci. **8**, 299 (1958).
[3] F. Käppeler, H. Beer and K. Wisshak "s-process nucleosynthesis – nuclear physics and the classical model," Rep. Prog. Phys. **52**, 945 (1989).
[4] C.E. Rolfs and W.S. Rodney, "Cauldrons in the Cosmos," The University of. Chicago Press, (1988).
[5] http://physci.llnl.gov/Research/RRSN/semr/30kev.html, LLNL.
[6] J. K. Tuli: Nuclear Wallet Cards 2005, BNL.
[7] http://wwwndc.tokai-sc.jaea.go.jp; JENDL 3.3, Japan Atomic Energy Agency.
[8] D. Arnett, "Supernovae and Nucleosynthesis", Princeton Series in Astrophysics, (1996).

Journal of Physics: Conference Series **202** (2010) 012025 doi:10.1088/1742-6596/202/1/012025

# Carbon burning in stars – Prospects for underground measurements of the $^{12}C+^{12}C$ fusion reactions

**Frank Strieder**

Institut für Experimentalphysik, Ruhr-Universität Bochum, Germany

strieder@ep3.rub.de

**Abstract**. The $^{12}C+^{12}C$ fusion reactions are together with the reaction $^{12}C(\alpha,\gamma)^{16}O$ the most important nuclear processes in the late stellar evolution. These fusion reactions play a key role in the understanding of various types of astrophysical objects. Thus, a measurement of the $^{12}C+^{12}C$ cross section at very low energies can serve as flagship experiment for a future underground accelerator laboratory. It is hoped that an appropriate facility for such a study will be created in the near future somewhere in the world. The prospects for the measurement of the carbon fusion reactions will be discussed in the present work.

## 1. Introduction

The LUNA facility [1] at the Laboratori Nazionali del Gran Sasso, Italy, is yet a unique underground accelerator laboratory for nuclear astrophysics. The availability of high intensity proton and alpha beams and the maximum accelerator voltage of 400 kV makes it an ideal tool for the study of low-energy (p,γ) and (α,γ) capture reactions. Since there are many reactions to be studied and/or reinvestigated the facility will continue for several more years. However, for an improved measurement of some important reactions induced by charged particles the present LUNA accelerator is not sufficient either because experiments in inverse kinematics are favoured or because only heavy ions are involved. Such key reactions for a future underground accelerator laboratory are the $^{12}C+^{12}C$ fusion reactions.

## 2. The carbon fusion reactions

The reactions $^{12}C(^{12}C,\alpha)^{20}Ne$ (Q = 4.617 MeV) and $^{12}C(^{12}C,p)^{23}Na$ (Q = 2.241 MeV) are referred to as carbon burning in stars [2]. Carbon-burning represents the third stage of stellar evolution for massive stars ($M > 8M_{sun}$) and proceeds mainly through the $^{12}C + ^{12}C$ fusion reactions since the Coulomb barrier is lowest among the involved nuclei. Carbon and oxygen are produced as a result of helium burning. The stellar CO core contracts after the central He exhaustion and the core temperature increases. Only massive stars develop temperatures large enough for further burning phases. On the contrary, in low and intermediate mass stars, a high pressure from electron degeneracy and the cooling caused by the production of plasma neutrinos prevent the onset of the carbon burning. The minimum mass of a star for the carbon ignition, called $M_{up}$, is a fundamental astrophysical parameter that marks the upper limit for CO White dwarf progenitors and the lower limit for core-collapse Supernova progenitors [3]. Stars with an initial mass slightly larger than $M_{up}$ experience an off-centre carbon ignition under condition of high electron degeneracy. As a result, a relatively large degenerate core with a mass in the range of 1.1 – 1.37 $M_{sun}$ made of O, Ne and Mg is formed. These stars evolve through a super-AGB phase (**A**symptotic **G**iant **B**ranch) and end their lives as massive ONeMg White dwarfs or, eventually, as electron-capture supernovae [4]. Stars more massive than another mass limit $M'_{up}$ – the minimum mass for Ne ignition – ignite central carbon burning and proceed through neon, oxygen, and silicon burning, followed by the iron core collapse [5-7]. On the contrary, stars with $M < M_{up}$ cannot progress through carbon ignition and conclude their life after the AGB phase as CO White dwarfs. CO White dwarfs accreting mass in close binary systems are the progenitors of the majority of Novae [8] and Type Ia Supernovae [9]. Thus not only the frequencies of Novae and the various types of Supernovae depend on the mass limits $M_{up}$ and $M'_{up}$ but also a large fraction of the stellar nucleosynthesis, i.e. the synthesis of most elements with mass numbers larger than 20.

The determination of $M_{up}$ relies on a precise understanding of the carbon burning reaction rate under stellar conditions. However, at the relevant stellar energies the reaction rate of these $^{12}C+^{12}C$ fusion reactions is not very well known and provided only by extrapolations of high energy data [10]. The stellar temperature range corresponds to a Gamow peak energy for $^{12}C+^{12}C$ fusion reactions of about $E_G = 1.5 \pm 0.3$ MeV and thus the cross section of these key processes must be known with high accuracy down to this energy region. Previous experiments [11–16] using charged-particle or γ-ray spectroscopy obtained useful data over a wide range of energies down to the centre-of-mass energy E = 2.0 MeV. However, below E = 3.0MeV the reported cross sections are partly rather uncertain, because at these energies the presence of $^{1}H$ and $^{2}H$ contamination inside or on the surface of the C targets hampered in most cases the measurement of the $^{12}C+^{12}C$ processes. For example, in γ-ray spectroscopy the transitions from the first excited state in $^{20}Ne$ ($E_\gamma$ = 1634 keV) and $^{23}Na$ ($E_\gamma$ = 440 keV) were normally the prominent lines in the γ-spectra but at low energies their observation suffered from an intense background from the $E_\gamma \approx 2.36$ MeV line of $^{1}H(^{12}C,\gamma)^{13}N$ and the $E_\gamma = 3.09$ MeV line of $^{2}H(^{12}C,p\gamma)^{13}C$. In one of the studies [16] C targets with a low hydrogen contamination were produced allowing for reasonable precise measurements by γ-ray spectroscopy of the $\alpha_1$- and $p_1$-channels down to E = 2.0 MeV.

## 3. Future studies of the $^{12}C+^{12}C$ fusion reactions

As a key condition improved studies of the $^{12}C+^{12}C$ fusion reactions require C targets with an ultra-low hydrogen contamination. Moreover, a reduction of γ-ray background from natural radioactivity and cosmic-rays is also an essential step for a measurement towards a lower energy limit than presently available. The reduction of the γ-ray background from natural radioactivity can be achieved by conventional passive shielding such as lead placed around the γ-ray detectors and active shielding to reduce the influence of cosmic-rays. However, the passive shielding is limited to a certain thickness. One cannot add further shielding material since cosmic-ray muons interact with the material and create energetic neutrons which, in turn, create γ-rays in the lead. Clearly, all background components are dramatically reduced with the significantly suppressed muon flux in an underground laboratory. The best solution is to install an accelerator facility in a laboratory deep underground. In summary, only these two key requirements, i.e. ultra-low H contamination and a low-background environment, combined with a high intensity $^{12}C$ ion beam, e.g. a particle current of 300 μA, will allow for a γ-ray spectroscopy measurement of the $^{12}C+^{12}C$ fusion reactions down into the region of the Gamow peak.

The γ-ray spectroscopy has the disadvantage that the transitions $\alpha_0$ and $p_0$ directly into the ground state of $^{20}Ne$ and $^{23}Na$, respectively, cannot be observed. Direct particle identification would allow for measuring the total cross section for each channel. Such a measurement could be performed with standard silicon surface barrier detectors in close geometry to the target and under backward angles with respect to the beam direction. Furthermore, a compact ΔE-E Silicon detector telescope or a Bragg detector would allow for particle identification, e.g. α and p. In this geometry background problems arise mainly from elastically scattered $^{12}C$ nuclei and the reaction $^{12}C(d,p)^{13}C$. In addition, a close distance to the target leads to a heating of the detectors due to the heat radiation of the target under a high intensity $^{12}C$ beam. This effect influences drastically the performance of the particle detector. On the other hand an ion beam intensity of about 300 particle μA $^{12}C$ – a factor 20 to 50 more than in previous experiments, e.g. [13] – would allow to mount the detectors in a far geometry without a significant loss of count rate compared to previous experiments but avoiding the problems of detector heating. However, such a particle spectroscopy measurement might also benefit from an underground laboratory in particular in case of Si detectors if the two detectors of the ΔE-E telescope are very close to each other and the detectors have a large surface area. This technique was successfully exploited in the measurement of the reaction $^3He(^3He,2p)^4He$ at the LUNA I facility [17] and a clear advantage of the underground environment for the ΔE-E Si detector setup was observed [18]. An important issue in the particle spectroscopy is the background discrimination in particular for low reaction yield: a distinguished background detection for heavy ion reactions is difficult if not impossible. In case of background reactions emitting ejectiles of the same species – may be even with a comparable energy – as the expected experimental signal, the background cannot be identified with any detection technique. In turn, this is an advantage of the γ-ray spectroscopy which provides a clear and unique signal if the signal-to-noise ratio allows for a precise observation of the signal. Therefore, the $^{12}C+^{12}C$ fusion reactions are an excellent case for experimental studies using a future underground facility with a larger accelerator than presently available and an ion source able to provide high intensity heavy ion beams.

The possible improvement for a γ-ray spectroscopy measurement of the $^{12}C+^{12}C$ fusion reactions in an underground laboratory can be estimated from the recent $^3He(\alpha,\gamma)^7Be$ experiment at LUNA [19, 20]. The detection setup of this experiment, in particular the arrangement of the passive shielding, was adapted to meet the special requirements for a measurement of γ-ray energies below $E_\gamma = 2.5$ MeV. The achievements in terms of background reduction can be considered as a kind of standard for future nuclear cross section measurements. Hence, the natural γ-ray background near the relevant 1634 keV γ-ray line of the $^{12}C(^{12}C,\alpha)^{20}Ne$ reaction can be used to calculate the sensitivity limit. The background for the HPGe detector in the well-shielded setup of the LUNA experiment is reduced by a factor 800 compared to the shielded setup in a recent experiment at earth's surface [16].

A similar or improved γ-ray background can probably be expected for all future underground accelerator laboratories at depths comparable to Gran Sasso. Unfortunately, the environmental (unshielded) γ-ray background in this γ-ray energy region in an underground laboratory located in sedimentary rocks is relatively high if compared to the reduction of cosmic-ray background. An advantage for measurements in this particular γ-ray energy region could come from the potential of salt mines where typically the unshielded γ-ray background provided by the salt environment is significantly reduced with respect to sedimentary rocks, including the $^{40}K$ γ-ray line. This reduction approximately amounts to one order of magnitude, e.g. between the Boulby UK Dark Matter underground laboratory located in the Boulby Salt and Potash Mine, U.K., – where the laboratory is placed in a salt layer – and Gran Sasso (sedimentary rock). The reduction could be even larger for sites located in salt layers with a very low $^{40}K$ content. Unfortunately, most of such sites are at depths less than 300 m, e.g. in Romania [21], and thus with an insufficient shielding against cosmic-ray background.

The $^{12}C+^{12}C$ experiment could be performed in a similar way as previously with a single detector in close geometry. In a recent theoretical work [22] a hypothetical resonance at E = 1.5 MeV has been proposed based on the carbon ignition temperature in superburst models. The proposed experiment with a high intensity $^{12}C$ ion beam, reasonable experimental parameters, and the above background assumption would have a resonance strength detection limit of at least 1 μeV for a resonance at E = 1.5 MeV. Thus, the hypothetical resonance of [22] would be within reach in such a measurement. Unfortunately, a proper passive shielding similar to the LUNA $^3He(\alpha,\gamma)^7Be$ experiment for an improved detection system such as a Ge crystal ball would lead to handling problems and large experimental difficulties. This approach could take advantage of the potential of salt mines due to the lower unshielded γ-ray background and the shielding requirements may be reduced. However, even in a well-shielded detector setup the actual environmental and cosmic-ray induced backgrounds are only some of the components determining the final background in the detector. Other important components are the purity of the shielding material, and in particular beam-induced background. Very often the final detection limit can only be estimated through the measurement in the underground laboratory itself. While for most cases in low-energy cross section measurements the natural background is dominating the γ-ray spectra or at least the beam induced background can be reduced to the level of the other background sources that might be different in an underground measurement. In particular new beam induced background components might appear which at low-energies give rise to the same yield than expected from the actual measurement, but normally in earth's surface experiments covered by the natural background (see for example the case of the reaction $^{25}Mg(p,\gamma)^{26}Al$ as shown in [1]).

In summary, our knowledge of the reaction rate of the $^{12}C+^{12}C$ fusion reactions will certainly be improved very much by a measurement in an appropriate future underground accelerator laboratory. The final sensitivity limit depends not only on the reduction of the natural background but also on the purity of the C target, i.e. hydrogen contamination must be reduced to a minimum. Nevertheless, with an optimized experimental setup and an accelerator such as a 3 MV high-current, single-stage machine with an ECR ion source a measurement over the energy range of the Gamow peak for massive stars at E = 1.5 MeV appears feasible. Finally, both experimental approaches, i.e. particle and γ-ray spectroscopy, to measure the $^{12}C+^{12}C$ cross section would benefit from the underground environment.

**Acknowledgments**
The present work has been supported by the Deutsche Forschungsgemeinschaft (DFG, Ro 429/44). In addition, the author would like to thank O. Straniero for fruitful discussions.

**References**

[1] Costantini H, Formicola A, Imbriani G, Junker M, Rolfs C and Strieder F 2009 *Rep. Prog. Phys.* **72** 086301-25
[2] Rolfs C E and Rodney W S 1988 *Cauldrons in the Cosmos* (University of Chicago Press)
[3] Becker S A and Iben I 1979 *Bulletin of the American Astronomical Society* **12** 417
[4] Siess L 2007 *Astron. Astrophys.* **476** 893-909
[5] Chieffi A, Limongi M and Straniero O 1998 *Astrophys. J.* **502** 737-762
[6] Woosley S E, Heger A and Weaver T A 2002 *Rev. Mod. Phys.* **74**, 1015-1071
[7] El Eid M F, Meyer B S and The L S 2004 *Astrophys. J.* **611**, 452-465
[8] Kraft R P 1964 *Astrophys. J.* **139** 457-475
[9] Hillebrandt W and Niemeyer J 2000 *Annu. Rev. Astron. Astrophys.* **38** 191-230
[10] Caughlan GR and Fowler W A 1988 *Atomic Nucl.Data Tables* **40** 283-334
[11] Kettner K U, Lorenz-Wirzba H and Rolfs C E 1977 *Phys. Rev. Lett.* **38** 337-340
[12] High M D and Cujec B 1977 *Nucl. Phys.* A **282** 181-188
[13] Becker H W, Kettner K U, Rolfs C E and Trautvetter H P *Z. Phys. A* **303** 305-312
[14] Aquilera E F *et al* 2006 *Phys. Rev. C* **73** 064601-12
[15] Barron-Palos L *et al* 2006 *Nucl. Phys.* A **779** 318-332
[16] Spillane T *et al* 2007 *Phys. Rev. Lett.* **98** 122501-4
[17] Junker M *et al* 1998 *Phys. Rev. C* **57** 2700-10
[18] Junker M 1996 *Ph. D. thesis* (Ruhr-Universität Bochum, Germany)
[19] Confortola F *et al* 2007 *Phys. Rev. C* **75** 065803-4
[20] Caciolli A *et al* 2009 *Eur. Phys. J. A* **39** 179-186
[21] Bordeanu C, Rolfs C, Margineanu R, Negoita F and Simion C 2008 *J. Phys. G* **35** 014011-6
[22] Cooper R L, Steiner A W and Brown E F 2009 *Astroph. J* **702** 660-671

Journal of Physics: Conference Series **202** (2010) 012026 doi:10.1088/1742-6596/202/1/012026

# Forthcoming $(n, \gamma)$ measurements on the Fe and Ni isotopes at CERN n_TOF

**C. Domingo-Pardo[1*], S. Andriamonje[4], J. Andrzejewski[7], L. Audouin [21], V. Avrigeanu[40], F. Bečvář[12], F. Belloni[3], B. Berthier[21], E. Berthoumieux[4], M. Brugger[17], M. Calviani[26], F. Calviño[13], D. Cano-Ott[6], C. Carrapiço[16,4], F. Cerutti[17], F. Chiaveri[17], N. Colonna[19], G. Cortes[13], I. Dillmann[4]5, I. Duran[5], A. Ferrari[17], W. Furman[24], S. Ganesan[41], G. Giubrone[2], M.B. Gómez Hornillos[13], R. Gallino[25], I.F. Goncalves[18], E. Gonzalez-Romero[6], V. Gorlychev[13], F. Gramegna[26], C. Guerrero[6], F. Gunsing[4], S. Harissopulos[31], M. Heil[1], K. Ioannides[8], E. Jericha[10], F. Käppeler[9], Y. Kadi[17], D. Karadimos[8], M. Krtička[12], E. Lebbos[17], C. Lederer[43], H. Leeb[10], M. Lozano[15], R. Losito[17], J. Marganiec[7], S. Marrone[19], T. Martínez[6], C. Massimi[39], P. Mastinu[27], E. Mendoza[6], A. Mengoni[14,38], P.M. Milazzo[3], M. Mirea[40], M. Moinul[19], M. Mosconi[44], R. Nolte[44], C. Paradela[5], A. Pavlik[34], J. Perlowski[21], R. Plag[1], J. Praena[26], J.M. Quesada[15], R. Reifarth[1], C. Rubbia[22], R. Sarmento[16], G. Tagliente[19], J.L. Tain[2], L. Tassan-Got[21], D. Tarrio[5], G. Vannini[39], V. Variale[19], P. Vaz[16], A. Ventura[39], D. Villamarin[6], V. Vlachoudis[17], R. Vlastou[34], A. Wallner[43], and C. Wiess[10]**

**The n_TOF Collaboration**

[1]GSI Helmholtzzentrum für Schwerionenforschung GmbH, Darmstadt, Germany
[2]Instituto de Física Corpuscular, CSIC-Universidad de Valencia, Spain
[3]Istituto Nazionale di Fisica Nucleare, Trieste, Italy
[4]CEA/Saclay - DSM, Gif-sur-Yvette, France
[5]Universidade de Santiago de Compostela, Spain
[6]Centro de Investigaciones Energeticas Medioambientales y Technologicas, Madrid, Spain
[7]University of Lodz, Lodz, Poland
[8]University of Ioannina, Greece
[9]Forschungszentrum Karlsruhe GmbH (FZK), Institut für Kernphysik, Germany
[1]0Atominstitut der Österreichischen Universitäten,Technische Universität Wien, Austria
[11]Centre National de la Recherche Scientifique/IN2P3 - IReS, Strasbourg, France
[12]Charles University, Prague, Czech Republic
[13]Universitat Politecnica de Catalunya, Barcelona, Spain
[14]International Atomic Energy Agency - IAEA, Nuclear Data Section, Vienna, Austria
[15]Universidad de Sevilla, Spain
[16]Instituto Tecnológico e Nuclear(ITN), Lisbon, Portugal
[17]CERN, Geneva, Switzerland
[19]Istituto Nazionale di Fisica Nucleare, Bari, Italy
[21]Centre National de la Recherche Scientifique/IN2P3 - IPN, Orsay, France
[24]Joint Institute for Nuclear Research, Frank Laboratory of Neutron Physics, Dubna, Russia
[26]Istituto Nazionale di Fisica Nucleare(INFN), Laboratori Nazionali di Legnaro, Italy
[27]Centre National de la Recherche Scientifique/IN2P3 - CENBG, Bordeaux, France
[28]Los Alamos National Laboratory, New Mexico, USA

[29]Tokyo Institute of Technology, Tokyo, Japan
[30]Oak Ridge National Laboratory, Physics Division, Oak Ridge, USA
[31]NCSR Demokritos, Athens, Greece
[33]National Technical University of Athens, Greece
[34]Institut für Isotopenforschung und Kernphysik, Universität Wien, Austria
[38]ENEA, Bologna, Italy
[39]Dipartimento di Fisica, Università di Bologna, and Sezione INFN di Bologna, Italy
[40]National Institute of Physics and Nuclear Engineering - IFIN, Bucharest, Romania
[41]BARC, Mumbai, India [42]Argonne National Laboratory, Chicago, USA
[43]VERA Laboratory - Isotopenforschung & Kernphysik, Faculty of Physics, University of Vienna, Austria
[44]Physikalisch-Technische Bundesanstalt - PTB, Braunschweig, Germany
[45]Physik-Department E12, Technische Universität München, Beschleunigerlaboratorium, Garching, Germany

E-mail: c.domingopardo@gsi.de

**Abstract.** An overview of the past, present and future research activities at the CERN neutron time of flight facility n_TOF is given, with special focus on the astrophysical aspects. During the first campaign (Phase I), neutron capture cross sections of relevance for several aspects of the s-process nucleosynthesis have been measured. A second campaign has recently started (Phase II), aiming at the study of the weak s-process component via accurate $(n, \gamma)$ measurements of the Fe and Ni isotopes. Some changes in the facility will allow us to perform these measurements with improved experimental conditions.

## 1. Introduction

The n_TOF installation [1] aims at the measurement of accurate nuclear data for the Accelerator Driven System (ADS) project [2] and for nuclear astrophysics [3]. This facility combines a very large flight path of 185 m with an intense neutron spallation source, thus yielding an excellent energy resolution of 0.1% at 30 keV and a neutron intensity of more than $10^5$ neutrons/pulse in the 1 keV-100 keV range. More details about the performance and commissioning of the installation can be found in Refs. [4, 5, 6]. In n_TOF Phase I, the first beam time was devoted to the further characterization of the resolution function [1], the ToF-neutron energy calibration [7], the spatial beam profile [8] and the detailed study of the pulse height weighting technique [9, 10]. A summary of the astrophysically relevant measurements carried out in Phase I, between 2002 and 2004, can be found in Sec. 2. Due to radiation-safety issues at the spallation source area, a new target has been constructed and the facility has resumed its activity in 2009. This upgrade and the new experimental programme are briefly described in Sec. 3.

## 2. n_TOF Phase I: measurements of astrophysical relevance for the $s$ process

The list of isotopes measured in this phase at n_TOF is listed in Table 1. Particular emphasis deserves the measurement of a $^{151}$Sm ($t_{1/2}$ = 93 years) sample, with a radioactivity of 156 GBq, which was possible thanks to the high instantaneous neutron flux, low background and very small duty-cycle of n_TOF. Based on this experimental result, a detailed study of the thermal conditions during the recurrent thermal pulses of low mass AGB-stars could be performed [11, 12]. A detection system with reduced neutron sensitivity [13, 14] allowed the accurate measurement of low cross sections, i.e. the Mg isotopes and closed neutron shell nuclei (and neighboring isotopes) like Zr, La, Pb and Bi. The latters act as bottlenecks along the s-process path, hence accumulating large abundances.

**Table 1.** $(n, \gamma)$ measurements made during n_TOF Phase I.

| Isotope | Astrophysically relevant aspects | References |
|---|---|---|
| $^{24,25,26}$Mg | Isotopic anomalies in meteorites | [15] |
| | Strength of $^{22}$Ne$(\alpha, n)$ source | |
| $^{90,91,92,93,94,96}$Zr | Weak/Main $s$ process components | [16, 17, 18] |
| | Bottleneck at $N = 50$ | |
| | $^{95}$Zr branching sensitive to neutron flux | |
| $^{139}$La | Indicator of s-process activity | [19] |
| | Bottleneck at $N = 82$ | |
| $^{151}$Sm | Branching of s-process path | [11, 20, 12] |
| | Thermal conditions of AGB stars | |
| $^{186,187,188}$Os | Nuclear cosmo-chronology | [21] |
| $^{204,206,207}$Pb, $^{209}$Bi | Termination of s-process path | [22, 23, 24, 25] |
| | Constraint for r-process model calculations | |
| | Bottleneck at $N = 126$ | |

## 3. n_TOF Phase II: s process in massive stars

Some aspects have been improved in the facility, particularly concerning the spallation source. On one hand, the target is equipped now with the possibility to include a borated or heavy water moderation layer, which will allow one to highly reduce the $\gamma$-rays produced by neutron captures in the hydrogen of the moderator, thus improving the peak-to-background ratio in the keV-MeV region by more than a factor of 5. On the other hand, the new spallation target has been designed in such a way, that it may serve in the future as a common neutron source for a second experimental area at a distance of 20 m. The latter would provide an increase in sensitivity by two to three orders of magnitude.

Since May 2009 a series of measurements have been carried out for commissioning the upgraded facility. The neutron flux has been determined from measurements with a micromegas detector containing $^{10}$B and $^{235}$U deposits, from the $^{6}$Li$(n, \alpha)$ reaction measured with a silicon array (SiMon) surrounding a thin Li-foil, and from the $^{235}$U(n,f) measured with a fission chamber from PTB (Physikalisch-Technische Bundesanstalt) Braunschweig. An absolute value of the neutron fluence at 4.9 eV was obtained from the $^{197}$Au(n,$\gamma$) measurement with the Total Absorption Calorimeter [26]. A 2D-measurement of the spatial beam profile versus the neutron energy has been carried out using micromegas detectors. The background conditions in the experimental area will be characterized from measurements made with lead, graphite and empty samples. Finally, the resolution function will be characterized from the measurement of a $^{56}$Fe sample.

**Table 2.** $(n, \gamma)$ measurements planned for n_TOF Phase II.

| Isotope | Astrophysically relevant aspects |
|---|---|
| $^{54,56,57,58}$Fe | Prime s-process contributions to the galactic chemical |
| $^{58,60,61,62,(63),64}$Ni | enrichment and quantitative study of the weak s process |

The weak *s* process has been found to be extremely sensitive to the neutron capture cross sections of the isotopes near the seed distribution around Fe [27, 28]. Therefore, accurate $(n, \gamma)$ measurements on these nuclei will be the main goal of the present campaign (see Table 2). Indeed, the neutron exposure during the weak s process is too small to achieve flow equilibrium, which means that the cross section uncertainty on one isotope affects not only its own abundance, but also the production of the subsequent higher mass isotopes. Due to this propagation effect, it is important to measure the cross section of all involved Fe-group nuclei over a broad energy range and with high accuracy [29].

## References

[1] U Abbondanno *et al* 2002 (*Preprint* CERN n_TOF Facility: Performance Report, CERN/INTC-O-011 INTC-2002-037 CERN-SL-2002-053ECT)
[2] C Rubbia *et al* 2000 (*Preprint* 1st Management Report of the n_TOFND-ADS EC programme under contract no. FIKW-CT-2000-00107)
[3] Mengoni A 2002 *Astrophysics, Symmetries, and Applied Physics at Spallation Neutron Sources* ed PE Koehler *et al* p 25
[4] C Borcea *et al* 2003 *Nuclear Instruments and Methods in Physics Research A* **513** 524–537
[5] U Abbondanno *et al* 2001 (*Preprint* Tech. Rep. CERN-INTC-2001-038)
[6] A Ferrari *et al* 2002 (*Preprint* Tech. Rep. CERN-SL-2002-011 ECT)
[7] G Lorusso *et al* 2004 *Nucl. Instr. and Meth. Phys. Res. A* **532** 622–630
[8] J Pancin *et al* 2004 *Nuclear Instruments and Methods in Physics Research A* **524** 102–114
[9] JL Tain *et al* 2002 *Jour. of Nucl. Sci. and Tech.* vol 2 (Tsukuba) p 689
[10] U Abbondanno *et al* 2004 *Nuclear Instruments and Methods in Physics Research A* **521** 454–467
[11] U Abbondanno *et al* 2004 *Phys. Rev. Lett.* **93** 161103
[12] S Marrone *et al* 2006 *Phys. Rev. C* **73** 034604
[13] R Plag *et al* 2003 *Nuclear Instruments and Methods in Physics Research A* **496** 425–436
[14] S Marrone *et al* 2004 *Nuclear Instruments and Methods in Physics Research A* **517** 389–398
[15] E Strandberg 2007 *AGB Star nucleosynthesis of the Magnesium Isotopes* Ph.D. thesis University of Notre Dame http://physics.nd.edu/Pdf/Theses/StrandbergE042007.pdf
[16] G Tagliente *et al* 2006 *International Symposium on Nuclear Astrophysics - Nuclei in the Cosmos*
[17] G Tagliente *et al* 2008 *Phys. Rev. C* **77** 035802
[18] G Tagliente *et al* 2008 *Phys. Rev. C* **78** 045804
[19] R Terlizzi *et al* 2007 *Phys. Rev. C* **75** 035807 (*Preprint* arXiv:nucl-ex/0610034)
[20] S Marrone *et al* 2005 *Nucl. Phys. A* **758** 533–536
[21] M Mosconi *et al* 2007 *Prog. Part. and Nucl. Phys.* **59** 165–173
[22] C Domingo-Pardo *et al* 2006 *Phys. Rev. C* **74** 025807
[23] C Domingo-Pardo *et al* 2006 *Phys. Rev. C* **74** 055802
[24] C Domingo-Pardo *et al* 2007 *Phys. Rev. C* **75** 015806 (*Preprint* arXiv:nucl-ex/0610033)
[25] C Domingo-Pardo *et al* 2007 *Phys. Rev. C* **76** 045805 (*Preprint* arXiv:nucl-ex/0707.3679)
[26] C Guerrero *et al* 2009 *American Institute of Physics Conference Series* (*American Institute of Physics Conference Series* vol 1090) ed Jolie J, Zilges A, Warr N and Blazhev A pp 372–375
[27] CM Raiteri *et al* 1991 *The Astr. Jour.* **371** 665–672
[28] T Rauscher *et al* 2002 *The Astr. Jour.* **576** 323–348 (*Preprint* arXiv:astro-ph/0112478)
[29] M Heil *et al* 2008 *Phys. Rev. C* **77** 015808

Journal of Physics: Conference Series **202** (2010) 012027 doi:10.1088/1742-6596/202/1/012027

# Nuclear Astrophysics with Tagged Photons: NEPTUN @ S–DALINAC, Darmstadt

**L Schnorrenberger, K Sonnabend, J Glorius, B Löher, N Pietralla, D Savran, V Simon, C Wälzlein**

Institut für Kernphysik, Technische Universität Darmstadt, Germany

E-mail: schnorrenberger@ikp.tu-darmstadt.de

**Abstract.** Tagged photons can be used to study astrophysically relevant cross sections with highest energy resolution. The tagging facility NEPTUN at the S–DALINAC, Darmstadt, Germany, is presented and it is demonstrated how NEPTUN can be used to study short-lived branching nuclei of s-process nucleosynthesis.

## 1. Motivation

The nucleosynthesis of the elements heavier than iron is today mainly explained by three processes: *s*, *r*, and *p* process. The *s* and the *r* process are based on neutron-capture reactions with adjacent $\beta$ decays while the *p* process is governed by photodisintegration reactions such as ($\gamma$,n), ($\gamma$,p), or ($\gamma$,$\alpha$).

The study of the neutron capture cross section of the so-called branching nuclei of the *s* process is mandatory to understand the produced abundance patterns. However, direct measurements were only performed for long-lived branching nuclei so far (e.g. $^{151}$Sm [1, 2] or $^{147}$Pm [3]). For short-lived ones the measurement of the inverse ($\gamma$,n) reaction yields constraints for the theoretical prediction of the neutron capture cross section (e.g. [4, 5, 6]). This is realized using the activation technique with continuous-energy bremsstrahlung sources like e.g. the High Intensity Photon Setup (HIPS) at Darmstadt, Germany [7]. However, the energy resolution is quite poor because the cross section has to be deconvolved from the energy-integrated cross section $I_\sigma$

$$I_\sigma = N_\mathrm{t} \cdot \int_{S_\mathrm{n}}^{E_\mathrm{max}} \sigma(E) \cdot N_\gamma(E, E_\mathrm{max}) \cdot \mathrm{d}E \qquad (1)$$

with $N_\mathrm{t}$ being the number of target atoms, $\sigma(E)$ the ($\gamma$,n) cross section, and $N_\gamma(E, E_\mathrm{max})$ the photon flux, respectively.

To avoid this problem, quasi-monoenergetic photon beams from Laser Compton Backscattering facilities like e.g. HI$\gamma$S at Duke University [8] can be used. The energy range of the produced photon spectrum is much smaller compared to HIPS, however, the energy resolution at these facilities is still limited to values on the order of 100 keV.

To measure the energy dependence of the photodissociation cross section with better accuracy – especially in the astrophysically relevant region close above the threshold in the case of ($\gamma$,n) reactions – tagged photons can be used. The NEPTUN tagger setup [9, 10, 11] that is constructed at the S-DALINAC, Darmstadt, Germany, is suitable for high resolution studies of astrophysically relevant cross sections.

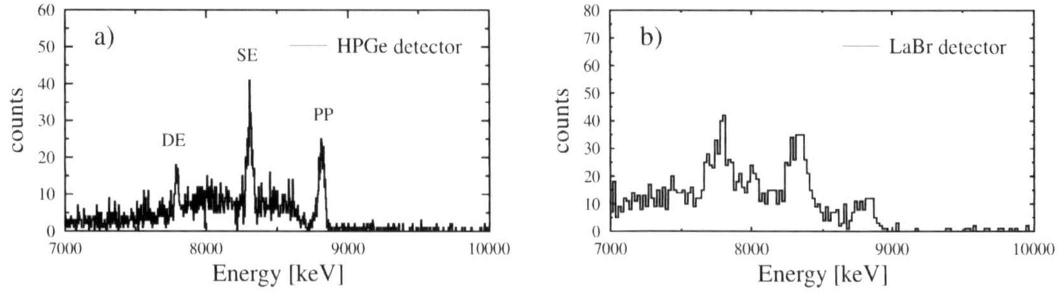


**Figure 1.** a) Spectrum of an in-beam HPGe detector at a tagged photon energy of about 8.8 MeV. The spectrum shows the typical detector response to monoenergetic photons, i.e. the photopeak (PP), single-escape (SE), and double-escape (DE) peaks, as well as a Compton continuum. b) Same spectrum for a $LaBr_3$ detector under indentical experimental conditions. Both spectra were taken within about 2 hours of beamtime.

## 2. The NEPTUN photon tagging facility

The basic concept of photon tagging is to select photons of a given energy out of a continuous-energy bremsstrahlung spectrum. The Superconducting DArmstadt LINear ACcelerator S-DALINAC delivers a monoenergetic electron beam. To produce bremsstrahlung the beam strikes a thin radiator target. Typically, a gold target with a thickness of about 10 $\mu$m is used. Thus, it is guaranteed that 99% of the electrons produce at most one photon. This is crucial to guarantee the correlation between electron and photon energies described further down. Downstream the radiator target a large dipole magnet is located to analyze the *electrons*. Electrons that produced a photon are scattered into a cone such that 99% of them are located within an angle of 8°. The magnet's acceptance has been chosen accordingly.

The position of the electrons in the focal plane of the magnet is determined by a detector array that is built of 64 scintillating fibers attached to photomultipliers. The granularity of the array has been chosen to be 1 mm to meet the resolution power of the magnet. An upgrade to 128 fibers is in progress.

The energy of the scattered electrons deduced from the position in the focal plane has a nominal resolution of 25 keV (at an energy corresponding to 10 MeV tagged photons). Neglecting the recoil momentum, energy conservation yields

$$E_\gamma = E_0 - E_e \tag{2}$$

whereas $E_\gamma$ is the photon energy, $E_0$ the energy of the primary electron beam, and $E_e$ the energy of the scattered electron.

To be able to connect the photons to the corresponding electrons a coincidence setup is necessary. Data on the electron hits in the focal plane detector array are stored digitally in the ring buffer of a multihit-capable TDC. The photons or photon-induced reaction products trigger the readout of the ring buffer. Timing correlations between signals from these detectors and the focal plane fibers within up to 2 $\mu$s are possible which is favorable for experiments close to particle thresholds where reaction products need relatively long times to reach the detectors due to their low kinetic energy.

With the full focal plane setup the energy window to be measured in one run without a change of the magnetic field will be about $\Delta E_\gamma = 2 - 3$ MeV depending on the settings. NEPTUN has been designed to tag photons in the energy range from 6 – 20 MeV [9, 10, 11].

## 3. NEPTUN as a testing device for the characterization of photon detectors

NEPTUN has been shown to deliver a tagged photon beam of 35 keV energy resolution,already. [11]. This is suited to explore the characteristics of detector materials and shapes. NEPTUN can be used to determine the energy resolution and single-escape probabilities up to 20 MeV. These properties are compared to results of simulations being the basis for ion beam experiments involving photon detection.

In a recent experiment the detector response of a $LaBr_3$ scintillator (1.5 inch diameter and length) with a photomultiplier has been studied. A first spectrum is shown in figure 1. For comparison, the detector response of a HPGe detector (100%) with an energy resolution exceeding the one of NEPTUN is shown.

## 4. The s-process branching nucleus $^{95}$Zr

The s-process branching nucleus $^{95}$Zr is located near the borderline between the weak and the main component of the *s* process [12]. Therefore, the prediction of the elemental abundance patterns corresponding to this branching is a crucial test for the validity of a full stellar model as described e.g. in [13].

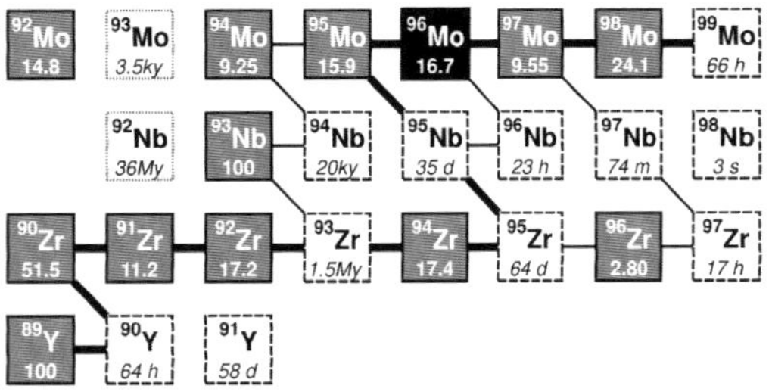


**Figure 2.** The *s*-process path around the branching nucleus $^{95}$Zr. The main flow of the *s*-process is marked with thick lines, thinner lines indicate branches. Grey shaded boxes illustrate stable isotopes and their natural abundances. The s-only isotope $^{96}$Mo is indicated by a black box. The white boxes display $\beta$-instable nuclei and their ground-state half-lives (from [14]).

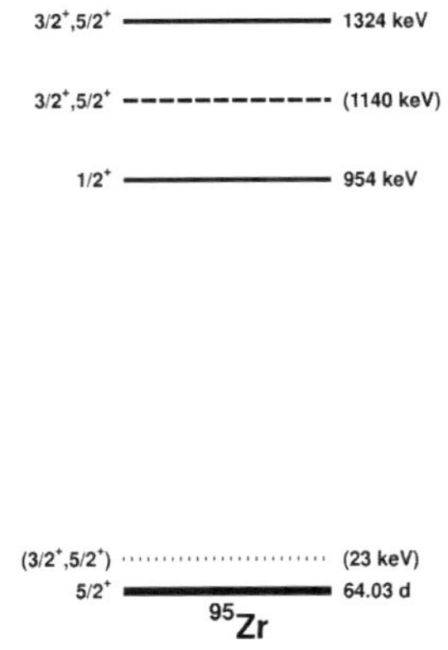


**Figure 3.** Complete level scheme of $^{95}$Zr up to 1.5 MeV [16]. The level at 23 keV was only mentioned by [17] and could be excluded by [18].

Several problems concerning the prediction of the Zirconium abundance patterns measured in SiC grains are reported in [19]. The uncertainty in the predicted Maxwellian average cross section (MACS) of $^{95}$Zr is mentioned as one of the possible error sources due to the wide range of predicted values. Another reason for the deviation between the measured and predicted abundance patterns might be the uncertainty in the half-life $T_{1/2}$ of $^{95}$Zr at *s*-process temperatures. However, as confirmed by a recent measurement [18] the first excited level is at $E = 954$ keV and, thus, is not significantly populated at the mean temperature $kT = 30$ keV. Hence, the half-life of $^{95}$Zr does not depend on temperature under *s*-process conditions and can be omitted as an error source because of its small uncertainty: $T_{1/2}(^{95}\text{Zr}) = (64.032 \pm 0.006)$ d [16].

The reaction $^{95}$Zr(n,$\gamma$) is not accessible to a direct measurement. New high-intense neutron sources like e.g. FRANZ at Frankfurt, Germany, will enable measurements with smallest amounts of target material. However, the reaction product $^{96}$Zr is stable and, thus, prohibits using activation techniques. Therefore, either the neutrons or the $\gamma$ cascades in $^{96}$Zr have to be measured to determine the reaction yield. But this is also hampered due to the high intrinsic activity of a $^{95}$Zr target. Although the cross section of the inverse reaction $^{96}$Zr($\gamma$,n) was already studied with bremsstrahlung [14] and Laser Compton Backscattered photons [15] an enhanced energy resolution is still desirable.

A study with NEPTUN would allow to distinguish between the population of the ground state and first excited state in $^{95}$Zr. Thus, the energy dependence of both de-excitation channels can be developed separately. In addition the measurement of the angular distribution of the emitted neutrons as planned with the neutron detector array of NEPTUN [20] yields further information. A precise and detailed understanding of the cross section of $^{96}$Zr($\gamma$,n) is mandatory to extract reliable constraints on the neutron capture cross section of the *s*-process branching nucleus $^{95}$Zr.

## 5. Outlook

The photon-tagging facility NEPTUN will enable high-resolution cross section measurements of importance in astrophysics like the investigation of neutron capture cross sections of short-lived *s*-process branching nulcei via the inverse reaction. Additionally, it is perfectly suited to characterize the response function of photon detectors as it can be treated like a monoenergetic photon source with energies between 2 MeV and 20 MeV. E.g. the CsI cristalls to be used in the CALIFA array of the future R$^3$B setup at GSI/FAIR will be investigated in near future.

## Acknowledgments

We thank the accelerator group of the S-DALINAC and all the members of AG Pietralla for their help during beam-time. This work is supported by DFG (contract SFB 634), BMBF (contract 06DA9040I) and by the Helmholtz International Center for FAIR within the framework of the LOEWE program (Landesoffensive zur Entwicklung Wissenschaftlich-Ökonomischer Exzellenz) launched by the State of Hesse.

## References

[1] Abbondanno U *et al.* (n_TOF collaboration) 2004 *Phys. Rev. Lett.* **93** 161103
[2] Wisshak K *et al.* 2006 *Phys. Rev. C* **73** 015802
[3] Reifarth R *et al.* 2003 *Astrophys. J.* **582** 1251
[4] Hasper J *et al.* 2008 Phys. Rev. C **77** 015803
[5] Müller S *et al.* 2006 *Phys. Rev. C* **73** 025804
[6] Sonnabend K *et al.* 2003 *Astrophys. J.* **583** 506
[7] Mohr P *et al.* 1999 *Nucl. Instr. Meth. Phys. Res. A* **423** 480
[8] Weller H R *et al.* 2009 *Prog. Part. Nucl. Phys.* **62** 257 – 303
[9] Elvers M *et al.* 2008 *J. Phys. G* **35** 014027
[10] Lindenberg K 2007 Design and construction of the low-energy photon tagger NEPTUN
[11] Savran D *et al.* The low-energy photon tagger NEPTUN *Nucl. Instr. Meth. Phys. Res. A*, submitted
[12] Käppeler F 1999 *Prog. Part. Nucl. Phys.* **43** 419
[13] Cristallo S *et al.* 2009 *Astrophys. J.* **696** 797–820
[14] Sonnabend K *et al.* 2004 vol 704 ed Arnould M, Lewitowicz M, Münzenberg G, Akimune H, Ohta M, Utsunomiya H, Wada T and Yamagata T (AIP) pp 463–472
[15] Sonnabend K *et al.* 2009 vol 1090 ed Jolie J, Zilges A, Warr N and Blazhev A (AIP) pp 481–485
[16] NNDC Online Data Service, ENSDF database, http://www.nndc.bnl.gov/ensdf/
[17] Frota-Pessôa E and Joffily S 1986 *Nuovo Cimento A* **91** 370
[18] Sonnabend K *et al.* 2003 *Phys. Rev. C* **68** 048802
[19] Lugaro M *et al.* 2003 *Astrophys. J.* **593** 486
[20] Simon V 2009 Development and Construction of a detector array for ($\gamma$,n) Experiments at NEPTUN, unpublished

Nuclear Physics in Astrophysics IV IOP Publishing
Journal of Physics: Conference Series **202** (2010) 012028 doi:10.1088/1742-6596/202/1/012028

# Understanding the Sun: Borexino

**Aldo Ianni**
INFN Laboratori Nazionali del Gran Sasso, SS 17 bis Km 18+910, 67010 Assergi (AQ), Italy
E-mail: `aldo.ianni@lngs.infn.it`

**Abstract.** Solar neutrinos have been detected over the past 40 years providing the first evidence of physics beyond the Standard Model for elementary particle interactions. The hypothesis of neutrino oscillations has offered a solution to the long-standing solar neutrino problem (missing solar neutrinos). The Solar Standard Model is a fundamental ingredient in the interpretation of the solar neutrino measurements. New determinations of solar metal abundances caused the Solar Standard Model predictions to be in conflict with helioseismological measurements. The direct measurement of $^7$Be solar neutrinos performed by the Borexino experiment at the Gran Sasso Underground Laboratory allows to probe the Solar Standard Model assuming the neutrino oscillation scenario. At present Borexino has measured $^7$Be solar neutrinos at the level of 10%. In the near future a measurement at the level of 5-3% could be achieved. This accuracy could allow to determine the metal abundances at the center of the Sun. In this framework solar neutrinos offer an independent method to determine the source of the conflict between Solar Standard Model and helioseismology.

The Borexino results are presented and discussed in the framework of the Solar Standard Model and neutrino oscillations.

## 1. Solar neutrino measurements, Solar Standard Model and helioseismology

The Sun is a huge source of low energy (<15 MeV) electron neutrinos, $\nu_e$. Fig. 1 shows the solar neutrino spectrum at Earth with the different components. The main production process is the fusion of two protons; the average energy is about 0.3 MeV. Detection of these low energy neutrinos is a challenge. In spite of this solar neutrinos have been detected over the past 40 years. In Tab. 1 we summarize results of measurements. Collected data can be understood in the framework of neutrino oscillations [1]: from their origin, in the core of the Sun, to Earth solar electron neutrinos get transformed to other flavours. A fundamental ingredient in the interpretation of solar neutrino measurements is the so-called Solar Standard Model (SSM) [2]. The SSM has been updated with new values of two important astrophysical factors, $S_{34}$ and $S_{1,14}$ from LUNA [3]. At present. the SSM predicts the $^7$Be($^8$B) solar neutrino flux at the level of 6%(11%). The SSM can be tested by means of helioseismology [4]. New determination of solar metal abundances [5], [6] caused the SSM predictions to be in conflict with helioseismology. Recently, a new determination of solar abundances has been proposed [6] by using an improved solar atmosphere model and atomic data, and a better selection of spectroscopic lines. In Tab. 2 we report measurements and predictions for the surface helium abundance and the depth of the convective zone. In Fig 2 we plot data against measurements including correlations. From Fig 2 it is evident that at present a strong conflict exists between the prediction of the SSM with new metal abundances (low metallicity, LZ) and observations. On the contrary previous abundances (high metallicity, HZ) were well in agreement with the model. The point we want to make in this paper is how solar neutrino measurements could help solving this controversy. The impact of new metal abundances with respect to older ones on solar neutrino fluxes is as follows [7], [8]: 1) the flux from $^7$Be decreases by 10.6%(8.7%) for LZ05(LZ09); 2) the flux of $^{13}$N decreases by 34.4%(26.6%) for LZ05(LZ09); 3) the flux of $^{15}$O decreases by 38.3%(29.7%) for LZ05(LZ09).

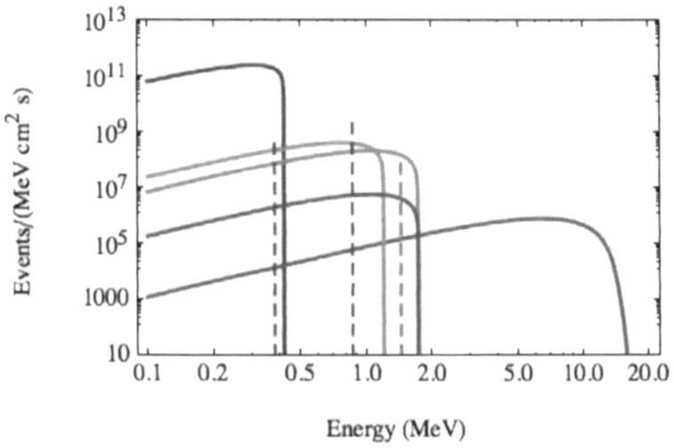


**Figure 1.** Spectrum of solar neutrinos at Earth. Solid lines from left to right: pp, $^{13}$N, $^{15}$O, $^{17}$F and $^8$B. Dashed lines from left to right: $^7$Be(0.38 MeV), $^7$Be(0.862 MeV) and pep.

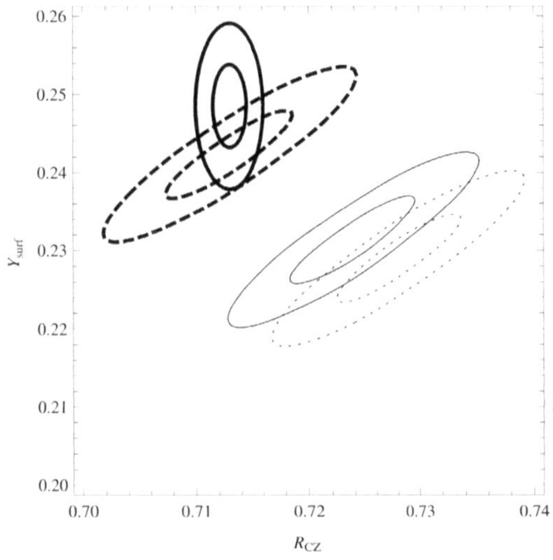


**Figure 2.** High metallicity against Low metallicity as from data reported in Tab. 2. Solid thick line: data. Dashed thick line: HZ. Solid line: LZ09. Dotted line: LZ05. Contours correspond to 68% and 99% C.L.

**Table 1.** Summary of results from solar neutrino measurements. CC = charged-current detection interaction. NC = neutral-current detection interaction. ES = elastic scattering detection interaction.

| Experiment | $\nu_e$ sources detected | data/model |
|---|---|---|
| Homestake(CC) | $^7$Be(13.1%)+pep(2.7%)+CNO(2.4%)+$^8$B(81.8%) | 0.31±0.03 |
| Gallium exps(CC) | pp(55%)+$^7$Be(28.3%)+pep(2.3%)+CNO(3.4%)+$^8$B(11%) | 0.53±0.05 |
| Super-Kamiokande(ES) | $^8$B(100%) | 0.451±0.017 |
| SNO(CC) | $^8$B(100%) | 0.28±0.01 |
| SNO(NC) | $^8$B(100%) | 0.88±0.05 |
| Borexino(ES) | $^7$Be(100%) | 0.65±0.07 |

## 2. Detecting solar neutrinos and probing the Solar Standard Model with Borexino

Borexino has started taking data in May 2007. For the first time sub-MeV solar neutrinos can be detected in real-time. Borexino is an organic liquid scintillator with a target mass (for solar neutrinos) of 100 tons. Fig 3 shows a sketch of the detector which is based on several layers of

**Table 2.** The Sun's $^4$He surface abundance and depth of the convective zone (in units of solar radius) from Solar Standard Models and from helioseismology. LZ09 is from [7] using metal abundances from [6]. LZ05 makes use of metal abundances from [5]. Correlations are taken from [8]

| | data | SSM(HZ) | SSM(LZ09) | SSM(LZ05) |
|---|---|---|---|---|
| $Y_{surf}$ | 0.2485±0.0035 | 0.242±0.0037 | 0.231±0.0037 | 0.229±0.0037 |
| $R_{CZ}$ | 0.713±0.001 | 0.713±0.0037 | 0.724±0.0037 | 0.728±0.0037 |

passive shielding material with high radiopurity to reduce external background. More details on the detector can be found in [9].

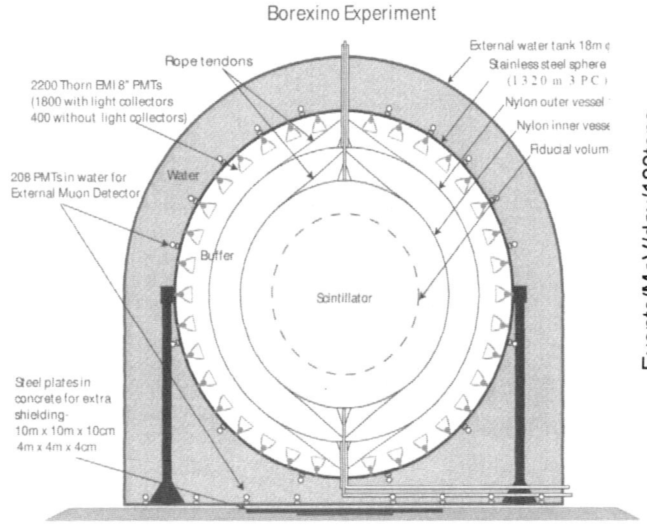


**Figure 3.** Sketch of the Borexino detector (taken from [9]).

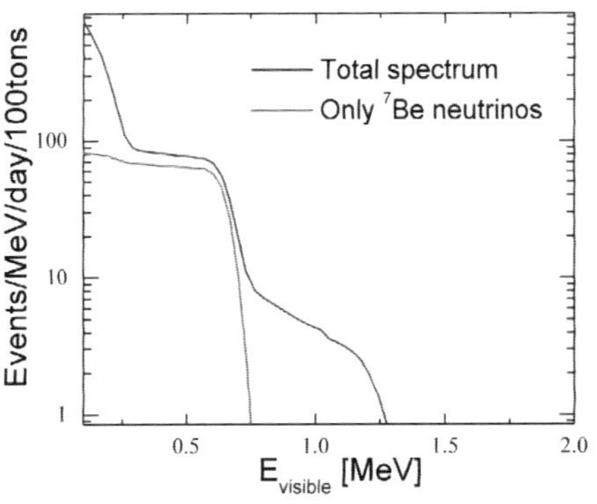


**Figure 4.** Spectrum of recoil electrons from solar neutrino interactions in Borexino.

Borexino detects solar neutrinos by elastic scattering. In Fig 4 we show the expected spectrum of recoiled electrons. The main contribution comes from $^7$Be solar neutrinos which are monoenergetic. About 74 counts/day/100tons are expected according to the HZ SSM. In Fig 5 we show the measured spectrum of solar neutrinos and backgrounds in Borexino (taken from [10]). For more details on the analysis see [10]. As reported in Tab 1 Borexino measures about 48 counts/day/100tons instead of 74. This is consistent with the hypothesis of neutrino oscillations. At present the $^7$Be rate is given with a 10% uncertainty. In Fig 6 we show iso-rate

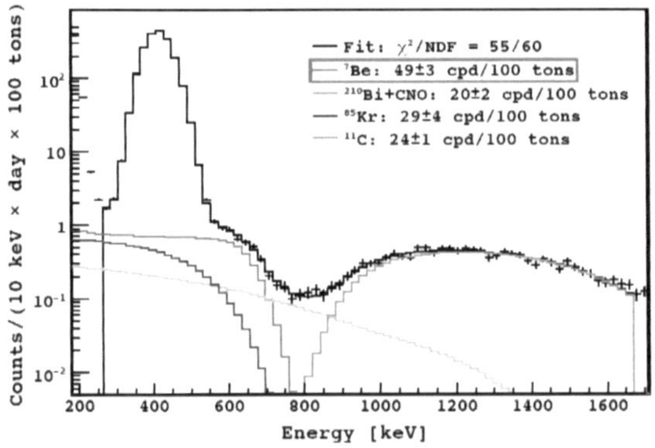


**Figure 5.** Borexino data fitted with the solar neutrino signal and backgrounds [10]. When systematic uncertainties are taken into account $R_{7Be} = 49 \pm 3 \pm 4$ cpd/100 tons.

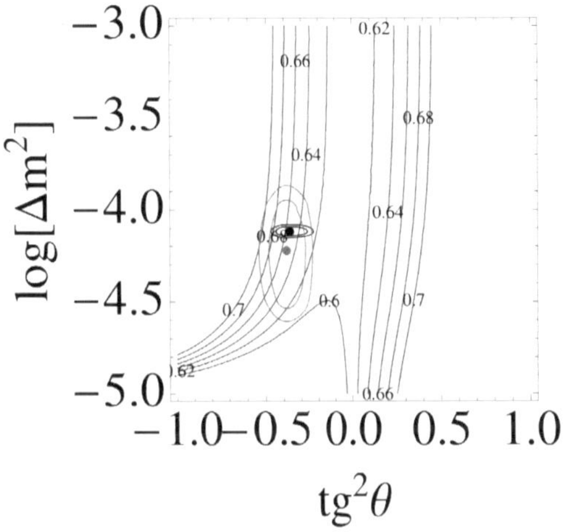


**Figure 6.** Iso-rate, $R_{7Be}(\theta_{12}, \delta m^2)/R_{7Be}^{SSM}$, contours superimposed to the allowed regions at 90% and 99% C.L. from a global analysis of other solar and reactor experiments.

contours normalized to the SSM(HZ) prediction for Borexino. In the same figure we report the allowed regions from a global analysis of solar (larger region) and solar+reactor (smaller region) data. Due to the fact that at present Borexino reports a value of 0.65±0.07, no improvements on the oscillation parameters can be extracted from the data. Future results could reach a 5-3% uncertainty due to performed calibration runs which aimed to reduce the systematic errors.

Present Borexino data however already give important results on SSM parameters, when we make a comparison with previous (that is prior to Borexino) analyses. As a matter of fact, using SNO and Super-Kamiokande data it is possible to determine the flux of $^8$B solar neutrinos without using the SSM. This is shown in Fig 7. The $^8$B flux normalized to SSM(HZ) is found to be 0.90±0.05. Using this result and data from other solar neutrino experiments (see Tab 1) and assuming neutrino oscillations one can determine the flux of pp and CNO neutrinos with and without the luminosity constraint [11]. This is shown in Fig 8. In particular, it can be found that the pp flux normalized to the SSM(HZ) is $1.04^{+0.16}_{-0.19}(1.02\pm0.10)$ without(with) the luminosity constraint. This is a great improvement with respect to the result before Borexino. More can be done by using the correlation between $^8$B and $^7$Be solar neutrinos as suggested by C. Pena-Garay [12]. In Fig 9 we show data against predictions: in particular, for the data we report $f_B$ from SNO (lower value) and from SNO+Super-Kamiokande; for predictions we report the SSM(LZ05) lower left ellipse and the SSM(HZ) upper right ellipse at 68% C.L. From this figure it is evident that: 1) the $^8$B solar neutrino flux measurement sits in between the two models; 2) uncertainties are too large at present too try to disentangle the two models with this method. In the future this method can be used with improved results from SNO (after the low threshold analysis will be completed) and with smaller errors from Borexino. Other possibilities using CNO neutrinos have been proposed [13].

By assuming the SSM(HZ) fluxes the Borexino data can be used to determine the solar electron neutrino survival probability at Earth. Due to the high radiopurity achieved Borexino has been able to detect $^8$B neutrinos above 3 MeV [14]. In Fig 10 we report the survival

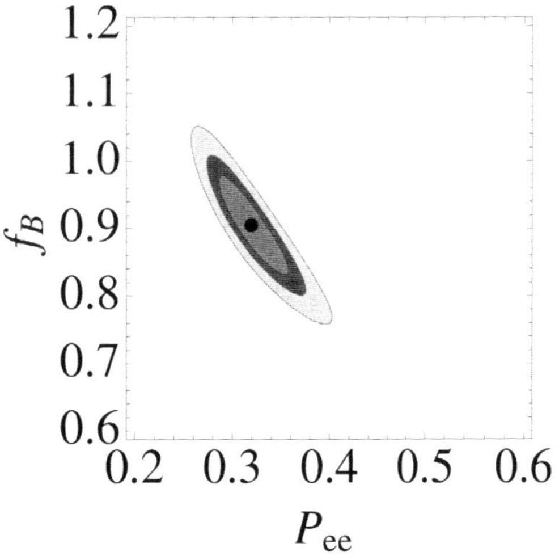


**Figure 7.** Measurement of the $^8$B solar neutrino flux with SNO and Super-Kamiokande. $f_B$ is the $^8$B solar neutrino flux normalized to the SSM(HZ); $P_{ee}$ is the electron neutrino survival probability at Earth. Contours are given at 68%, 90% and 99% C.L.

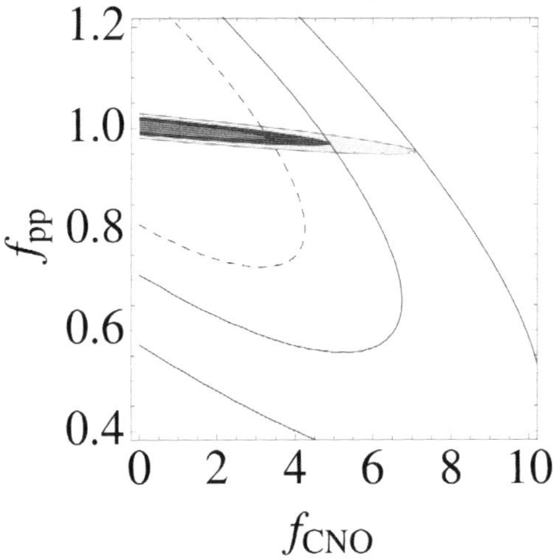


**Figure 8.** Measurement of the pp vs CNO solar neutrino fluxes with Borexino with and without the luminosity constraint. Contours are given at 68%, 90% and 99% C.L.

probability from Borexino and SNO together with the one which can be calculated for pp neutrinos. This figure shows also the Large Mixing Angle prediction (solid line). The plot shows the agreement between data and model for neutrino oscillations. Future Borexino measurements of $^7$Be and $^8$B solar neutrinos will confirm better this agreement or show possible sub-leading effects.

## 3. Conclusions

First results from Borexino on $^7$Be and $^8$B solar neutrinos, with less than one year of livetime, have already shown a great potentiality for understanding the Sun and the hypothesis of neutrino oscillations (see Fig 8, 10). At present important results have been obtained for the indirect measurement of the pp solar neutrino flux, a fundamental astrophysical parameter, and for the electron neutrino survival probability. Without a calibration campaign with external sources (in order to not contaminate the liquid scintillator) Borexino has performed a 10% measurement of the $^7$Be(0.862 MeV) solar neutrino flux. Future measurements with reduced systematics (5-3%), after a full calibration campaign with sources deployed inside the liquid scintillator, will allow to probe better the SSM helping to solve *the solar abundance problem.*

## Acknowledgments

The author wishes to acknowledge the organizers of the NPA4 for their assistance and the effort made during the conference.

## References

[1] G.L. Fogli et al., Prog.Part.Nucl.Phys.**57**, (2006) 742-795. T. Schwetz et al., New J.Phys.**10**, (2008) 113011.

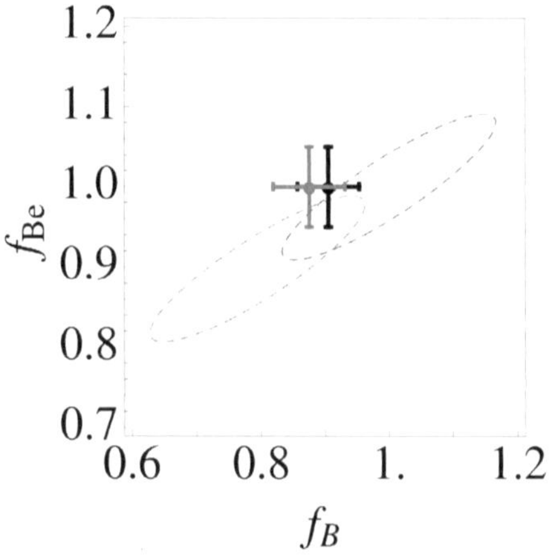


**Figure 9.** $^7$Be against $^8$B from Borexino and SNO/Super-Kamiokande. Upper(lower) ellipse is for the high(low 2005 [5]) metallicity model. 68% C.L. are shown. Left experimental point is from SNO only; right point is from SNO+Super-Kamiokande.

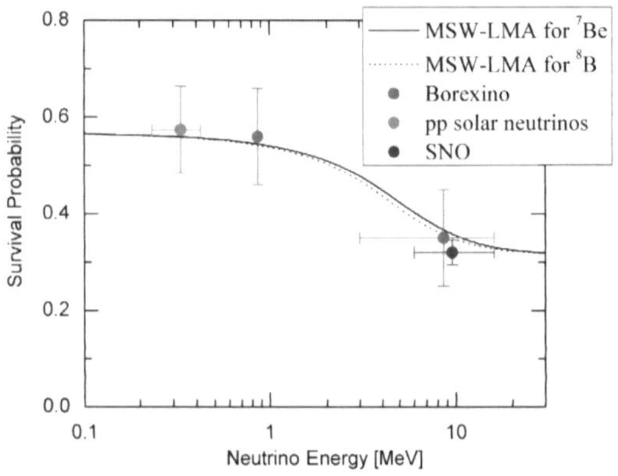


**Figure 10.** Survival probability for $\nu_e$ at Earth assuming the SSM(HZ).

[2] J.N. Bahcall et al., Astrophys.J.Suppl.**165**, (2006) 400-43. J.N. Bahcall, Phys.Scripta T **121**, (2005) 46-50.
[3] H. Costantini et al., Nuclear Phys. A **814**, (2008) 144. M. Marta et al., Phys. Rev. C **78**, (2008) 022802.
[4] S. Basu et al., Astrophys.J. **655** (2007) 660-67.
[5] M. Asplund, N. Grevesse and J. Sauval, Nucl. Phys. A 777, 1 (2006).
[6] M. Asplund, N. Grevesse, J. Sauval and P. Scott, 2009, ARA&A, in press.
[7] A.M. Serenelli, B. Basu and J. Ferguson, *New solar composition: the solar abundance problem revisited*, in press.
[8] C. Pena-Garay and A.M. Serenelli, arXiv:0811.2424.
[9] G. Alimonti et al., Borexino collaboration, NIM A **600** (2009) 568-593.
[10] C. Arpesella et al., Borexino collaboration, Phys. Rev. Lett. **101**, 091302 (2008).
[11] J.N. Bahcall, Phys. Rev. C **65**, (2002) 025801.
[12] C. Pena-Garay, PHYSUN, Gran Sasso Laboratory, Oct 16-17, 2008. http://www.mi.infn.it/PHYSUN/.
[13] W.C. Haxton and A.M. Serenelli, Astrophys.J. **687**, (2008) 678-691.
[14] G. Bellini at al., Borexino collaboration, arXiv:0808.2868.

Nuclear Physics in Astrophysics IV | IOP Publishing
Journal of Physics: Conference Series **202** (2010) 012029 | doi:10.1088/1742-6596/202/1/012029

# Simulated analysis of low-flux solar neutrinos and their terrestrial detection

**G. I. Karathanou[1], V. Tsikoudi[1], T.S. Kosmas[2]**

[1] University of Ioannina, Division of Astrophysics, Ioannina 45110 GREECE

[2] University of Ioannina, Theoretical Physics Division, Ioannina 45110 GREECE

**Abstract.** The energy spectra of hep and pep solar neutrinos produced by pp-chain and those produced from the CNO-cycle reactions are analyzed for various nuclear detectors in current terrestrial experiments. We intend to investigate the differential and total cross section of the coherent scattering of neutrinos off nuclear isotopes like $^{28}Si$. In such experiments the relevant quantity to measure is the kinetic energy distribution of the recoiling nucleus. This type of experiment has been proposed in order to observe the coherent elastic neutrino-nuclear scattering. We focus on the differential cross section $d\sigma/dT$ as a function of the kinetic energy T of the promising nuclear isotope $^{28}Si$.

## 1. Introduction

More than forty years ago, neutrinos were conceived as a way to test the validity of the standard solar model which tells that stars are powered by nuclear fusion reactions. The first experiment built to detect solar neutrinos was performed by Raymond Davis in the late 60s in the Homestake mine in South Dakota and detected only one third of the expected flux, deficit known as the Solar Neutrino Problem. Different experiments were built in order to understand the origin of this discrepancy. Now we know that neutrinos undergo oscillations, i.e. they change flavor travelling from the core of the Sun to the terrestrial detectors [1].

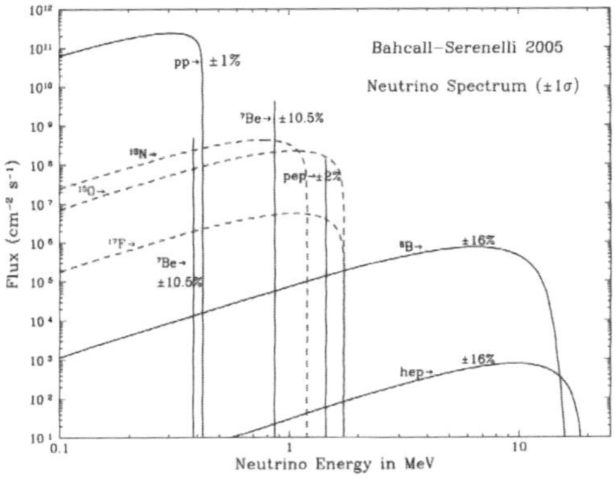


**Figure 1.** Neutrino energy spectrum as predicted by the Solar Standard Model (SSM).

Recently SNO has finished its final phase as a solar neutrino detector, and has accumulated over neutrino data covering the energy range from approximately 5 – 20 MeV. Almost all the observed neutrinos are due to the $^8B$ reaction in the Sun. Neutrinos produced through the hep process, $^3He+p \to ^4 He+e^+ + \nu_e$, also in the Sun, has not yet been observed. The hep neutrino energy endpoint extends above the $^8B$ spectrum. The most recent data analysis by SNO has also lowered the threshold for the detection of $^8$B solar neutrinos using neutral and charged-current reactions of neutrinos on deuterium while recent results from the Borexino experiment measured the rate of $^7$Be solar neutrinos that are in lower energy than $^8$B solar neutrinos (see Fig.1) [2].

Several future solar neutrino experiments have been proposed. Measurements of the pep and CNO solar neutrinos are a goal of the SNO+ experiment [3]. The flux of pep solar neutrinos is calculated to 1.5% uncertainty in the Standard Solar Model [4].

In the elastic neutral current (NC) neutrino-nucleus scattering, a neutrino scatters off a nucleus at low momentum transfer Q such that the nucleon wavefunction amplitudes are in phase and add coherently. For these reactions is very useful to calculate the differential cross section as a function of recoil nucleus kinetic energy (T), instead of scattering angle ($\Phi$) while many of the neutrino detectors have no directional sensitivity. The maximum recoil kinetic energy is directly related to the energy (E) of the incoming solar neutrino.

Since the coherent neutrino-nucleus scattering has not been experimentally proven, the development of new detectors (cryogenic, etc.) are planned for future applications in low energy neutrino physics. Such studies could be probed in conjunction with Cold Dark Matter (e.g. CRESST) and double beta decay experiments (e.g. those using $^{150}$Nd and $^{116}$Cd absorbers).

In the present work, we study coherent neutrino-nucleus elastic scattering from a theory point of view. We currently concentrate on calculations for reaction cross sections of the form $d\sigma/dT$ (see Sect. 2) for coherent neutrino-nucleus processes. Such processes could be, furthermore, used in the future to search for new physics like neutrino magnetic moment, non-standard neutral current interactions, etc.

## 2. Coherent Neutrino-Nucleus Scattering

Coherent neutrino scattering on a nucleus (A,Z),

$$\nu + (A, Z) \to \nu + (A, Z)' \tag{1}$$

is a neutral current interaction (i.e. sensitive to all active neutrino flavors) and should occur for processes with small momentum transfer (A represents the mass number and Z the atomic number, A=N+Z with N the neutron number). In these reactions the nuclear wavefunction remains unchanged i.e. we have $gs \to gs$ transition. Only the nucleus experiences a recoil as a whole.

The maximum recoil kinetic energy in neutrino-nucleus coherent scattering is

$$T_{max} = \frac{2E^2}{M + 2E}, \tag{2}$$

where E is the incident neutrino energy, and M is the mass of the target nucleus [5].

The four-momentum transfer is related to the recoil kinetic energy by $Q^2 = 2MT$, and the three-momentum transfer q is approximately $\sqrt{2MT}$. For neutrino energies below 20 MeV and nuclear targets such as $^{28}Si$ and $^{32}S$ , the maximum recoil kinetic energy is approximately around 30 keV, and therefore the maximum possible q is quite small, $< 1fm^{-1}$.

The dependence of the cross section on the scattering angle means that solar neutrino elastic scattering events will, in principle, point back to the Sun. However, many of neutrino detectors do not have directional sensitivity, and so it is most useful to calculate event rates as a function of recoil nucleus kinetic energy [6]. The scattering angle and the recoil kinetic energy are related via 2- body kinematics and the cross section can be expressed in terms of the kinetic energy T of the recoiling nucleus as

$$\frac{d\sigma}{dT} = \frac{G_F^2}{2\pi} M[1 + (1 - \frac{T}{E})^2 - \frac{MT}{E^2}] \frac{Q_w^2}{4} F^2(Q^2) \tag{3}$$

In the latter expression, $G_F$ denotes the Fermi constant, $G_F^2 M/2\pi = 5.14 \times 10^{-41} cm^2/MeV$ and $Q_w$ which is given by

$$Q_w = N - Z(1 - 4\sin^2\theta_w) \tag{4}$$

is the weak charge ($\sin^2\theta_w \approx 0.231$) with $\theta_w$ the Weinberg angle. The quantity $Q^2$ is the squared momentum transfer

$$Q^2 = \frac{2E^2TM}{(E^2 - ET)}, \tag{5}$$

and $F(Q^2)$ is the nuclear form factor. Denoting the neutron and proton densities as $\varrho_{n,p}(r)$ the form factor is

$$F(Q^2) = \frac{1}{Q_w} \int [\varrho_n(r) - (1 - 4\sin^2\theta_w)\varrho_p(r)] \frac{\sin(Qr)}{Qr} r^2 dr \tag{6}$$

From Eq.(6) we see that, since $(1 - 4/sin^2\theta_w)$ is small, a neutrino scattered elastically off a spin-zero nucleus couples mostly to the neutron distribution. A measurement of the cross section for this process would, at some level, provide a determination of the neutron form factor. This would be complementary to the parity violating experiment with lead because it would provide additional data, obtained at different energy ranges and with different nuclei that could be used to calibrate nuclear structure calculations [7].

In the present work we are treating the proton and neutron distributions separately, as in Eq.(6). The form factor we use is

$$F(Q^2) = \frac{1}{Q_w}[NF_n(Q^2) - Z(1 - 4\sin^2\theta_w)F_p(Q^2)] \tag{7}$$

where $F_n(Q^2)$ and $F_p(Q^2)$ are the nuclear neutron and proton (electromagnetic) form factors, respectively.

**Table 1.** The differential cross section $d\sigma/dT$ as a function of the kinetic energy T using theoretical nuclear form factors F(Q) [9] for the nuclear isotope $^{28}Si$.

| Q ($fm^{-1}$) | F(Q) | T (MeV) | $\frac{d\sigma}{dT}(\times 10^{-39} cm^2 MeV^{-1})$ |
|---|---|---|---|
| 0.14 | 0.9705 | $6.4216 \times 10^{-5}$ | 4.050 |
| 0.15 | 0.9662 | $2.7547 \times 10^{-4}$ | 4.013 |
| 0.16 | 0.9616 | 0.0018 | 3.967 |
| 0.17 | 0.9667 | 0.0050 | 3.912 |
| 0.18 | 0.9516 | 0.010 | 3.846 |
| 0.19 | 0.9462 | 0.017 | 3.768 |
| 0.20 | 0.9405 | 0.027 | 3.677 |
| 0.21 | 0.9346 | 0.040 | 3.573 |
| 0.22 | 09284 | 0.056 | 3.453 |
| 0.23 | 0.9220 | 0.075 | 3.319 |

In the case of $^{28}Si$ and $^{32}S$ we have performed explicit calculations for $d\sigma/dT$ and received preliminary results in the following cases:

- We employ experimental nuclear form factors obtained from electron scattering experiments [8].
- We utilized the theoretical nuclear form factors obtained in the frame note of the deformed BCS calculations for a set of light nuclei in Ref. [9].

The results for the nuclear isotope $^{28}Si$ are listed in Table 1.

The values of $d\sigma/dT$ in Table 1 are in agreement with similar calculations performed by Vergados et al. (Ref. [5]) where examined the coherent contribution of all neutrons in neutrino nucleus scattering due to the neutral current considering the boron solar neutrinos which could potentially become a source of background in the future dark matter searches.

## 3. Summury and Conclusions

Current proposed and ongoing solar neutrino experiments like BOREXINO and SNO+, expect to observe low-rate solar neutrinos as pep and CNO neutrinos significant to study solar physics. Neutrinos will help in understanding the metallicity in the solar core that appears to contrast with recent interpretations of solar surface chemical abundances.

Furthermore, other type of experiments (cryogenic, etc.) are designed with the aim to study coherent neutral-current scattering of neutrinos on nuclei. Such experiments may provide a potential to detect low-energy neutrinos with an excellent energy resolution. Neutrino-induced recoil events may constitute a background to direct dark matter searches.

Our aim in this work, is to provide the theoretical background of such searches by evaluating differential cross sections $d\sigma/dT$ as functions of the kinetic energy $T$ for promising nuclear detectors. We presented results for the $^{28}Si$ isotope (one of the best choices for coherent neutrino-nucleus studies) and we are currently working in evaluating event rates for a set of nuclear isotopes in the area of $^{28}Si$ and $^{32}S$ isotopes.

## 4. Acknowledgments

The present work was partially supported by the ΠΕΝΕΔ No 03ΕΔ807 project of the General Secretariat for Research and Technology of the Hellenic Ministry of Development.

## References

[1] Lino Miramonti, 2009, http://arxiv.org/abs/0901.3443v1.
[2] C. Arpesella et al. (Borexino Collaboration), 2008, Phys. Rev. Lett. 101, 091302.
[3] Chris Howard (SNO Collaboration), 2009, 0906.0040v1 [ hep-ex].
[4] Bahcall J. N., Serenelli A.M. and Basu S., 2005, Astrophys. J. 621 L85.
[5] J.D. Vergados and H. Ejiri, 2008, hep-ph/0805.2583v2.
[6] J. Monroe, P. Fisher, 2007, Phs. Rev. D 76, 033007
[7] P. S. Amanik and G. C. McLaughlin, 2009, J. Phys. G: Nucl. Part. Phys. 36 015105 (9pp).
[8] H.De Vries, C.W.De Jager, and C.De Vries, 1987, Atomic data and Nuclear data tables 36, 495-536.
[9] Ch.C. Moustakidis, T.S. Kosmas, F. Simkovic, Amand Faessler, 2005, nucl-th/0511013v1.

# Extra-mixing in red giant stars: challenges for nuclear physics

**Sara Palmerini and Enrico Maiorca**
I.N.F.N. sezione di Perugia
Dipartimento di Fisica Universitá degli Studi di Perugia,
via Pascoli, 06123, Perugia, Italy

E-mail: sara.pamerini@fisica.unipg.it

**Abstract.** The existence of extra-mixing phenomena has been often invoked as a possible solution for the Li-abundance puzzle in low-mass red giant stars. In particular, [1] have shown that extra-mixing phenomena induced by stellar magnetic fields can justify the surface Li enrichment as well as its depletion in low mass giants. In the framework of this model, we test here how sensitive is the Li production to the reaction rate for the $^7$Be electron capture, in order to establish whether the presence of intense magnetic fields can alter the Li yield.

## 1. Introduction

Gratton et al. (2000) [2] set to $\log\epsilon(Li) \leq 0$[1] the upper limit for the Li abundance of low mass and low metallicity stars immediately after the main sequence. The concentration of $^7$Li (which is by far the most abundant lithium isotope) is determined by the interplay between its synthesis by electron capture on $^7$Be and its destruction by proton capture, taking place at relatively low temperatures (T$\leq 2.5 \cdot 10^6 K$). Due to the prevailing destruction effect, $^7$Li is almost totally consumed during the pre-main sequence and main sequence stages. However, red giants with an enhanced Li abundance, sometimes higher than the interstellar medium one [$\log\epsilon(Li) \leq 3.3$], have been observed [3]. This shows that an enrichment should have taken place after the main sequence. Several viable explanations have been proposed, including accretion of Li from outside, such as in the engulfment of a giant planet [4]. Since a thin radiative layer free of entropy barriers is present in the Red Giant Branch (hereafter RGB) stars, between the H-burning shell and the convective envelope, the most common hypothesis has been to assume that some fast extra-mixing should have affected this radiative layer and triggered a Cameron-Fowler mechanism [5]. This a transport of $^7$Be from above the H-burning shell up to the envelope at a rate larger than p-captures along the path. Sackmann & Boothroyd (1999) have shown that it is possible to increase or deplete the Li supply in RGB envelopes by a deep mass circulation mechanism, depending on the mixing depth and velocity. This phenomenon was recognized to be related also to CNO isotopic ratio anomalies in stars along the RGB and the following Asymptotic Giant Branch (AGB) phase.

Several works have investigated causes and mechanisms of extra-mixing (either fast or slow). In particular, processes induced by rotation, with shear instabilities, diffusion and meridional

[1] Throughout the paper, the Li abundance is given on the scale $\log\epsilon(\mathrm{Li}) = \log \mathrm{N(Li)/N(H)} + 12$. The solar photospheric Li abundance on this scale is $\log\epsilon(\mathrm{Li}) = 1.1$.

circulation were presented to reproduce the Li enrichment during the initial part of the RGB phase [7],[8]. These possibilities were, to a large extent, frustrated by the discovery that rotation would not yield sufficiently extended extra-mixing processes to account for the observations [9]. More recently, two other extra-mixing models have been proposed as possible sources of deep circulation in stars. The first is the so-called thermohaline diffusion, a double diffusive process of sinking of heavy envelope matter, due to the molecular weight inversion induced by the reaction $^3$He+$^3$He→$^4$He + 2p [10]. However, this model provides a slow mixing velocity, while fast a transport is requested to provide for a significant Li enrichment.

The other suggested process is the buoyancy of magnetized H-burning ashes [11], [12]. This second mechanism was shown to be able to lead both to fast and to slow mixing episodes, depending on the buoyant structure geometry [13] and on its heat exchange rate with the surrounding material [14]. More recently Guandalini et al. (2009) [1] presented an analysis of Li abundances (based on the interplay of fast and slow extra-mixing) and proposed a possible evolution of it during the RGB and AGB phases of low mass stars. If the mixing is operated by magnetized structures, the presence of strong fields (up to $10^5$ Gauss) is required in the inner stellar regions, where the Lorentz force could then generate a non-homogeneous charge distribution at microscopic scale. If this were not the case, the magnetic field may affect the nucleosynthesis experienced by mixed materials through changes in the otherwise Maxwellian distribution of the gas velocities, which might in turn modify the effects of electron screening and electron captures.

As an example, in the ppIII-chain network:

$$^3He + ^4He \longrightarrow ^7Be + \gamma \tag{1}$$
$$^7Be + H \longrightarrow ^8B + \gamma \longrightarrow ^4He + ^4He + \gamma \tag{2}$$
$$^7Be + e^- \longrightarrow ^7Li + \nu \tag{3}$$
$$^7Li + H \longrightarrow ^4He + ^4He + \gamma \tag{4}$$

reactions (1) and (3) play a key role in determining the $^7$Be (and so the $^7$Li) abundance. In particular, at high temperature (T$> 2 \cdot 10^7 K$) $^7$Be is completely ionized and the electron captures take place via free electrons in the plasma. If the electron velocities were not following a Maxwellian distribution, for the presence of a non-kinetic term in their internal energy, introduced by electromagnetic forces, this could significantly modify the efficiency of reaction (3) .

In this work, we simulate what could be the resulting Li abundance in red giant stars if the electron capture on $^7$Be were inhibited in the material affected by extra-mixing. This is done by repeating the same computation proposed by [1] (for a $2M_\odot$, solar-metallicity star) for different values of this reaction rate, from the nominal accepted value to zero.

## 2. Magnetic extra-mixing and resulting Li abundances

In RGB stars the hydrogen burning shell surrounds the partly degenerate He-rich core, while in AGB stars there is also a second helium rich shell (below the hydrogen one) surrounding the CO inactive core. Apart from this, the structure of the radiative layer at the base of the convective envelope in a RGB star and in an AGB, during hydrogen shell burning, is almost the same. Fig.1 shows the typical abundances of $^3He$, $^7$Be, $^7$Li and a few CNO isotopes in the radiative region of a $2M_\odot$, solar-metallicity star on the RGB and on the AGB. In order to increase the envelope abundance of $^7$Li the extra-mixing mechanism has to carry rapidly $^7$Be from its production zone to cool regions, where p-captures on $^7$Li are suppressed because of the low temperatures.

Following the mixing scenario suggested by [1], we consider an initial mixing phase at the RGB luminosity bump[2], where only small portions of magnetized material (hereafter called 'bubbles')

[2] The luminosity bump corresponds to the point when the outgoing hydrogen burning shell encounters and erases

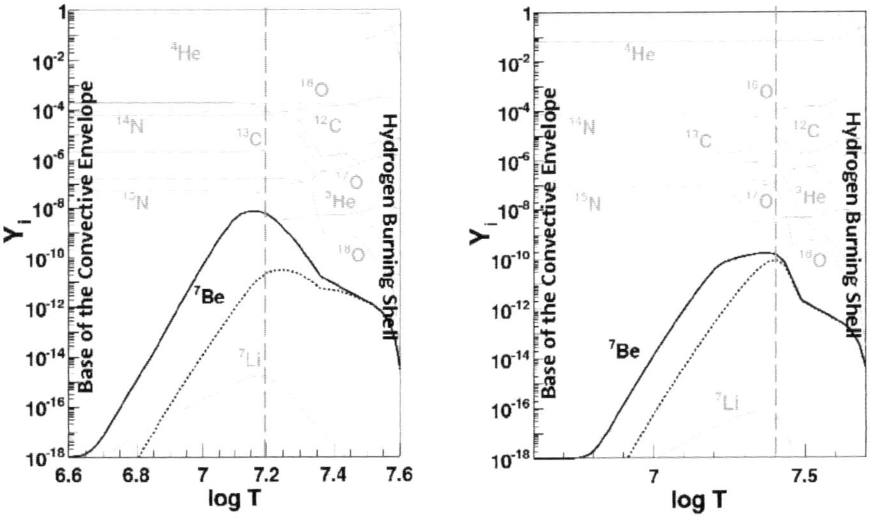


**Figure 1.** Left panel: Abundances in the radiative region between the H-burning shell and the convective envelope as a function of the temperature (in Kelvin) for a $2M_{\odot}$ star of solar-metallicity during RGB phase. The dashed black line shows the $^7$Be profile obtained using reaction rates from [15]. Instead, the solid black line represents the result obtained by setting to zero the rate of the electron capture on $^7$Be. (A recent reaction rate measurement of (1) was presented by [16]and it is included in our calculation). The dashed grey vertical line indicates the maximum temperature and hence the deepest layer reached by the extra-mixing. Right panel: Same plot as in the left panel but dealing with the inner structure during the tenth inter-pulse period of the AGB phase.

detach, driven by local magnetic instabilities, from regions near the H shell and travel toward the convective envelope at the Alfvén speed, of a few m/s (case A). Since the motion is very fast, the hypothesis that no burning occurs in this mass circulation is likely. The resulting effect is the mixing of $^7$Be (produced near the H shell) into the envelope, where it decays to Li. This first fast mixing phase may be followed by a slow circulation regime [1], where larger structures become buoyant (case B). The mixing velocity is then reduced to values of a few cm/s, due to the gradual heat exchange between the magnetized zones and the surrounding environment [14] and this slow mixing is responsible for Li destruction. This slow mass circulation is assumed to work during the last part of the RGB phase ($\sim 4.7Myr$) and during AGB hydrogen burning periods (the so call 'inter-pulses'). In both cases, A and B, an equally fast downflow of envelope material guarantees mass conservation.

First, we performed a calculation using the reaction rate from [15] and [16], and the mixing parameters from [1]. Then, we repeated the calculation using the same choices for the stellar model and mixing parameters but under the hypothesis that no electron capture on $^7$Be takes place in the radiative zone and inside the magnetic structures.

Fig. 2 shows the temporal evolution of the surface Li abundance of the RGB and AGB stars under study. If electron captures are suppressed, the enrichment of Li in case A is more efficient by a factor of 10 (from $\log\epsilon = 1.3$ to $\log\epsilon = 2.4$). In the same way, the Li depletion obtained in case B is strongly reduced for the RGB star, and can actually be reversed into a production

the mean molecular weight discontinuity left by the penetration of the convective envelope at the beginning of the RGB phase (the so-called first dredge-up).

(from $\log \epsilon = -2$ to $\log \epsilon = 2.4$) depending on the mixing rate. Changes in the electron captures induce smaller variations on the AGB. However, also here the rate of Li destruction is reduced and can actually be converted into a (small) production (see Fig.2 right panel).

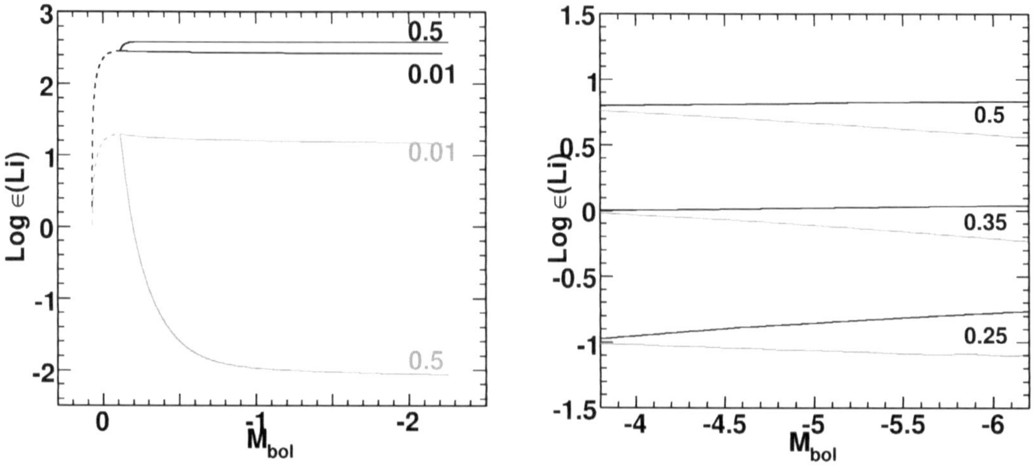


**Figure 2.** Left panel: Evolution of the envelope Li abundance as a function of the stellar luminosity, during the RGB phase. The dotted lines indicate the Li enrichment due to fast magnetic instabilities (case A). Solid lines show the effects of slower transport phenomena at different mixing rates (case B). These might be induced by the buoyancy of large structures, exchanging heat with the environment. Results presented in Fig.5 by [1], for a mixing rate of $5 \cdot 10^{-7}$ and $10^{-8} M_{\odot}/yr$ (grey curves), are compared with our results obtained considering no electron captures on $^{7}$Be (black curves). Right panel: Evolution of Li abundance due to slow extra-mixing during the AGB phase, for different mixing rates and different initial abundance of $^{7}$Li in the envelope. As in the left panel, grey curves report results by [1], instead black curves show results obtained considering no electron captures on $^{7}$Be.

## 3. Conclusions

Magnetic buoyancy, inducing slow and fast mass circulation, might offer an useful tool to understand Li enrichment and destruction in low mass red giants. However, if the presence of strong magnetic fields (up to $10^5$ Gauss) modified the charge distribution at microscopic scale, this would affect the results of the extra-mixing nucleosynthesis. In particular, we have tested that reduction of electron captures on $^{7}$Be may increase the Li abundance, in some cases by orders of magnitude. Because of its high sensitivity to the temperature and velocity experienced by circulating materials, the stellar Li abundance provides constraints to the extra-mixing model stronger than ones yielded by CNO isotopes. For these reasons, it is crucial to understand whether the magnetic forces can modify the distribution of charges at the microscopic level in the transported materials.

## 4. Acknowledgments

We are indebted to the Italian Ministry of Research for a PRIN grant (n.2006/022731) and to the National Institute of Nuclear Physics (Section of Perugia, ERNA experiment) for providing support and computing facilities.

## 5. References

[1] Guandalini, R., Palmerini, S., Busso,M., Uttenthaler, S., (2009), PASA, in press.
[2] Gratton, R. G., Carretta, E., Matteucci, F., Sneden, C., (2000), A&A, 358, 671.
[3] de La Reza, R., Drake, N.A., da Silva, L., Torres, C.A.O., Martin, E.L., (1997), ApJ Lett, 482, 77.
[4] Denissenkov, P. A., VandenBerg, D. A., (2003), ApJ, 593, 509.
[5] Cameron, A. G. W., Fowler, W. A., (1971) , ApJ, 164, 111.
[6] Sackmann, I. J., Boothroyd, A. I., (1999), ApJ, 510, 217.
[7] Charbonnel, C., Do Nascimento, J. D. Jr., (1998), A&A, 336, 915.
[8] Palacios, A., Talon, S., Charbonnel, C., Forestini, M., (2003), A&A, 399, 603.
[9] Palacios, A., Charbonnel, C., Talon, S., Siess, L., (2006), A&A, 453, 261.
[10] Eggleton, P. P., Dearborn, D. S. P., Lattanzio, J. C., (2006), Science, 314, 1580.
[11] Busso, M,Wasserburg, G. J., Nollett, K. M., Calandra, A., (2007), ApJ, 671, 802.
[12] Nordhaus, J., Busso, M., Wasserburg, G. J., Black- man, E. G., Palmerini, S., (2008), ApJ, 684, L29.
[13] Palmerini, S., Busso, M., (2008), NewAR, 52, 412.
[14] Denissenkov, P.A., Pinsonneault, M., Mac Gregor K.B., (2009), ApJ, in press.
[15] Adelberger, E. G., et al., (1998), Rev. Mod. Phy., 70, 1265.
[16] Di Leva, A., et al., (2009), PRL 102, 232502.

# Progress of LUNA

**Alessandra Guglielmetti for the LUNA collaboration**

Università degli Studi di Milano and INFN, Milano, Italy

alessandra.guglielmetti@mi.infn.it

**Abstract**. At H-burning temperatures in the Sun, nuclear cross sections are extremely small and experimental measurements in a laboratory at the Earth's surface are hampered by the cosmic background. The LUNA collaboration has exploited the unique features of LNGS underground laboratory in terms of background reduction, to study very important H-burning reactions at astrophysically relevant energies. This paper will focus on the results recently obtained for the $^{3}He(^{4}He,\gamma)^{7}Be$ reaction using two different experimental techniques and with a close eye on the minimization of the systematical uncertainties. On going and future measurements will also be presented.

## 1. Introduction

It is well known [1] that stars generate energy and produce elements by means of thermonuclear fusion reactions, which start from the most abundant and lightest element, hydrogen, and gradually synthesize heavier elements. The hydrogen burning can proceed either through the p-p chain or through the more efficient CNO cycle or even through the NeNa and MgAl cycles in stars more evoluted and hotter than our Sun. The net result is the transformation of 4 protons into a $^{4}He$ nucleus with an energy release of about 27 MeV. The fusion of hydrogen into helium occurs during the longer part of the star's life (main sequence) and is responsible for the prodigious luminosity of the star itself.

All these fusion reactions occur in a very well defined energy range, the so-called Gamow peak, which is the result of the overlapping between the energetic distribution of nuclei in stars at a given temperature and the energetic dependence of the reaction cross-section. The first, given by the Maxwell-Boltzmann distribution, has a maximum for E=kT (where T is the star temperature and k the Maxwell-Boltzmann constant) and then decreases exponentially for increasing energy while the second, given by the tunneling probability through a Coulomb barrier, decreases exponentially for decreasing energy. In particular the reaction cross-section for charged particles can be written as:

$$\sigma(E) = (S(E)/E) \exp(-2\pi\eta) \qquad (1)$$

where S(E) is the astrophysical S factor which contains the pure nuclear behavior of the cross section and $\eta$ is the Sommerfeld parameter given by:

$$\eta = (31.29/2\pi)\, Z_1\, Z_2\, (\mu/E)^{1/2} \qquad (2)$$

being $Z_1$ and $Z_2$ the charges of the interacting nuclei, $\mu$ the reduced mass and E the interaction energy in the center of mass system in keV. As an example, in our Sun, whose temperature is 1.5 $10^{7}$ °K, the maximum of the Maxwell-Boltzmann distribution occurs at about 1 keV, the Coulomb barrier for most of the reactions of the p-p cycle is between 0.5 and 2 MeV and the Gamow peak for the same reactions is below 30 keV. At such low energy, the cross section is extremely low, down to the femtobarn level in "the worst" cases, due to the already mentioned exponential decrease of the cross section with decreasing energy. Therefore, direct measurements are mostly hampered by the cosmic background in a laboratory at the earth's surface where the signal to background ratio would be too small. The usual approach is to extrapolate the observed energy dependence of the cross section at high energies down to the low and significant energy region leading to substantial uncertainties. In particular a possible narrow resonance in the unmeasured region, the tail of a broad or sub-threshold resonance or even a change in the reaction mechanism can completely change the reaction rate.

## 2. Why underground: the LUNA project

The LUNA (Laboratory for Underground Nuclear Astrophysics) collaboration has exploited the silent environment of the underground laboratory under the Gran Sasso Mountain in Italy (LNGS) to overcome this difficulty by performing direct measurements at the relevant energies. The rock cover of about 1500 m (3400 m water equivalent) reduces the muon component of the cosmic background of a factor $10^6$, the neutron component of a factor $10^3$ and the gamma component of a factor 10 with respect to the earth's surface. As a result, the gamma background above 3 MeV in an HPGe detector placed underground at LNGS is reduced of a factor ~2500 with respect to the same detector placed over-ground. Moreover, for lower energy gammas where the background is dominated by environmental radioactivity, the effect of passive shielding is enhanced by going underground. Indeed, in a laboratory on the Earth's surface, a passive shielding can be built around the detector but the shield efficiency cannot be increased by making it thicker with the addition of further material since cosmic muons interact with it creating background signals in the detector itself. This problem is of course dramatically reduced in an underground laboratory.

The LUNA collaboration has installed two accelerators underground: a compact 50 kV "home-made" machine [2] and a commercial 400 kV one [3]. Common features of the two are the intense current production, the long term stability and the precise energy determination. The first two features are essential to maximize the reaction rate, while the third is important in view of the exponential energy dependence of the cross section. With the first machine, operating between 1992 and 2001, two fundamental reactions of the p-p chain were studied at solar Gamow peak energies: the $^3$He($^3$He,2p)$^4$He [4] and d(p,$\gamma$)$^3$He [5] and the screening effect, consisting in a lowering of the Coulomb barrier for a nucleus surrounded by electrons with respect to a bare one, was investigated through the d($^3$He, p)$^4$He reaction [6]. The 400 keV machine, started operations in the year 2000 and is still operating. The main results obtained are the cross sections of the $^{14}$N(p,$\gamma$)$^{15}$O reaction [7], which represents the bottle-neck of the CNO cycle, and of the $^3$He($^4$He,$\gamma$)$^7$Be one. In this paper I will concentrate on the latter result and on the ongoing and future measurements at LUNA.

## 3. The $^3$He($^4$He,$\gamma$)$^7$Be reaction

The $^3$He($^4$He,$\gamma$)$^7$Be reaction is the onset of the $^7$Be and $^8$B branches of the p-p chain from which the $^7$Be and $^8$B neutrinos are generated. Solar neutrino fluxes depend on both astrophysical and nuclear physics inputs: in the case of $^7$Be and $^8$B neutrinos, one of the most important parameters is the cross section of the $^3$He($^4$He,$\gamma$)$^7$Be reaction. After the successful measurements of the $^8$B neutrino flux with a 3.5% uncertainty performed by SNO [8] and SuperKamiokande [9] and in view of a similar result on the $^7$Be neutrino flux from the Borexino experiment [10], a comparable uncertainty on the $^3$He($^4$He,$\gamma$)$^7$Be cross section is demanded. Only under these circumstances, the neutrinos will give reliable information on the interior of the Sun and on its physical and chemical properties

Moreover, the $^3$He($^4$He,$\gamma$)$^7$Be reaction is fundamental for the Big-Bang nucleosynthesis (BBN) being the main source for $^7$Li production. The predicted abundance of such isotope is a factor 2-3 higher than the observed one. Even if it is unlikely that the explanation of this discrepancy could come from the $^3$He($^4$He,$\gamma$)$^7$Be cross section, a better knowledge of its value represents the necessary basis for the search of alternative solutions to the $^7$Li problem.

In the last forty years, the $^{3}He(^{4}He,\gamma)^{7}Be$ reaction has been measured using two different approaches: either the detection of direct α-capture γ-rays (prompt γ method) or the detection of the $^{7}Be$ radioactivity through the electron capture of $^{7}Be$ into $^{7}Li$ and its subsequent de-excitation from the first excited to the ground state (activation method). Results obtained with the two techniques show, on average, a discrepancy of about 15% which is the origin of the large uncertainty on the $^{3}He(^{4}He,\gamma)^{7}Be$ cross section [11].

To solve this problem, the LUNA collaboration planned and performed an experiment with both techniques at the same time and with a careful eye on reducing the systematical uncertainties. A re-circulating $^{3}He$ extended windowless gas target was coupled to a 137% HPGe detector in close view of the interaction chamber. The $^{4}He$ intense beam of the LUNA 400 kV machine was focused on a removable copper beam-stop which served both as the end cap of a calorimeter (to measure the beam current) and as a catcher for the $^{7}Be$ nuclei allowing for a delayed off-line measurement of their activity. Being the Q-value of the reaction about 1.5 MeV, the prompt γ-rays are in the region of natural background: a massive shielding (0.3 $m^3$) was therefore built around the target chamber and the HPGe detector. This consisted in several layers of lead bricks and an inner layer of Oxygen Free High Conductivity (OFHC) copper bricks. Moreover, the target chamber and the calorimeter were constructed using OFHC copper and no welding materials were used. All this system was enclosed in an anti-radon envelope consisting in a Plexiglas box flushed with $N_2$ gas to avoid $^{222}Rn$ accumulation. As already underlined, the efficiency of the shielding was increased by the underground location and a background suppression of five orders of magnitude was obtained for γ-rays below 2 MeV with respect to a background spectrum measured underground with no shielding [12]. The typical α-beam current was high (about 250 μA), thus causing a reduction of the target density due to the so-called beam-heating effect. This was measured through the detection of the beam particles elastically scattered on the target atoms [13]. Moreover, in order to reduce the systematic error coming from the uncertainty on the angular distribution, a lead collimator was positioned inside the target chamber to collect mostly γ-rays emitted at 55°, where the contribution of the second Legendre polynomial vanishes. After each run, the calorimeter cap was dismounted and its activity counted off-line with a 125% HPGe detector, shielded by 15 cm of lead and 10 cm of copper on each side and placed in the low activity laboratory of LNGS [14].

The experiment was divided into two phases: in the former the cross section was measured at E=127 and E=148 keV (CM reference frame) with the activation technique only [15], in the latter both techniques were used to measure the cross section at E= 93, 106 and 170 keV [16]. A comparison of the results obtained with the two techniques was therefore possible and reduced significantly the systematic uncertainties. Moreover, the cross sections measured with the two methods were consistent at the level of 4% [16], as can be seen by looking at the inset of figure 1.

While the LUNA data partly covered the energy window interesting for BBN (80-400 keV), an extrapolation is still needed for reaching the solar Gamow peak at 22 keV. The most recent results [17, 18] were taken into account together with the LUNA data [16]: two different theoretical approaches were followed [19, 20] to fit each of the three experiments and the weighted average of the extrapolated S(0) values was considered as the "best" value. This resulted in:
$S(0) = 0.567 \pm 0.018 \pm 0.004$ keV· b where the last error accounts for the uncertainty on the adopted theoretical model. The most recent results [17, 18] and the LUNA data [16] for the astrophysical S-factor as a function of the center of mass energy are shown in figure 1, together with the theoretical extrapolations [19, 20] of the S(0) value.

The goal of the 3% uncertainty was therefore reached but a unique set of data exploring the entire interesting energy range is highly desirable to precisely determine the S factor energy dependence.

## 3. Ongoing and future measurements

After the completion of the $^{3}He(^{4}He,\gamma)^{7}Be$ experiment, LUNA started to measure the $^{25}Mg(p,\gamma)^{26}Al$ reaction; the slowest of the Mg-Al cycle which operates in massive stars like AGB's. The $\beta^+$ decay of the ground state of $^{26}$Al ($T_{1/2}$ = 7 $10^5$ y) to the first excited state of $^{26}$Mg gives rise to a 1.8 MeV γ-ray which is very important for γ-astronomy [21]. Indeed, the presence of $^{26}$Al in the interstellar medium has been determined from observation of this γ-line from satellites like COMPTEL [22] and INTEGRAL [23] and from the observation of $^{26}$Mg isotopic enrichment in carbonaceous meteorites (extinct $^{26}$Al ) [24]. Satellite observations provide evidence that $^{26}$Al nucleosynthesis is still active on a large scale while meteorites enrichment in $^{26}$Mg demonstrates that $^{26}$Al must have been produced not later than 4.6 billion years ago (time of the condensation of the solar-system material). Any astrophysical scenario for $^{26}$Al nucleosynthesis must be concordant with both observations. The $^{25}$Mg(p,γ)$^{26}$Al reaction is the most important source of $^{26}$Al production. The reaction is dominated by narrow resonances which have been measured down to 190 keV. However, from the known level scheme of $^{26}$Al, one expects low-lying resonances at E = 93, 109 and 130 keV, among which the lowest energy one appears to be the most important. The LUNA collaboration used enriched $^{25}$Mg solid targets and two different detection set-ups: a high efficiency but low resolution BGO detector to measure the strength of the lower energy resonances (93, 109, 130 and 190 keV) and a high resolution but low efficiency HPGe to determine cascades and branching ratios for the higher energy ones (190 and 317 keV). Moreover, an AMS independent measurement of the same higher energy (190 and 317 keV) resonances was used for comparison. The data analysis of this experiment is still ongoing.

At the beginning of year 2007, the LUNA collaboration prepared a detailed proposal, resulting from a two-year study effort, containing a list of reactions of astrophysical relevance which could be studied with the 400 kV machine and which could profit from the underground location. Among these, reactions of the CNO cycle, NeNa cycle and BBN are present. This proposal was approved by LNGS and funded by INFN and the experimental program started last year. One of these reactions, namely the $^{15}$N(p,γ)$^{16}$O, has been indeed orally presented during this conference [25] being at the ultimate stage of data taking.

The next reaction to be studied will be the $^{2}$H(α,γ)$^{6}$Li, the main $^{6}$Li production source during BBN. $^{6}$Li abundance in metal poor stars is three orders of magnitude higher than BBN predictions. Existing direct data reach E=650 keV while indirect measurements at lower energies are in strong disagreement. The region of interest for BBN is between 50 and 300 keV and could be reached with the 400 kV machine. Presently, the beam induced-background due to the d(d,n)$^{3}$He and d(d,p)$^{3}$H has been deeply investigated both experimentally and with simulations leading to encouraging results. A set-up similar to the one used for the $^{3}$He($^{4}$He,γ)$^{7}$Be experiment (the Q-values of the two reactions are almost identical), but with an eye on the beam-induced background minimization, will be designed and built and data taking will start soon.

In addiction to the already cited proposal, the LUNA collaboration prepared a Letter of Intent (LOI) with the ambitious program of installing a 3 MV machine at LNGS. This could allow the study of He burning reactions among which the $^{12}$C(α,γ)$^{16}$O "holy grail" of nuclear astrophysics and of several (α,γ) and (p,γ) processes having deep consequences in nucleosynthesis, stellar evolution, supernova mechanism, etc. The proposed LOI received very positive response but the decision of LNGS is still pending, major problems being space requirements and possible interferences with other low-background experiments.

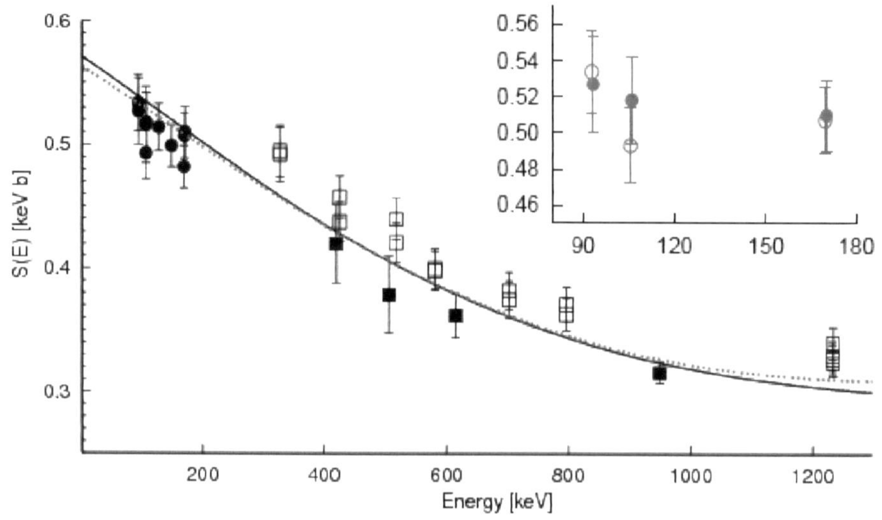


**Figure 1**. Astrophysical S factor for the $^3He(^4He,\gamma)^7Be$ reaction obtained from recent experiments. The filled squares are from reference [17], the open squares from reference [18] while the filled circles are the LUNA data [16]. The solid and dotted curves are the theoretical predictions by [19] and [20] respectively, obtained considering the weighted average of S(0) of the different experiments (see text). The inset shows a detailed comparison of the prompt (filled circles) and activation (open circles) LUNA results.

**References**

[1] Rolfs C and Rodney W 1988 *Caluldrons in the Cosmos* (Chicago: University of Chicago Press)
[2] Greife U et al. 1994 *Nucl. Instr. and Meth. A* **350** 327
[3] Formicola A et al. 2003 *Nuc.l Instr. and Meth.* A **507** 609
[4] Bonetti R et al. 1999 *Phys. Rev. Lett.* **82** 5202
[5] Casella C et al. 2002 *Nucl. Phys.* A **706** 203
[6] Costantini H et al. 2000 *Phys. Lett.* B **482** 43
[7] Formicola A et al. 2004 *Phys Lett.* B **591** 61
Lemut A et al. 2006 *Phys. Lett.* B **634** 483
[8] Aharmin B et al. 2007 *Phys. Rev.* C **75** 045502
[9] Hosaka J et al. 2006 *Phys. Rev.* D **73** 112001
[10] Arpesella C et al. 2008 *Phys. Lett.* B **658** 101
[11] Adelberger E et al. 1998 *Rev. Mod. Phys.* **70** 055502
[12] Caciolli A et al. 2009 *Eur. Phys. J.* A **39** 179
[13] Marta M et al. 2006 *Nucl. Instr. and Meth.* A **569** 729
[14] Arpesella C et al. 1996 *Appl. Radiat. Isot.* **47** 991
[15] Bemmerer D et al. 2006 *Phys. Rev. Lett.* **97** 122502
[16] Confortola F et al. 2007 *Phys. Rev.* C **75** 065803
[17] Sing B N et al. 2004 *Phys. Rev. Lett.* **93** 262503
[18] Brown T A D et al 2007 *Phy.s Rev.* C **76** 055801
[19] Descouvemont P et al. 2004 *At. Data Nucl. Data Tables* Sect. A **88** 203
[20] Kajino T, Toki H and Austin S M 1987 *Astrophys J.* **319** 531
[21] Diehl R et al. 1995 *Astronomy and Astrophysics* **298** 445
[22] Knodlseder J et al. 1999 *Astronomy and Astrophysics* **345** 813
[23] Winkler C et al. 2003 *Astronomy and Astrophysics* **411** L1
[24] Wasserbur G J 1985 *Protostars and Planet II* (Tucson: University of Arizona Press) 703
[25] Caciolli A for the LUNA collaboration 2009 *this volume*

Nuclear Physics in Astrophysics IV IOP Publishing
Journal of Physics: Conference Series **202** (2010) 012032 doi:10.1088/1742-6596/202/1/012032

# Observations of cosmic nuclear gamma-ray lines

**Michael G Lang[1,2], R Diehl[1], W Wang[3,1], P Martin[1], R Voss[1,2] and K Kretschmer[1]**

[1] Max Planck Institute for Extraterrestrial Physics, Giessenbachstraße, D-85748 Garching, Germany

[2] Excellence Cluster Universe, Technische Universität München, Boltzmannstr. 2, D-85748 Garching, Germany

[3] National Astronomical Observatories, Chinese Academy of Sciences, Beijing 100012, China

E-mail: mll@mpe.mpg.de

**Abstract.** Nucleosynthesis in stars and supernovae creates radioactive nuclei, which are ejected into the interstellar medium by stellar winds and explosive events. Their radioactive decay is accompanied by the emission of gamma-ray lines that can be observed by space observatories. Different isotopes are identified and their distribution and kinematics is studied using spectroscopic information. With more than five years of INTEGRAL/SPI observations we can spatially resolve the $^{26}$Al emission along the Galactic disk, derive the scale height of the Galactic $^{26}$Al distribution and provide constraints on $^{26}$Al kinematics for different regions. Groups of massive stars are detected in $^{26}$Al and thus can serve as laboratories to study the feedback of the freshly synthesised nuclei into the surrounding interstellar medium. We predict the yields of massive star clusters using population synthesis models. Additionally, INTEGRAL has observed the decay lines of $^{44}$Ti, constraining the dynamics of $^{44}$Ti within the Cassiopeia A supernova remnant.

## 1. Production of long-lived isotopes in massive stars

Gamma-ray telescopes observe the decay of radioactive elements which are produced during stellar and explosive nucleosynthesis [1]. The long-lived isotopes $^{26}$Al, $^{60}$Fe and $^{44}$Ti with half-lifes of 0.717 My, 2.62 My and 58.9 y, respectively [2–4], are produced in massive stars, supernovae and novae.

$^{26}$Al is mainly produced by proton capture on $^{25}$Mg in the context of the Ne-Na cycle and the Mg-Al chain in massive stars, in the AGB phase of intermediate mass stars and in nova explosions [5]. $^{26}$Al decays into $^{26}$Mg, emitting a $\gamma$-ray photon of 1808.65 keV [2].

Massive stars are as well known to produce $^{60}$Fe during the final stages of stellar evolution via successive neutron captures on stable iron nuclei [6]. The decay of $^{60}$Fe into $^{60}$Co is accompanied in 2% of the decays by the emission of a 58.6 keV photon [2]. This line has not been detected up to now. In a successive decay into $^{60}$Ni two additional photons at 1173.23 keV and 1332.50 keV are released [2].

The decay time scales of these two elements are similar to the timescales of massive star evolution. After the first detection of $^{26}$Al by HEAO 3 [7] it was shown with COMPTEL that $^{26}$Al is distributed mainly along the Galactic disk [8, 9], supporting a massive star origin [10]. $^{60}$Fe was detected from the Galaxy with RHESSI [11] and INTEGRAL [12, 13]. Predictions of the yields of massive stars depend strongly on the prescription of nuclear reaction rates, stellar

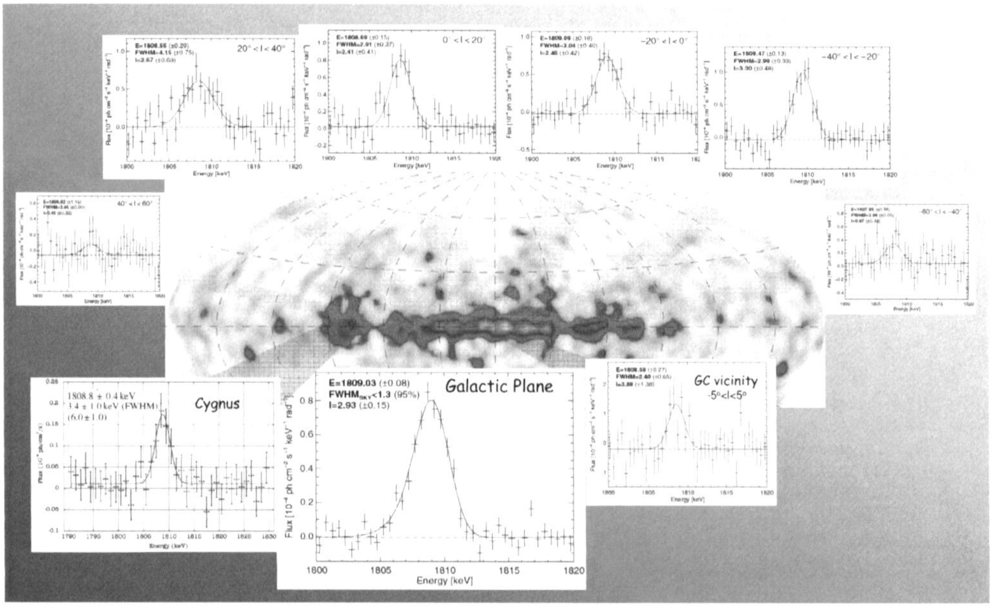


**Figure 1.** SPI spectra for different regions of the Galaxy [22, 23]. (INTEGRAL picture of the month on integral.esa.int)

winds, convection, mixing and rotation [6, 14, 15]. Since both elements are created by massive stars, the ratio of $^{60}$Fe to $^{26}$Al provides a good test for stellar models.

In core-collapse supernova explosions, conditions suited for the formation of $^{44}$Ti during explosive Si-burning with $\alpha$-rich freeze-out can be also found. The yields now strongly depend on the dynamics and on the conditions close to the mass cut [16, 17]. Detections of $^{44}$Ti from Cassiopeia A have been reported by COMPTEL, BEPPOSAX/PDS and INTEGRAL/IBIS [18–20]. $^{44}$Ti decays via $^{44}$Sc into $^{44}$Ca, emitting photons at 67.87 keV, 78.34 keV and 1157.03 keV, respectively. Spectroscopic measurement of the decay lines of the $^{44}$Ti decay chain with SPI can constrain properties of the explosion such as velocities of the ejecta.

## 2. Observations of Galactic $^{26}$Al and $^{60}$Fe

The International Gamma-Ray Astrophysics Laboratory INTEGRAL is in orbit since 2002. The spectrometer onboard INTEGRAL (SPI) [21] provides an excellent spectral resolution of $E/\delta E = 500$ at 1.3 MeV and offers the possibility to spectroscopically resolve Doppler velocities down to few tens of km s$^{-1}$.

### *2.1. Large-scale properties*

Diehl et al. have shown that the $^{26}$Al line is intrinsically narrow [24], thus ruling out a result obtained earlier with GRIS [25]. Due to continuous observations we could gain in sensitivity and, after 5 y of observations, improve the upper limit for the line width from 2.8 keV to less than 1.3 keV ($2\sigma$), corresponding to Doppler velocities of $\approx 150$ km/s. This result is consistent with the expectations from Galactic rotation and modest turbulence of the interstellar medium surrounding the $^{26}$Al sources.

The $^{26}$Al scale height of the inner Galaxy is $130^{+120}_{-70}$ pc, larger than the scale height of the molecular disk or the massive stars, but smaller than the outer thick disk component of Taylor and Cordes' model for the density distribution of free electrons [26]. We showed that the spiral

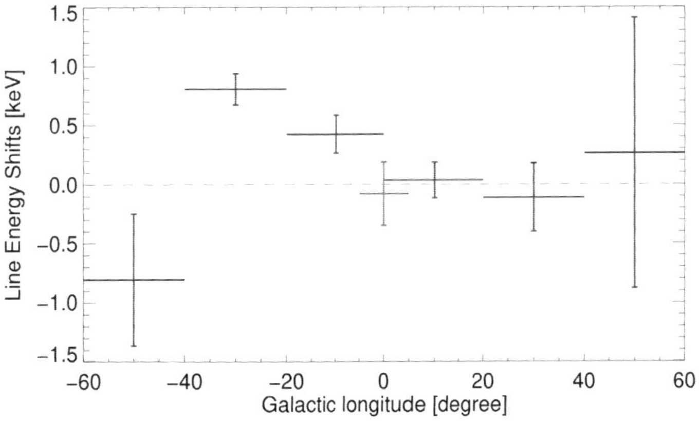


**Figure 2.** Doppler shifts of the 1809 keV line for different Galactic longitudes. The horizontal error bars represent the longitude interval.

arm component of this model describes the distribution of $^{26}$Al in the inner Galaxy better than a simple exponential disk model. This is consistent with massive stars located in the spiral arms and ejecting $^{26}$Al.

### *2.2. Spatially resolved spectroscopy*

We have further continued our spectroscopic study in the inner Galaxy [24, 27] and increased the number of spectra to more than 6 from different sky regions [22, see Fig. 1]. The spectral parameters intensity, line shift and width have been obtained for 20° wide longitude bins and a 10° wide bin centered at the direction towards the Galactic Center. The intensity distribution along the Galactic plane is in agreement with the intensity derived from the COMPTEL $^{26}$Al map. The obtained widths of the spectral lines can be explained by the instrumental spectral resolution except for longitude bin $l \in [20°, 40°]$. The line broadening corresponds to velocities of $\approx 480 \pm 170$ km/s. Figure 2 shows the shifts in line position along the Galactic plane. The overall pattern reflects the Doppler shifts expected from Galactic differential rotation. A possible contribution from a Galactic bar is being investigated.

### *2.3. Nearby star forming regions*

In addition to the properties of the Galactic disk as a whole, we investigate $^{26}$Al in specific nearby star forming regions. A prominent region in COMPTEL's $^{26}$Al map is the Cygnus region with an assembly of well-known nearby OB associations. With SPI, the $^{26}$Al line is detected and an upper limit for $^{60}$Fe emission is given [23].

In order to interpret our experimental results with respect to yields calculated from stellar models we have updated our population synthesis model of massive stars including rotation, supernova yields, wind prescription and the new value for the $^{60}$Fe lifetime in order to predict the time dependence of the $^{26}$Al and $^{60}$Fe yields [28]. In this way we can calculate the time evolution of the $^{60}$Fe/$^{26}$Al ratio. For a typical size of a stellar cluster of 100 stars, statistical fluctuations play an important role. We assess the non-gaussianity of the uncertainties by Monte-Carlo simulations.

For Cygnus, the predicted fluxes for $^{26}$Al and $^{60}$Fe are in agreement with the observations if the age of the $^{26}$Al emitting association is younger than 5 My [23]. In future studies we will apply the code as well to other nearby star forming regions such as Scorpius and Centaurus (Sco-Cen) (Diehl et al, in preparation) or Orion, where we have a hint for $^{26}$Al emission.

## 3. $^{44}$Ti sources

Using 4 y of SPI data we derived upper limits for the 1157 keV decay line of $^{44}$Ti for Cas A. A minimum line broadening that translates into a lower limit for the expansion velocity of the $^{44}$Ti ejecta of $\approx 500$ km/s is required to explain our non-detection, given the detection of $^{44}$Ti with the hard X-ray imager IBIS onboard INTEGRAL [29]. As typical velocities within the remnant are all larger than this limit, this result does not allow to constrain the location of $^{44}$Ti.

Up to now no other direct detection of a $^{44}$Ti source has been reported [30]. This remains a puzzle as supernova models predict yields that should be reflected as $^{44}$Ti sources in the Galactic disk observable with INTEGRAL (see [31] for a discussion).

## Acknowledgments

This research was supported by DLR project 50 OR 0405 and the DFG cluster of excellence 'Origin and Structure of the Universe'. Wang is also supported by the National Natural Science Foundation of China under grant 10833003. The INTEGRAL project is supported by the German government through DLR grant 50 OG 0502 and by grants in all member states of the hardware teams. The SPI project has been completed under responsibility and leadership of CNES/France. We are grateful to ASI, CEA, CNES, DLR, ESA, INTA, NASA and OSTC for support.

## References

[1] Leising M and Diehl R 2009 *ArXiv e-prints* (*Preprint* 0903.0772)
[2] Firestone R B and Ekström L P 99 Table of Radioactive Isotopes URL http://ie.lbl.gov/toi/index.htm
[3] Rugel G *et al.* 2009 *Physical Review Letters* **103** 072502–+
[4] Ahmad I *et al.* 2006 *Phys. Rev.* C **74** 065803–+
[5] Prantzos N and Diehl R 1996 *Phys. Rep.* **267** 1–69
[6] Limongi M and Chieffi A 2006 *Astrophys. J.* **647** 483–500 (*Preprint* arXiv:astro-ph/0604297)
[7] Mahoney W A, Ling J C, Jacobson A S and Lingenfelter R E 1982 *Astrophys. J.* **262** 742–+
[8] Diehl R *et al.* 1995 *Astron. Astrophys.* **298** 445–+
[9] Plüschke S *et al.* 2001 *Exploring the Gamma-Ray Universe* (*ESA Special Publication* vol 459) ed Gimenez A, Reglero V and Winkler C pp 55–58
[10] Knödlseder J 1999 *Astrophys. J.* **510** 915–929
[11] Smith D M 2004 *5th INTEGRAL Workshop on the INTEGRAL Universe* (*ESA Special Publication* vol 552) ed Schoenfelder V, Lichti G and Winkler C pp 45–+
[12] Harris M J *et al.* 2005 *Astron. Astrophys.* **433** L49–L52 (*Preprint* arXiv:astro-ph/0502219)
[13] Wang W *et al.* 2007 *Astron. Astrophys.* **469** 1005–1012 (*Preprint* 0704.3895)
[14] Woosley S E and Heger A 2007 *Phys. Rep.* **442** 269–283 (*Preprint* arXiv:astro-ph/0702176)
[15] Tur C, Heger A and Austin S M 2009 *ArXiv e-prints* (*Preprint* 0908.4283)
[16] Woosley S E and Weaver T A 1995 *Astrophys. J. Supplement* **101** 181–+
[17] Diehl R, Prantzos N and von Ballmoos P 2006 *Nuclear Physics A* **777** 70–97
[18] Iyudin A F *et al.* 1994 *Astron. Astrophys.* **284** L1–L4
[19] Vink J *et al.* 2001 *Astrophys. J.* **560** L79–L82 (*Preprint* arXiv:astro-ph/0107468)
[20] Renaud M *et al.* 2006 *Astrophys. J.* **647** L41–L44 (*Preprint* arXiv:astro-ph/0606736)
[21] Vedrenne G *et al.* 2003 *Astron. Astrophys.* **411** L63–L70
[22] Wang W *et al.* 2009 *Astron. Astrophys.* **496** 713–724 (*Preprint* 0902.0211)
[23] Martin P 2008 *La vie et la mort des étoiles massives révélées par l'observation des raies gamma nucléaires grâce au spectromètre INTEGRAL/SPI* Ph.D. thesis Université Toulouse III – Paul Sabatier
[24] Diehl R *et al.* 2006 *Astron. Astrophys.* **449** 1025–1031 (*Preprint* astro-ph/0512334)
[25] Naya J E *et al.* 1996 *Nature* **384** 44–46
[26] Taylor J H and Cordes J M 1993 *Astrophys. J.* **411** 674–684
[27] Diehl R *et al.* 2006 *Nature* **439** 45–47 (*Preprint* arXiv:astro-ph/0601015)
[28] Voss R *et al.* 2009 *ArXiv e-prints* (*Preprint* 0907.5209)
[29] Martin P, Knödlseder J, Vink J, Decourchelle A and Renaud M 2009 *Astron. Astrophys.* **502** 131–137
[30] Renaud M *et al.* 2004 *5th INTEGRAL Workshop on the INTEGRAL Universe* (*ESA Special Publication* vol 552) ed Schoenfelder V, Lichti G and Winkler C pp 81–+
[31] The L S *et al.* 2006 *Astron. Astrophys.* **450** 1037–1050 (*Preprint* arXiv:astro-ph/0601039)

Nuclear Physics in Astrophysics IV IOP Publishing
Journal of Physics: Conference Series **202** (2010) 012033 doi:10.1088/1742-6596/202/1/012033

# The long-standing problem of $^{176}$Lu/$^{176}$Hf branching: a new approach with full stellar evolutionary models

**S. Cristallo[1,2], L. Piersanti[2], R. Gallino[3], O. Straniero[2], F. Käppeler[4], I. Domínguez[1] & P. Mohr[5]**

[1] D.pto Física Teorica y del Cosmo, Universidad de Granada, 18071 Granada, Spain
[2] Osservatorio Astronomico di Teramo - INAF, 64100 Teramo, Italy
[3] D.pto Fisica Generale, Universitá di Torino, 10125 Torino, Italy
[4] Forschungszentrum Karlsruhe, Institut für Kernphysik, D-76021 Karlsruhe, Germany
[5] Diakonie-Klinikum Schwäbisch Hall, D-74523 Schwäbisch Hall, Germany

E-mail: cristallo@ugr.es

**Abstract.** We present an analysis of the s-process branching at $A$=176, which is determined by the two s-only isotopes $^{176}$Lu and $^{176}$Hf. This work is motivated by the discrepancy between the current theoretical expectations and solar-system abundances of the two isotopes involved in the branching. Asymptotic Giant Branch models have been calculated with the FRANEC evolutionary code, coupling a full nuclear reaction network with the equation describing the physical evolution of the stellar structure. With respect to previous studies, we can better follow the complex interplay between mixing and burning, which both strongly influence this branching. We show that a substantial reduction of the mixing velocity within the convective zone generated by a He-shell flash (or thermal pulse) may alleviate the discrepancy between the theoretical expectations and the observed solar abundances. Nonetheless, we reckon that a definitive solution of this problem has to be searched in the nuclear coupling scheme between the thermally populated levels in $^{176}$Lu.

## 1. Introduction

The s-process branching at mass number A=176 determines the abundances of the s-only isotopes $^{176}$Lu and $^{176}$Hf, which are produced by the reaction sequences

$$^{175}\text{Lu} \overset{(n,\gamma)}{\rightarrow} {}^{176}\text{Lu and } {}^{175}\text{Lu} \overset{(n,\gamma)}{\rightarrow} {}^{176}\text{Lu}^{*} \overset{\beta^-}{\rightarrow} {}^{176}\text{Hf}.$$

Due to its long half-life of ~40 Gyr, $^{176}$Lu was initially considered as a potential cosmochronometer [1]. However, it turned out that the half-life is dramatically shortened at the temperatures of the s-process site in low mass AGB stars due to a coupling of the long-lived ground state with the short-lived isomer at 123 keV with $t_{1/2}$ = 3.68 h. Because direct transitions between the two states are forbidden by K-selection rules, the coupling can only occur via thermal population of intermediate states at much higher excitation energies, in particular through a state at 839 keV (hereafter IS@839). Accordingly, the branching at A=176 has been interpreted as a sensitive s-process thermometer (see [2] and references therein) in the temperature range above $1.5 \times 10^8$ K, where the coupling between ground state and isomer becomes relevant, leading to a drastic reduction of the effective half-life of $^{176}$Lu and, therefore, to a larger $^{176}$Hf production.

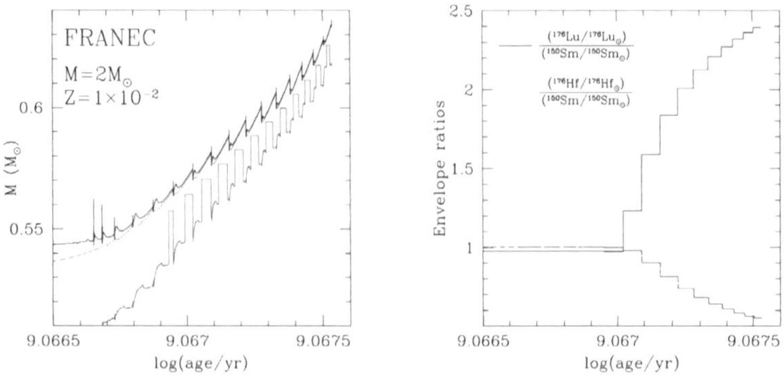


**Figure 1.** *Left Panel*: Physical evolution of (top to bottom): inner border of the convective envelope, H-shell and He-shell of a star with $M = 2M_\odot$ and $Z = 10^{-2}$. *Right Panel*: Behavior of the surface $^{176}$Lu/$^{150}$Sm and $^{176}$Hf/$^{150}$Sm ratios, normalized to their solar abundances.

The main neutron source in low mass AGB stars (the major contributors to the s-process) is the $^{13}C(\alpha, n)^{16}O$ reaction, which works under radiative conditions within the so-called $^{13}$C-pocket [3]. In this case, temperatures are too low ($T < 1.5 \times 10^8$ K) to couple $^{176}$Lu isomer and ground state and, therefore, they behave as two separate species. However, an additional minor neutron exposure comes from the $^{22}Ne(\alpha, n)^{25}Mg$ reaction, which is active during He shell flashes (Thermal Pulses, TPs). In this regime the maximum temperature at the bottom of the convective He shell reaches about $3 \times 10^8$ K, where the population of the two states depends sensitively on the temperature and the neutron density. The probability that the s-process reaction flow passes through the neutron capture channel (to the detriment of the $\beta^-$ decay channel) is described by the branching factor $f_n = \lambda_n/(\lambda_n + \lambda_-)$[1]. As a consequence of the coupling between ground state and isomer, the beta decay rate $\lambda_-$ is therefore a complex function of the local temperature and neutron density [2,5]:

$$\lambda_- = \lambda_n \cdot \frac{C_1 I_\sigma exp[-C_2/KT] + (1 - B)\lambda_n}{I_\sigma exp[-C_3/KT] + C_1 B} \ . \tag{1}$$

Various symbols are defined as:

- $\lambda_n = n_n v_T \langle\sigma\rangle$ is the neutron capture rate ($n_n$ denotes the neutron density, $v_T$ the mean thermal velocity and $\langle\sigma\rangle$ the $(n, \gamma)$ cross section averaged over the stellar neutron spectrum);
- $I_\sigma \sim V \times (1 - V)/\tau_i$ is the energy-integrated cross section ($\tau_i$ is the underlying lifetime, while $V$ is the decay branching of the IS@839 toward the isomer);
- B = 1-IR (IR being the isomeric ratio in the $^{175}Lu(n, \gamma)^{176}Lu$ reaction);
- $C_1$, $C_2$ and $C_3$ are constant factors.

Basing on an updated set of nuclear input parameters[2], [4] performed a sequence of post-process calculations and, by varying the two fundamental parameters $I_\sigma$ and $B$, found a $^{176}$Lu/$^{176}$Hf ratio compatible with the observed solar system abundances. This agreement was, however, challenged by new experimental data [5] leading to a lower lifetime of the intermediate state $\tau_i$ of 6.6 ps compared to the previously adopted 80 ps and to an increase of $B$ from 0.14 to 0.20. In the following the parameter sets of Refs. [4] and [5] are labelled HEIL and MOHR, respectively.

The stellar evolutionary code used to perform our theoretical calculations is described in Section 2 and the results are discussed in Section 3.

[1] see formula 25 in [2].

[2] thus well constraining the value of $\lambda_n$

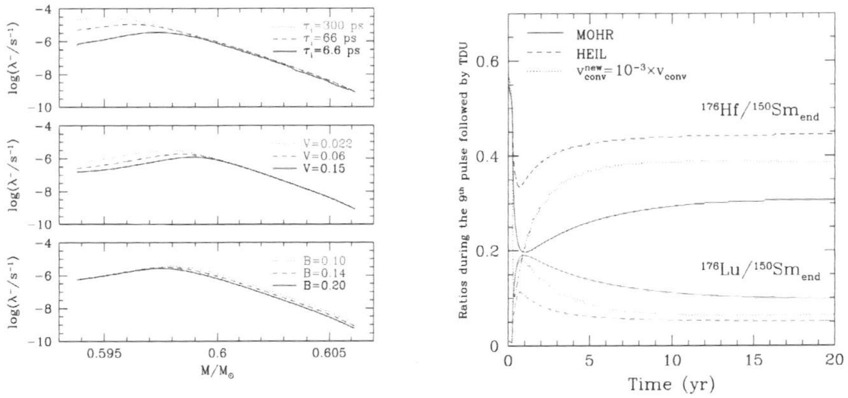


**Figure 2.** *Left Panel*: behavior of the $\beta$-decay channel within the convective shell during the $9^{th}$ TP as a function of different parameter sets (see text for details). *Right Panel*: Evolution of the $^{176}$Lu/$^{150}$Sm and $^{176}$Hf/$^{150}$Sm ratios within the convective shell for different choices of the parameter set and for a reduced mixing efficiency.

## 2. The FRANEC code

We revisit the Lu/Hf problem by using an up-to-date version of the FRANEC stellar evolutionary code [6,7,8]. Owing to the particular relevance for the present paper, our models have been calculated by fully coupling the physical evolution of the stellar structure with an extended nuclear reaction network involving about 500 isotopes (from H to Bi), linked by about 1000 reactions. Such a coupling allows for a proper treatment of all those situations when the nuclear burning time scales are comparable to those of convective mixing. The $^{176}$Lu/$^{176}$Hf ratio, mainly determined during the freeze-out phase of each TP, can therefore be evaluated with a larger degree of accuracy than could be obtained in previous works. Models presented here have been calculated using the MOHR parameters.

## 3. Results

With the aim of reproducing the solar main s-process component (i.e. to obtain a flat distribution of the isotopes synthesized by the s-process only, see [9]), we follow the AGB phase of a star with initial mass $M = 2M_{\odot}$ and $Z = 1 \times 10^{-2}$ (see left panel of Fig. 1)[3]. This model is characterized by 12 TPs followed by Third Dredge Up (TDU) episodes, which carry the synthesized s-process elements from the stellar interior to the surface. Since $^{176}$Lu and $^{176}$Hf are both shielded by their stable isobar $^{176}$Yb, we would expect for these isotopes the same overabundances (scaled to their solar value) as for $^{150}$Sm[4]. Instead, we obtain a strong overproduction of $^{176}$Lu (2.40) and a strong underproduction of $^{176}$Hf (0.55) with respect to $^{150}$Sm (see right panel of Fig. 1), thus confirming previous results by [5], who obtained 1.80 and 0.61, respectively[5]. In order to understand the origin of this disagreement between stellar models and theoretical expectations we start a deeper analysis of this model focusing on the $9^{th}$ TP followed by TDU. In left panel of Fig. 2 we report the behavior of the $\beta$-decay channel as a function of the three nuclear parameters ($\tau_i$, $B$ and $V$) in the convective shell. We confirm that major variations derive from the choice of $\tau_i$, pointing out that the maximum of the decay channel moves outward (increasing M/M$_{\odot}$ or stellar radius) with decreasing $\tau_i$. In contrast, the neutron capture channel has its maximum

[3] Note that a different metallicity or a different initial mass would imply minor changes to the results.
[4] This isotope is the best known case of an un-branched s-only isotope among the Rare Earth Elements.
[5] Note that differences with respect to $^{150}$Sm up to 10% can be reasonably tolerated.

always at the base of the convective shell, where the neutron density attains its maximum value. This fact suggests that the $^{176}$Lu/$^{150}$Sm and $^{176}$Hf/$^{150}$Sm ratios could depend on the mixing efficiency, since the two channels reach their maximum in different regions of the convective shell. We test this hypothesis during the evolution of the 9th TP by artificially reducing the mixing velocity by a factor of 1000 with respect to the value estimated by means of the mixing length theory. The evolution of the intershell abundances of $^{176}$Lu and $^{176}$Hf normalized to the final abundance of $^{150}$Sm are shown in right panel of Fig. 2. Solid curves refer to ratios obtained with the MOHR parameters, while dotted curves represent the ratios obtained with the same parameters but with reduced convective velocities. For comparison, we also run an additional model by using the HEIL parameters (dashed curves).

Right panel of Fig. 2 shows that, with respect to the MOHR results, a reduction of the convective velocities leads to a lower Lu production and to a lower Hf destruction (thus mimicking the HEIL results). In order to check the consistency of this test model, we look at the abundances of stable isotopes in the He-intershell at the end of the convective episode and find appreciable differences only for a restricted number of cases. Apart from the differences for $^{176}$Lu and $^{176}$Hf, major changes occur for r-only isotopes (such as $^{110}$Pd, $^{150}$Nd, and $^{160}$Gd), which are enhanced up to a factor 2, for $^{96}$Zr (enhanced by about 40%) and for the s-only $^{152}$Gd (enhanced by about a factor 2, due to the temperature dependent $\beta$ decay of its unstable isobar $^{152}$Eu). From this chemical analysis, therefore, there seem to be no arguments against an extreme reduction of the convective velocities. Note that, even if convective velocities have been reduced for a factor 1000, we obtain the same $^{128}$Xe/$^{130}$Xe ratio of the standard case, due to the accuracy characterizing the adopted temporal step (see [10] for details).

Nevertheless, the better agreement with theoretical expectations obtained with reduced convective velocities does not allow us to state that the problem of the $^{176}$Lu/$^{176}$Hf ratio has been solved. Our doubts arise mainly from the fact that recent 3D hydrodynamical calculations of shell He flashes [11] tend to exclude that velocities within He-shell convective zones differ strongly from the ones obtained in the framework of the mixing length theory. While one could accept uncertainties of a factor 2-3 in the determination of convective velocities, we find it hard to justify differences up to a factor 1000. In our opinion, it is more likely that the solution to this problem has to be searched in a better comprehension of the nuclear coupling scheme between the populated levels in $^{176}$Lu. In this context, [12] has recently hypothesized that the rate of equilibration between the two isotopes could be further enhanced via $K$-mixing. This new analysis may lead to a strong revision of the nuclear parameters adopted in the present work.

[1] Audouze J, Fowler W A and Schramm D N 1972 *Nature Phys. Sci.* **238** 8
[2] Klay N, Käppeler F, Beer H and Schatz G 1991 *Phys. Rev.* C **44** 2839
[3] Gallino R, Arlandini C, Busso M, Lugaro M, Travaglio C, Straniero O, Chieffi A and Limongi M 1998 *ApJ* **497** 388
[4] Heil M, Winckler N, Dababneh S, Käppeler F, Wisshak K, Bisterzo S, Gallino R, Davis A M, and Rauscher T 2008 *ApJ* **673** 434
[5] Mohr P, Bisterzo S, Gallino R, Käppeler F, Kneissl U and Winckler N 2009 *Phys. Rev.* C **79** 045804
[6] Chieffi A, Limongi M and Straniero O 1998 *ApJ* **502** 737
[7] Straniero O, Gallino R and Cristallo S 2006 *Nucl. Phys.* A **777** 311
[8] Cristallo S, Straniero O, Gallino R, Piersanti L, Domínguez I and Lederer M T 2009 *ApJ* **696** 797
[9] Arlandini C, Käppeler F, Wisshak K, Gallino R, Lugaro M, Busso M and Straniero O 1999, *ApJ* **525** 886
[10] Reifarth R, Käppeler F, Voss F, Wisshak K, Gallino R, Pignatari M and Straniero O 2004, *ApJ* **614** 363
[11] Herwig F, Freytag B, Hueckstaedt R M and Timmes F X 2006, *ApJ* **642** 1057
[12] Gintautas V, Champagne A E, Kondev F G and Longland R 2009, *Phys. Rev.* C **80** 015806

# Preparation of $^{60}$Fe, $^{7}$Be, $^{44}$Ti and other samples for nuclear physics experiments

**Dorothea Schumann and Marin Ayranov**

Paul Scherrer Institute Villigen, 5232 Villigen PSI, Switzerland

dorothea.schumann@psi.ch

**Abstract**. Chemical separation procedures were developed aimed to isolate exotic long-lived radionuclides from proton-irradiated material for use in experiments in the fields of basic nuclear physics and astrophysics. Some examples for separation as well as results of first applications are presented.

## 1. Introduction

Several long-lived exotic radionuclides like $^{44}$Ti, $^{60}$Fe and others play an important role in nuclear astrophysics topics. Experimental studies of processes and reactions in this scientific field require these isotopes to be used as targets, beams and other samples. In this context, the availability of sufficient amounts of these very rare isotopes is a precondition. One of the few possibilities for their production is the exploitation of activated components available from accelerator facilities at the Paul Scherrer Institute. There, the most powerful proton accelerator in Europe and also world-wide is operated with a proton energy of 590 MeV and a beam intensity of around 2.5 mA. In spallation reactions, a broad variety of exotic radionuclides is produced with high yield.

In the frame of the ERAWAST initiative [1], chemical separation procedures were developed which allow isolating the desired isotopes from several suitable exposed materials like beam dumps, targets, shielding equipment and cooling water. Due to the high radiation doses, usually the procedures have to be carried out using radio-protection equipment. The separation techniques are therefore very expensive and time-consuming.

Three selected examples concerning the separation and application of exotic radionuclides are presented in this report. One source for gaining these isotopes are samples from an irradiated copper beam dump, containing $^{44}$Ti, $^{60}$Fe and others as well as $^{60}$Co as the main contaminant. For the separation of $^{7}$Be the cooling water of the spallation neutron source (SINQ) of PSI was used.

## 2. High-activity separation of $^{44}$Ti, $^{60}$Fe, $^{26}$Al and others from a copper beam dump

The BMA station for pions cancer therapy was operated at PSI from 1980 to 1992. For this period the copper beam dump of the facility received a total dose of approximately 0.16 Ah of 590 MeV protons. After dismantling in 1993, it was cut and samples were taken for analytical purposes. The sampling collected 500 grams of high active copper chips that can be used for separation of exotic radionuclides [2]. The overall gamma analyses showed main nuclides present to be $^{60}$Co, $^{54}$Mn, $^{22}$Na, $^{65}$Zn and long lived $^{44}$Ti with a daughter nuclide $^{44}$Sc [3]. Further analyses by LSC and AMS demonstrated that significant amounts of $^{55}$Fe, $^{63}$Ni, $^{26}$Al, $^{53}$Mn, $^{59}$Ni and $^{60}$Fe are present in the copper beam dump. The analytical results estimate that about 100 MBq $^{44}Ti$, 500 MBq $^{53}Mn$, 7 kBq $^{26}Al$, 8 MBq $^{59}Ni$ and 5 kBq $^{60}Fe$ are available in the collected copper chips.

Due to the high activity of $^{60}Co$, approx. 4 GBq in total, the separation should be implemented using appropriate shielding, e.g. in a hot cell. The purpose of this work is to develop a simple, selective, efficient and easy to accommodate for remote manipulation procedure for the separation of $^{26}Al$, $^{59}Ni$, $^{53}Mn$, $^{44}Ti$ and $^{60}Fe$ from gram amounts of the copper beam dump.

The separation procedure, a combination of precipitation and ion exchange, is presented in Figure 1. Copper as the main matrix element interferes with the separation of all elements of interest. For this reason, after the copper dissolution in 7 M $HNO_3$ the solution is conditioned to 1 M $HNO_3$ and Cu(II) is precipitated by saturation with $H_2S$. At these conditions the copper precipitates selectively, while the exotic radionuclides remain in the solution. Further the solution is transferred in 12 M HCl and passed on a Dowex 1x8 anion exchange column. Al and Ni are not retained and are washed out of the column with 12 M HCl. The remaining ions are eluted consequently in the following way, Mn – 10 M HCl, Ti – 8 M HCl, Co – 5 M HCl, Cu – 2.5 M HCl and finally Fe with 0.5 M HCl [4]. Further Al is separated from Ni on a Dowex 50x4 cation exchange column. The Ni is purified with Eichrom Ni resin based on the traditional dimethylglyoxime precipitation chemistry.

The proposed separation procedure is easy for remote controlled implementation in a hot cell. The ion exchange separation of Ni, Al, Mg, Ti and Fe is complete and high decontamination factors for copper and cobalt were achieved.

A setup in a hot cell of a remotely controlled system for consecutive separation exotic radionuclides from gram amounts of copper beam dump is ongoing.

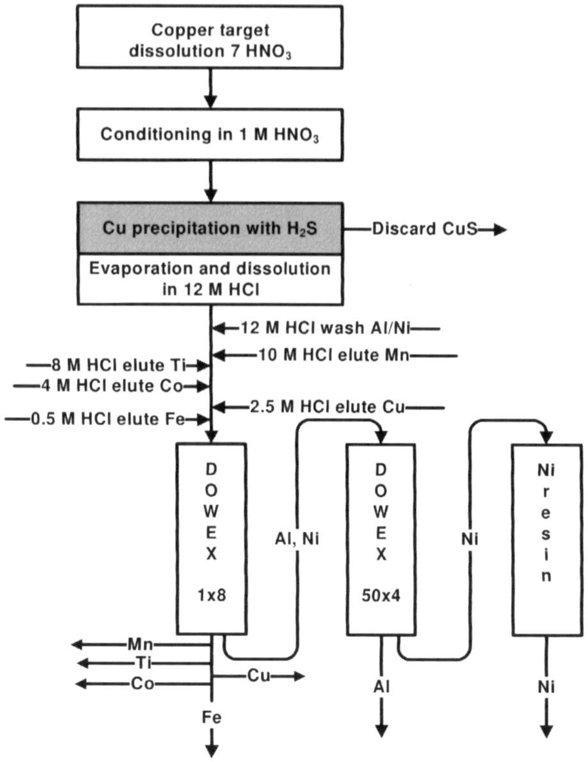


Figure 1. The flowchart of the exotic radionuclides separation procedure.

## 3. Laboratory-scale separation of $^{60}$Fe from the copper beam dump and first applications

The long-lived radionuclide $^{60}$Fe plays a key role in tracing the history of the Early Solar System (ESS) [5]. Attempts at reconstructing the inventory of radioactivity between Fe and Pb require, among others, the exact knowledge of the half-life and experiments aimed to determine the neutron capture cross section.

As a first attempt, two $^{60}$Fe samples were prepared from the copper beam dump in laboratory scale using around 3 g of material, respectively. Liquid-liquid extraction, precipitation and ion exchange are applicable to separate the impurities with a decontamination factor of $\sim 3 \cdot 10^8$ for $^{60}$Co and $\sim 5 \cdot 10^6$ for $^{44}$Ti. The chemical separation procedures to be applied depend on the specific requirements of the envisaged experiment.

The first sample was foreseen as target for the determination of the neutron capture cross section at stellar energies. For this application, the content of stable isotopes has to be as low as possible in order to avoid the production of the long-lived $^{59}$Fe. Therefore, precipitation was not possible; all separation steps were performed using liquid-liquid extraction. A detailed description of the target preparation is given in [6].The results of the neutron capture experiment can be found in [7].

The second sample was used for a re-measurement of the half-life of $^{60}$Fe [8]. Here, the presence of stable carrier is mandatory because isotope ratios are measured by Inductively Coupled Plasma Mass Spectrometry (ICP-MS). On the other hand, for this method it is important to reduce the content of isobaric elements as much as possible. So, an additional purification from Ni had to be performed.

Parts from this very pure $^{60}$Fe sample were used to prepare standards for AMS measurement. Several grams of samples with isotopic ratios of $I=10^{-12}$, $10^{-10}$ and $10^{-8}$, respectively, are now available and can be provided.

## 4. $^7$Be from SINQ cooling water

$^7$Be as well as $^{10}$Be are key radionuclides for investigations of several astrophysical processes and phenomena.

One of the "hot topics" is the half-life of $^{10}$Be, where the literature values differ from 1.34 to 1.51 My [9,10]. Additional measurements are, therefore, urgently needed. One possibility is the use of Liquid Scintillation Counting (LSC) for the determination of the activity and ICP-MS for measuring the number of atoms. The calibration of the ICP-MS requires at least 2 mass points in known amounts, and since Be has only one stable isotope ($^9$Be), $^7$Be can serve as the second marker.

Another application of $^7$Be is the study of key reactions concerning the solar neutrino flux, in particular the reaction $^7$Be(p,γ)$^8$B [11]. Highly-active $^7$Be targets in the range of several 100 GBq are required for such studies.

$^7$Be is produced in considerable amounts in the cooling water ($D_2O$) of the SINQ facility at PSI by spallation reactions on $^{16}$O with the generated fast neutrons. By-products can be nearly neglected, so that this cooling water establishes an ideal source for highly active $^7$Be-samples.

A filtration device shown in Fig.2. was installed as a by-pass for the cooling water into the cooling circle of SINQ for 6 weeks. The apparatus contains 1 litre of the mixed-bed ion exchanger LEWATIT (grain size 400 μm), which absorbs cations and anions as well.

## 5. Conclusions

First successful experiments with laboratory-scale produced amounts of exotic radionuclides like $^{60}$Fe or $^7$Be show the potential of the EAWAST project. With the development of a remote-controlled incell-equipment we will be able to provide a bright variety of these urgently needed rare isotopes in higher amounts.

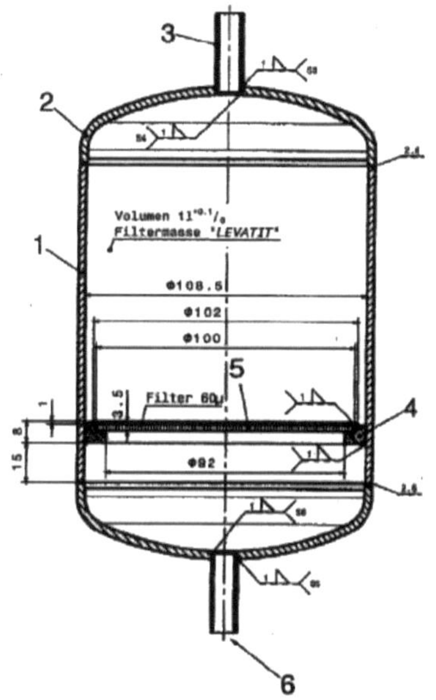

**Fig. 2:** Drawing of the adsorption device
1,2 -stainless steel vessel; 3 - inlet; 4 - ring for filter fixing; 5 - filter; 6 - outlet

**References**

[1] http://lch.web.psi.ch/webcontent/research/radwaste/workshop/index.html
[2] D. Schumann, J. Neuhausen, Accelerator waste as a source for exotic radionuclides, J Phys. G 35 (2008) 014046
[3] D. Schumann, J. Neuhausen, J. Eikenberg, M. Ruthi, M. Wohlmuther, P. W. Kubik, H. A. Synal, V. Alfimov, G. Korschinek, G. Rugel and T. Faestermann; Radiochemical Analysis of a Copper Beam Dump Irradiated with High-Energetic Protons, Radiochimica Acta, 97 (3) 2009
[4] M. Marhol, Ion Exchangers in Analytical Chemistry; Elsevier Scientific Publishing, Amsterdam, Oxford, New York, 1982, p. 330
[5] R. Diehl, N. Prantzos, and P. von Ballmoos, Nucl. Phys. A777, 70 (2006).
[6] D. Schumann, J. Neuhausen, I. Dillmann, C. Domingo Pardo, F. Käppeler, J. Marganiec, F. Voss, S. Walter, M. Heil, R. Reifarth, J. Goerres, E. Uberseder, M. Wiescher, M. Pignatari; Preparation of a $^{60}$Fe Target for Nuclear Astrophysics Experiments; NIM B (2009), accepted
[7] E. Uberseder, R. Reifarth, D. Schumann, I. Dillmann, C. Domingo Pardo, J. Görres, M. Heil, F. Käppeler, J. Marganiec, J. Neuhausen, M. Pignatari, F. Voss, S. Walter, and M. Wiescher, Measurement of the $^{60}$Fe($n$,$\gamma$)$^{61}$Fe Cross Section at Stellar Temperatures, PRL 102, 151101 (2009)
[8] G. Rugel, T. Faestermann, K. Knie, G. Korschinek, M. Poutivtsev; D. Schumann, N. Kivel, I. Günther-Leopold, R. Weinreich, M. Wohlmuther, On the Half-Life of $^{60}$Fe, PRL (2009), accepted
[9] Middleton, R; Brown, L; Dezfouly-Arjomandy, B; Klein, J: On the $^{10}$Be standards and the half-life of $^{10}$Be. Nucl.Instr.Methods B 82 p. 399-403 1993
[10]Fink, D; McKelvey, B; Hannan, D; Newsome, D: Cold rocks, hot sands: In-situ cosmogenic applications in Australia at ANTARES. NIM B 172 p. 838-846 2000
[11]L. T.Baby, C. Bordeanu, G. Goldring, M. Hass, L. Weissman, V. N. Fedoseyev, U. Köster, Y. Nir-El, G. Haquin, H. W. Gäggeler, and R. Weinreich , Precision Measurement of the $^{7}$Be($p$,$\gamma$)$^{8}$B Cross Section with an Implanted $^{7}$Be Target, PRL 90, 022501 (2003)

Journal of Physics: Conference Series **202** (2010) 012035 doi:10.1088/1742-6596/202/1/012035

# Target dependence in the study of collective modes in stable and exotic Ni nuclei

**T. Le Bleis[1,2,3], D. Rossi[4], A. Klimkiewicz[1,5], P. Adrich[1], K. Boretzky[1], F. Aksouh[1], H. Alvarez-Pol[6], T. Aumann[1], J. Benlliure[6], M. Boehmer[7], E. Casarejos[6], M. Chartier[8], A. Chatillon[1], D. Cortina-Gil[6], U. Datta Pramanik[9], H. Emling[1], O. Ershova[1,3], B. Fernandez-Dominguez[8], H. Geissel[1], M. Gorska[1], M. Heil[1], H. Johansson[1,10], A. R. Junghans[11], O. Kiselev[1,4], J.V. Kratz[4], N. Kurz[1], M. Labiche[12], R. Lemmon[13], Y. Litvinov[1], K. Mahata[1], P. Maierbeck[7], T. Nilsson[10,14], C. Nociforo[1], R. Palit[15], S. Paschalis[8], R. Plag[1,3], R. Reifarth[1,3], H. Simon[1], K. Sümmerer[1], A. Wagner[11], W. Walus[5], H. Weick[1], M. Winkler[1].**

[1] *GSI, Darmstadt, Germany,*
[2] *IPHC, Strasbourg, France,*
[3] *Uni. Frankfurt, Germany.*
[4] *University of Mainz, Germany,*
[5] *Jagelonian Uni., Krakow, Poland,*
[6] *Uni. Santiago de Compostela, Spain,*
[7] *Tech. Uni. Munich, Germany,*
[8] *Uni. Liverpool, United Kingdom,*
[9] *SINP Kolkata, India,*
[10] *Chalmers Göteborg, Sweden,*
[11] *FZ Dresden Rosendorf,*
[12] *Uni. Paisley, United Kingdom,*
[13] *Daresbury, United Kingdom,*
[14] *Tech. Uni. Darmstadt, Germany,*
[15] *TIFR Mumbai, India,*

E-mail: `t.lebleis@gsi.de`

**Abstract.** The appearance of the pygmy-dipole-resonance is a recently observed phenomenon that can be related to neutron-matter properties. Its study can be a tool to determine the nuclear symmetry-energy parameters and thus can contribute constraining neutron star models. We present the $(\gamma, n)$ cross sections for different Ni isotopes obtained from a measurement in inverse kinematics at about 500 MeV/u in the LAND reaction setup at GSI. The question of the disentanglement of the Coulomb and nuclear contributions is addressed.

## 1. Introduction

In the early 1990s, a new collective dipole mode, usually referred to as pygmy-dipole-resonance (PDR), was predicted for neutron-rich nuclei [1, 2]. Random-phase-approximation calculations for medium-heavy and heavy nuclei [3–6] attribute the concentration of strength located below the giant dipole resonance (GDR) region to the formation of a neutron or a proton skin in neutron or proton-rich nuclei, respectively. Since such a skin depends on the nuclear symmetry energy [7, 8], experimental observation of the PDR strength can lead to valuable information about the parameters of the symmetry energy and on the evolution of the neutron-skin thickness. Experimental PDR strength measurements using Coulomb dissociation of exotic secondary beams at kinetic energies of about 500 MeV/u were reported in the $^{132}$Sn region [9, 10] and compared with available data for stable nuclei from [11–15] and $(\gamma,\gamma')$ measurement for $^{68}$Ni from [16].

From that experiment, neutron skin thicknesses of $0.23 \pm 0.04$ fm and $0.24 \pm 0.04$ fm were determined for $^{130}$Sn and $^{132}$Sn, respectively [10]. Values for the symmetry-energy in pure neutron matter $a_4 = 32.0 \pm 0.04$ fm and the symmetry energy pressure $p_0 = 2.3 \pm 0.8$ MeV/fm$^3$, both at saturation densities, were obtained from the determination of the PDR strength.

Neutron stars are described as composed of a neutron solid crust that surrounds liquid neutronic matter. The solid-liquid phase transition depends on the pressure of the neutron-rich matter, which led Horowitz and Piekarewicz [17] to propose a study of neutron star observable by means of studying neutron-rich matter pressure and neutron skin. The study of PDR is then an effective tool to perform such studies.

Another experiment in stable and exotic Ni isotopes has been performed to study the PDR and is presented in Section 2. First results are presented and further discussed, particularly in the context of the target-charge and target-mass dependence of the excitation processes in Section 3.

## 2. Experimental observation

We carried out an experiment using beams of stable and exotic $^{56-72}$Ni isotopes at about 500 MeV/u delivered from the GSI facility. The secondary beams were selected after fragmentation reactions using the Fragment Separator (FRS). The dipole response of the beam projectiles is studied via relativistic heavy-ion induced electromagnetic excitation at beam energies around 500 MeV/u, where excitation energies up to about 15 MeV can be reached effectively [18]. Both dipole modes, the PDR and GDR are excited with relatively large cross sections. Other multipolarities are excited with much smaller cross sections. After interaction in the reaction target, the decay products from the excited nuclei are detected using the kinematically complete LAND reaction setup. The neutron detector LAND, the large acceptance dipole ALADIN, and the CsI crystals-based $\gamma$ detector are the most notable. We study the neutron-decay channels, which are the dominant channels for excitations above the neutron threshold, in particular for neutron-rich nuclei. For $^{68}$Ni, for instance, the proton threshold is much higher at 15.5 MeV, which is additionnaly hindred due to the Coulomb barrier.

One of the main difficulties in such measurements is to distinguish between the electromagnetic dissociation (ED) and the very peripheral nuclear dissociation (ND). For that purpose, we have used three different targets with different charges namely C (187 mg/cm$^2$), Sn (1515 mg/cm$^2$) and Pb (519 mg/cm$^2$).

The data are under analysis and the results presented in the following are preliminary.

## 3. Discussion

For the study of PDR and GDR, we are interested in the ED part of the dissociation cross sections and thus need to subtract the ND contribution. The dependence of the neutron-removal cross section on the charge and the mass of the nuclei have previously been studied [19–21]. From these one can conclude that (a) the nuclear and electromagnetic interaction can be added incoherently as interference effects are small. We can thus write $\sigma = \sigma_{ND} + \sigma_{ED}$. (b) the separation depends highly on a critical impact parameter $b_c$. For impact parameters $b < b_c$ the nuclear contribution is predominant, whereas for $b > b_c$, only the electromagnetic interaction applies.

Since the ED grows with the charge of the target, we want to extract our values from the measurements with the Pb target. Assuming the ED cross section is very small (negligible) with a C target, we are looking for a scaling factor $\alpha = \sigma_{ND}^{Pb}/\sigma^{C}$, in order to describe the nuclear contribution to the cross section in Pb using the total cross section measured in C. We illustrate this study using the case of the one-neutron removal from $^{68}$Ni as presented in Figure 1. ED cross sections were calculated in a semiclassical approach [22] for the three targets, using minimum impact parameters as proposed in [21]. An effective Z-dependence of $\sigma_{ED} \propto Z^{1.57}$ is found for the one-neutron removal.

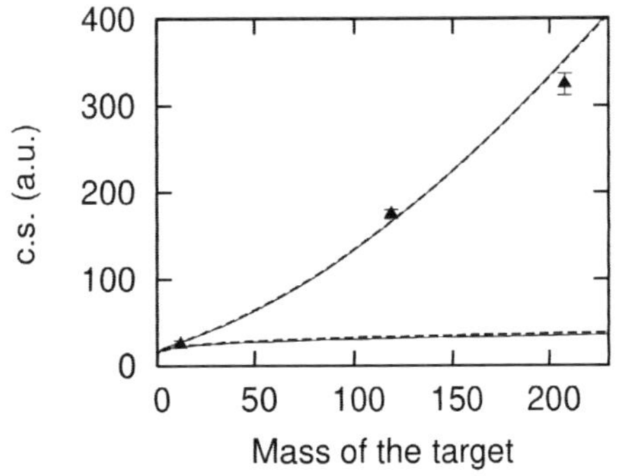


**Figure 1.** Integrated cross sections for one-neutron removal from $^{68}$Ni. The curves represent the fit results as described in the text for the black disk approach (solid lines) and the soft-sphere approach (broken lines). The upper curves represent the sum of the ED and ND contributions whereas the lower curves show the ND cross section only.

However, the nuclear part is more complex. We will consider two different approaches to represent the nuclear contribution. The first one is the black disk method: the nuclei are not transparent to each other. Therefore the interaction is simplified to a geometrical picture of two disks interacting with each other. It then only depends on the radii of the nuclei, giving $\sigma_{ND} \propto (A_T^{1/3} + A_P^{1/3})$. A second approach is the soft-sphere model. It is detailed in [21] and represents the fact that the interaction occurs at the surface of the nuclei. It is then proposed to be of the form $\sigma_{ND} = 2\pi \left[b_c - \frac{\Delta b}{2}\right] \Delta b$, where $\Delta b$ represents the width of impact parameters for which the reaction occurs on the surface of the nuclei. Benesh *et al.* suggest, following Glauber calculations over a rather wide range of nuclei, to use $\Delta b = 0.5$ fm. They also propose $b_c = r_0(A_T^{1/3} + A_P^{1/3} - x(A_T^{-1/3} + A_P^{-1/3}))$ with $r_0 = 1.34$ fm and $x = 0.75$.

We have fitted our integrated cross sections for the one-neutron removal from $^{68}$Ni with the functions $\sigma = p_1\sigma_{ED}+p_2\sigma_{ND}$, where the mass and charge dependence of $\sigma_{ED}$ and $\sigma_{ND}$ are taken from semiclassical calculations and our two models for the nuclear dependence, respectively. $p_{1,2}$ are taken as free parameters for the fit. The results are shown in Figure 1. The solid line was obtained using $\sigma_{ND}$ from the black disk model and the broken line from the soft sphere model. The fits show remarkable agreement for both the total and the nuclear cross sections. From these fits, we determine the scaling factors $\alpha_{disk} = 1.29$ and $\alpha_{sphere} = 1.38$.

For previous experiments, we used a factorisation approach as presented by Mercier *et al.* [19], there we determined a factor of $\alpha = 1.38$, which is in good agreement.

Using the $\alpha_{sphere}$ parameter, an ED cross section (including $1n$ and $1n1p$) of $^{58}$Ni of 108(10) mb is extracted, which is in agreement with the published value of 127(12) mb derived from [23] by folding with the virtual photon spectrum.

## Conclusion

The low-lying dipole strength has been recently studied in both experimental and theoretical approaches. It appears to be a tool to investigate the nuclear symmetry energy as well as the neutron star structure. The LAND-FRS setup proves to be an effective means to study the dipole response of exotic nuclei in the energetic regime of the PDR and GDR. The difficulty to disentangle the nuclear from the electromagnetic contributions to the neutron removal cross section of neutron-rich Ni nuclei has been addressed. We have shown that the black disk and the soft-sphere description of the nuclear contribution to the integrated cross section lead to similar results, and allows us to determine a scaling factor from the cross section of carbon to the nuclear contribution to the cross section of lead of about 1.38.

As the analysis progresses we will be able to verify this scaling factor, investigating, e.g., decay channels with thresholds above the energetic range of ED, like the three-neutron removal channel.

## Acknowledgments

This work is supported by the HGF Young Investigators Project VH-NG-327.

## References

[1] Suzuki Y, Ikeda K and Sato H 1990 *Prog. Theor. Phys.* **83** 180
[2] Isacker P V, Nagarajan M A and Warner D D 1992 *Phys. Rev.* C **45** R13
[3] Vretenar D, Paar N, Ring P and Lalasizzis G A 2001 *Nucl. Phys.* A **692** 496
[4] Tsoneva N, Lenske H and Stoyanov C 2004 *Phys. Lett.* B **586** 213
[5] Paar N, Ring P, Nikšić T and Vretenar D 2003 *Phys. Rev.* C **67** 34312
[6] Paar N, Vretenar D, Khan E and Coló G 2007 *Rep. Prog. Phys.* **70** 691
[7] Brown B A 2000 *Phys. Rev. Lett.* **85** 5298
[8] Furnstahl R J 2002 *Nucl. Phys.* A **706** 85
[9] Adrich P *et al.* 2005 *Phys. Rev. Lett.* **95** 132501
[10] Klimkiewicz A *et al.* 2007 *Phys. Rev.* C **76** 051603(R)
[11] Zilges A *et al.* 2002 *Phys. Lett.* B **542** 43
[12] Volz S 2006 *Nucl. Phys.* A **779** 1
[13] Ryezayeva N *et al.* 2002 *Phys. Rev. Lett.* **89** 272502
[14] Govaert K *et al.* 1998 *Phys. Rev.* C **57** 2229
[15] Savran D *et al.* 2008 *Phys. Rev. Lett.* **100** 232501
[16] Wieland O *et al.* 2009 *Phys. Rev. Lett.* **102** 092502
[17] Horowitz C J and Piekarewicz J 2001 *Phys. Rev. Lett.* **86** 5647
[18] Bertulani C A and Baur G 1988 *Phys. Reports* **163** 299
[19] Mercier M T *et al.* 1986 *Phys. Rev.* C **33** 1655
[20] Aumann T, Bertulani C A and Sümmerer K 1995 *Phys. Rev.* C **51** 416
[21] Benesh C J, Cook B C and Vary J P 1989 *Phys. Rev.* C **40** 1198
[22] Winther A and Alder K 1979 *Nucl. Phys.* A **319** 518
[23] Fultz S C *et al.* 1974 *Phys. Rev.* C **16** 608

Nuclear Physics in Astrophysics IV IOP Publishing
Journal of Physics: Conference Series **202** (2010) 012036 doi:10.1088/1742-6596/202/1/012036

# Study of the $^{15}$N(p,$\gamma$)$^{16}$O reaction at the LUNA accelerator with a BGO detector

**Antonio Caciolli for the LUNA collaboration**
Università degli Studi di Padova and INFN, Padova, Italy
E-mail: `caciolli@pd.infn.it`

**Abstract.** The $^{15}$N(p,$\gamma$)$^{16}$O reaction has been intensively studied by the LUNA collaboration. In this contribution the measurements performed using the same BGO detector and two different setups, with gas and solid target, are described in details. The measurement of the cross section for this reaction allows to determine the astrophysical S-factor. The results of the gas target experiment will be discussed while the analysis of the solid target data is still in progress.

## 1. Introduction

The $^{15}$N(p,$\gamma$)$^{16}$O reaction ($Q$-value 12.127 MeV) links the first CNO cycle to the second one providing the path to produce the $^{16}$O in stellar hydrogen burning [1].

The excitation curve was measured in pioneering experiments by Hebbard [2] and Rolfs and Rodney [3] using NaI and Ge(Li) detectors respectively. The reported low energy data differ by more than a factor of 2 and the same discrepancy also appears in the extrapolated S-factor. Of the two data sets, only the more recent one [3] was used in the NACRE compilation [4]. However, the reason of the discrepancy between these two sets remains unsolved.

New results on the S-factor have been recently published by Mukhamedzanov et al. [5]. They measured the Asymptotic Normalization Coefficients (ANC) and performed an R-Matrix fit including the previous direct data. This work provides an S-factor lower by a factor of 2 than the one suggested by NACRE and in agreement with the Hebbard's extrapolated S-factor [2].

Another recent R-Matrix analysis, using again the available direct data but limited to the capture reaction to the ground state [6], also indicates a much lower S-factor.

It is therefore clear that a new direct measurement is needed to understand this discrepancy. LUNA collaboration decided to perform an extended set of measurements on the $^{15}$N(p,$\gamma$)$^{16}$O reaction using both gas and solid target setups. The LUNA facility, thanks to its underground position inside the Gran Sasso National Laboratory (LNGS), offers a favorable laboratory background for the $\gamma$-ray spectra with respect to analogue facilities on the Earth's surface, allowing to reach lower energies close to the Gamow peak.

Three different approaches were used to investigate this reaction: a gas target with a BGO detector, a solid target with the same BGO detector and a solid target with a High Purity Germanium detector. In this contribution only the experiments and the associated results involving the BGO detector will be discussed.

## 2. Experiments and Target analysis

The two experiments performed with the BGO detector at the LUNA accelerator exploited the high efficiency of the BGO to reach lower energies than feasible with a low efficiency HPGe detector, where statistic is limited.

The first measurement involved a windowless gas target cell filled with 1 mbar natural nitrogen (isotopic composition: 99.634% $^{14}$N, 0.366% $^{15}$N [7]). The main goal of this experiment was the study of the $^{14}$N(p,$\gamma$)$^{15}$O reaction at low energies [8, 9], but the presence of $^{15}$N in the target material allowed for a spin-off analysis of the $^{15}$N(p,$\gamma$)$^{16}$O reaction.

The cylindrical target chamber was surrounded by a $4\pi$-BGO summing detector that, according to GEANT4 simulations and measurements with radioactive sources, has an efficiency of 70% for 12 MeV $\gamma$-rays. The results from this experiment are already published [10] and all details are already presented in that paper and reference therein.

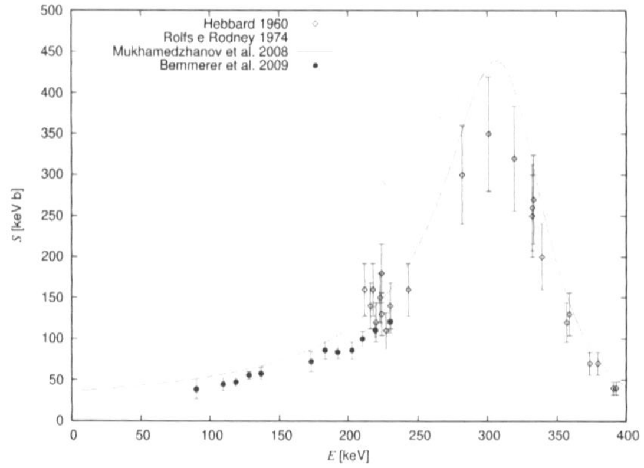


**Figure 1.** Astrophysical S-factor for the $^{15}$N(p,$\gamma$)$^{16}$O reaction as as function of the center of mass energy. The LUNA gas target are presented (blue circles) with data from Hebbard [2] (black diamonds) and from Rolfs and Rodney [3] (green triangles). The red curve represents the R-matrix fit from the ANC data [5].

In Figure 1 the S-factor obtained in the LUNA gas target measurements [10] is compared with the data from previous direct measurements [2, 3] and the results of the ANC analysis [5]. The LUNA results are lower than the adopted values in the NACRE database by a factor of 2 and lies below the ANC curve, which has 17% uncertainty [5]. However, all these results are in agreement within error bars. The uncertainties for the LUNA data [10] are smaller than in previous direct measurements with a main contribution from the beam induced background, as produced by the $^{11}$B(p,$\gamma$)$^{12}$C reaction.

The second experiment involved the same BGO detector setup and a solid state target. The main purposes of this new set of measurements at the LUNA accelerator was to extend the cross section to energies as close as possible to the Gamow peak and to further reduce the previous uncertainty.

The use of enriched solid targets, with a nominal amount of $^{15}$N of 98%, increased the statistic of two order of magnitude with respect to the natural nitrogen gas targets. The boron contamination was evaluated less then 1-2% in all $\gamma$-spectra. Only at the lowest measured beam energies ($E_p$ = 80 keV) the boron amount in the $\gamma$-spectra was about 3%. In all of the cases its contribution to the systematic uncertainty was negligible with respect to other systematic uncertainties.

The targets were made of TiN on a tantalum backing and were produced in Karlsruhe and

Legnaro laboratories using the reactive sputtering technique [11]. The target thickness was estimated to be about 100 nm (14 keV at $E_p$ = 278 keV) All targets suffered deterioration due to the intense beam current (typically 100 $\mu$A) that impinged on the target material for extended time, depositing several tens of Coulomb of charge on each target. Moreover, the target stoichiometry between titanium and nitrogen and the isotopic ratio of $^{14}$N and $^{15}$N are two important parameters in the data analysis. These target characteristics should be studied in depth when using enriched targets and their changes during the measurements should not be neglected in the data analysis.

The target characteristics were investigated with three different techniques in order to obtain the complete information on stoichiometry, isotopic ratio and the target deterioration:

- Resonance scan profiles performed at the Forschungszentrum Dresden-Rossendorf (FZD);
- Secondary Neutral Mass Spectrometry (SNMS) performed at ATOMKI in Debrecen;
- Elastic Recoil Detection (ERD) with high-Z beam performed in Munich.

All the target analysis were performed after the LUNA beam irradiation, since some of these technique are destructive. Two different regions of the target surface were analyzed with each technique in order to extrapolate the relevant information for the target area irradiated by the LUNA beam ("beamspot") and for a region not irradiated ("outside") that correspond to the conditions after and before the $^{15}$N(p,$\gamma$)$^{16}$O measurements. One technique, often used in this type of experiments with solid targets, is the scan of the target with a narrow resonance. The beam (1$\mu$A) was provided by the Tandetron Accelerator of the Forschungszentrum Dresden-Rossendorf to investigate the target using the resonance at 430 keV (in the laboratory system) of the $^{15}$N(p,$\alpha\gamma$)$^{12}$C reaction.

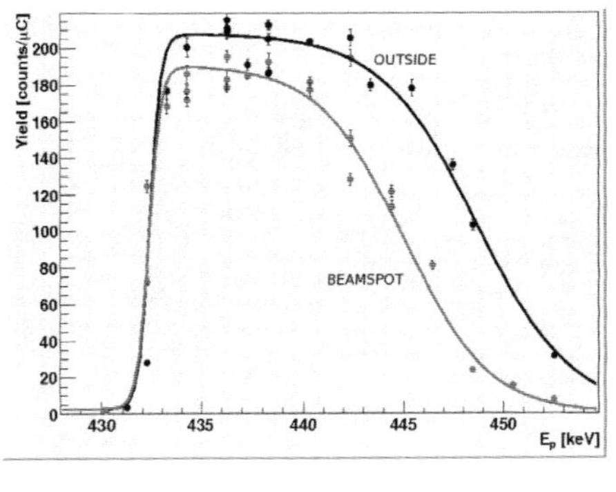


**Figure 2.** Resonance profile scan of the target used in the solid target measurements. The black points are related to the "outside" area and the red ones to the "beamspot" area. The points are fitted with a Fermi like function.

In Figure 2 the data for one of the target are displayed. By interpolating between the two profiles it is possible to extract information about the amount of $^{15}$N in the target during the measurements, as a function of the accumulated charge. From the leading edge in the two profiles situated at the same energy we can exclude any contamination of the target surfaces during the measurements at LUNA. The results obtained by the resonance scans have been used to monitor relative changes and not to obtain absolute values since the $\omega\gamma$ of the resonance and the angular distribution of the emitted $\gamma$-rays are affected by large uncertainties.

In order to know the exact stoichiometry, the targets were analyzed using SNMS and high-Z ERD techniques. While the high-Z ERD analysis is still in progress the SNMS results are shown in Figure 3 for one of the targets. Also in this case two regions of the target were analyzed: one related to the "beamspot" area and the other to the "outside" area.

SNMS uses rare gas ions at low energy (0.5 - 5 keV) to sputter the atoms from the target surface and obtain a depth profile detecting the sputtered atoms. As shown in Figure 3, the SNMS method provides not only results about the $^{15}$N concentration, but also about the stoichiometry and the isotopic concentration. The tantalum concentration was evaluated to determine the target thickness. All information collected on the target characteristics is necessary to evaluate the effective energy ($E_{eff}$) and effective stopping power ($\epsilon_{eff}$), since $E_{eff}$ and $\epsilon_{eff}$ are strictly dependent on the atomic composition and the atomic concentration in the targets. The results of SNMS and resonance scan analysis were compared and compatible values for the target deterioration and the profile shape were obtained.

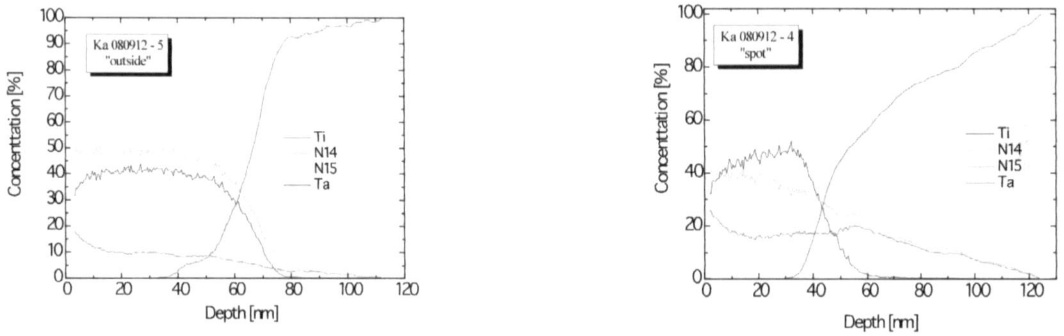


**Figure 3.** Target profile as obtained with the SNMS technique. The left and right panels correspond to the "outside" and "beamspot" areas, respectively

## 3. Summary and outlook

The LUNA collaboration has performed a detailed study of the $^{15}$N(p,$\gamma$)$^{16}$O reaction with natural $N_2$ gas target and enriched solid target. The data analysis of the solid target measurements is still in progress. To improve the uncertainty on the deduced S-factor, a precise knowledge of the target characteristics and their evolution after intense beam irradiation is required. The targets analysis seems to give promising results and a complete overview of the targets characteristic.

## 4. Acknowledgment

Financial support by INFN and in part by the European Union (TARI RII3-CT-2004-506222, AIM 025646 and SPIRIT 227012) and the Hungarian Scientific Research Fund (T49245 and K68801) is gratefully acknowledged.

[1] C. Iliadis, A. Champagne, J. Jose, S. Starrfield and P. Tupper, Astrophys. J. Suppl. **142**, 105-138 (2002); AIP Conf.Proc. 637 (2003) 114-118
[2] D.F. Hebbard, Nucl. Phys. **15** (1960) 289
[3] C. Rolfs and W. Rodney, Nucl. Phys. A **235**, 450 (1974).
[4] C. Angulo *et al.*, Nucl. Phys. A **656**, 3-187 (1999).
[5] A. M. Mukhamedzanov *et al.*, Phys. Rev. C **78**, 015804 (2008).
[6] F. C. Barker, Phys. Rev. C **78**, 044612 (2008).
[7] T. B. Coplen *et al.*, Pure Appl. Chem. **74**, 1987 (2002).
[8] A. Lemut *et al.*, Phys. Lett. B **634**, 483 (2006).
[9] D. Bemmerer *et al.*, Nucl. Phys. A **779**, 297 (2006).
[10] D. Bemmerer *et al.*, J. Phys. G: Nucl. Part. Phys. **36** 045202 (2009).
[11] V. Rigato *et al.*, Surface and Coatings Technology, **142-144** (2001) 943

Nuclear Physics in Astrophysics IV IOP Publishing
Journal of Physics: Conference Series **202** (2010) 012037 doi:10.1088/1742-6596/202/1/012037

# Double Beta Decay Searches

**Oliviero Cremonesi**

INFN - Sezione di Milano Bicocca, Piazza della Scienza, 3 I-20126 Milano, Italy

E-mail: oliviero.cremonesi@mib.infn.it

**Abstract.** Neutrinoless double beta decay is one of the most sensitive probes for the regime of physics beyond the standard model. It can provide fundamental information on the character of neutrinos and on their absolute mass scale. The present status of experiments searching for neutrinoless double-beta decay ($\beta\beta(0\nu)$) is reviewed. Given the observation of neutrino oscillations and the present knowledge of neutrino masses and mixing parameters, a possibility to observe $\beta\beta(0\nu)$ at a neutrino mass scale in the range 10-50 meV could actually exist. This is a real challenge faced by a number of new proposed projects. The most important parameters contributing to the experimental sensitivity are outlined. A short discussion on nuclear matrix element calculations is also given.

## 1. Introduction

First suggested by M.Goeppert-Mayer in 1935, Double Beta Decay (DBD) is a rare spontaneous nuclear transition in which the charge of two nuclei belonging to the same isobaric multiplet changes by two units with the simultaneous emission of two electrons. In order to avoid (or at least inhibit) the occurrence of the equivalent sequence of two single beta decays, it is generally required that both the parent and the daughter nuclei be more bound than the intermediate one. Because of the pairing term, such a condition is fulfilled in nature for a number of even-even nuclei. The decay can then proceed both to the ground state or to the first excited states of the daughter nucleus. Double beta transitions accompanied by positron emission or electron capture are also possible. However they are usually characterized by poorer experimental sensitivities and will not be discussed in the following. We refer to the most recent reviews on $\beta\beta$ for a more complete treatment on the subject[2].

Among the possible $\beta\beta$ modes two are of particular interest, the $2\nu$ mode ($\beta\beta(2\nu)$ ) ${}^A_ZX \rightarrow {}^A_{Z+2} X + 2e^- + 2\overline{\nu}$, which observes the lepton number conservation and it is allowed in the framework of the Standard Model (SM) of electro-weak interactions, and the $0\nu$ mode ($\beta\beta(0\nu)$ ) ${}^A_ZX \rightarrow {}^A_{Z+2} X + 2e^-$ which violates the lepton number by two units. A third decay mode ($\beta\beta(0\nu,\chi)$ ) in which one or more neutral bosons $\chi$ (Majorons) are emitted ${}^A_ZX \rightarrow {}^A_{Z+2} X + 2e^- + N\chi$ is also often considered. The interest in this decay is mainly related to the existence of Majorons, massless Goldstone bosons that arise upon a global breakdown of B–L symmetry.

From the point of view of Particle Physics $\beta\beta(0\nu)$ is of course the most interesting of the $\beta\beta$ decay modes. In fact, after 70 years from its introduction by W.H. Furry[3], $\beta\beta(0\nu)$ is still one of the most powerful tools to test neutrino properties: it can exist only if neutrinos are Majorana particles and it allows then to fix important constraints on the neutrino mass scale.

In all envisaged decay modes, $\beta\beta(0\nu)$ rate is usually expressed as $[T^{0\nu}_{1/2}]^{-1} = G^{0\nu}|M^{0\nu}|^2|\langle m_\nu\rangle|^2$, where $G^{0\nu}$ is the (exactly calculable) phase space integral, $|M^{0\nu}|^2$ is the nuclear matrix element and $\langle m_\nu\rangle$ is a combination of the neutrino mass eigenvalues $\langle m_\nu\rangle \equiv \sum_{k=1}^{3}|U^L_{ek}|^2 m_k e^{i\phi_k}$ which, in the limit of small neutrino masses becomes $\langle m_\nu\rangle = c^2_{12}c^2_{13}m_1 + s^2_{12}c^2_{13}e^{i\alpha_1}m_2 + s^2_{13}e^{i\alpha_2}m_3$.

Unfortunately, the presence of the Majorana phases $\alpha_k$ in the $\langle m_\nu\rangle$ expression implies that cancellations are possible. Such cancellations are complete for a Dirac neutrino since it is equivalent to two degenerate Majorana neutrinos with opposite CP phases. This stresses once more the fact that $\beta\beta(0\nu)$ can occur only through the exchange of Majorana neutrinos. It should be also pointed out that $\beta\beta(0\nu)$ represents the unique possibility to measure the neutrino Majorana phases.

Given the evidence for neutrino oscillations and the present knowledge of neutrino masses and mixing parameters, two possible orderings are possible (neutrino mass hierarchy problem). In the case that fortcoming $\beta\beta(0\nu)$ experiments would not observe any decay (and assuming that neutrinos are Majorana particles) the inverse ordering could finally be exluded thus fixing the problem of the neutrino absolute mass scale[4, 5].

$\langle m_\nu\rangle$ is actually the only $\beta\beta(0\nu)$ measurable parameter containing direct information on the neutrino mass scale. Unfortunaltely its derivation from the experimental results on $\beta\beta(0\nu)$ half-lifetimes requires a precise knowledge of the transition Nuclear Matrix Elements $M^{0\nu}$(NME). Many (often conflicting) evaluations are available in the literature leading to large uncertainty ranges for $\langle m_\nu\rangle$ . This has been recognized as a critical problem by the $\beta\beta$ community. New efforts have been therefore recently directed to its solution: new calculations based on different methods and various cross-checks to improve the reliability of these calculations are presently being undertaken[6, 7, 8]. A statystical analysis of the NME calculation results according to different methods and model parameters has also been considered[9].

## 2. Experimental approaches: present status

The only experimentally available informations in $\beta\beta(0\nu)$ are those carried by the daughter nucleus and the two emitted electrons. Different signatures depend therefore on the number of such informations which are actually measured: sum of the electron energies, single electron energy and angular distributions, identification and/or counting of the daughter nucleus. A better signature is often synonymous of a lower background and, definitely, of a better sensitivity. All experiments tend therefore to find a compromise between the desire to collect the maximum number of informations and the best way in which such a goal can be accomplished. Two main general approaches have been followed so far for $\beta\beta$ experimental investigation: i) indirect or inclusive methods, and ii) direct or counter methods. Inclusive methods are based on the measurement of anomalous concentrations of the daughter nuclei in properly selected samples, characterized by very long accumulation times. They include Geochemical and Radiochemical methods which, being completely insensitive to different $\beta\beta$ modes, can only give indirect evaluations of the $\beta\beta(0\nu)$ and $\beta\beta(2\nu)$ lifetimes. They have played a crucial role in $\beta\beta$ searches especially in the past.

Counter methods are based instead on the direct observation of the two electrons emitted in the decay. Different experimental parameters (energies, momenta, topology, etc) can then be registered according to the different capabilities of the employed detectors. These methods are further classified in *inhomogeneous* (when the observed electrons originate in an external sample) and *homogeneous* experiments (when the source of $\beta\beta$ 's serves also as detector).

In most cases the various $\beta\beta$ modes are separated just on the base of the different distribution expected for the electron sum energies: a continuous bell ditribution for $\beta\beta(2\nu)$ and $\beta\beta(0\nu,\chi)$ , and a sharp line at the transition energy for $\beta\beta(0\nu)$ . Direct counting experiments with very good energy resolution are presently the most attractive approach for $\beta\beta(0\nu)$ searches.

Experimental evidence for several $\beta\beta(2\nu)$ decays has been provided using the measured two-electron sum energy spectra, the single electron energy distributions and the event topology1. On the other hand, impressive progress has been obtained during the last years also in improving $\beta\beta(0\nu)$ half-life limits for a number of isotopes. The best results are still maintained by the use of isotopically enriched HPGe diodes for the experimental investigation of $^{76}$Ge (Heidelberg-Moscow[10] and IGEX[11]) but two other experiments have recently reached comparable sensitivities: NEMO3[12, 13] at LSM and CUORICINO at LNGS[14]. The former is a large inhomogeneous detector aiming at overcoming the intrinsic limits of the technique (relatively small active masses) by expanding the setup dimensions; the big advantage of the NEMO3 technique is the possibility to access single electron informations. CUORICINO is, on the other hand, a $TeO_2$ granular calorimeter based on the bolometric technique; it aims at exploiting the excellent performance of the bolometers (and the possibility they offer to be built with any material of practical interest[15, 16]) to scan the most interesting $\beta\beta(0\nu)$ active isotopes. NEMO3 will continue data taking until the end of 2010 while CUORICINO was stopped in June 2008 to be sustituted by CUORE-0, the first tower of CUORE. The NEMO3 effort to cover as many as possible $\beta\beta$ nuclei thus allowing a diret check for $\beta\beta(2\nu)$ NME elements is evident (Tab. 1).

The evidence for a $\beta\beta(0\nu)$ signal has also been claimed (and recently confirmed [17] by a small subset (KHDK) of the HDM collaboration at LNGS with $T^{0\nu}_{1/2} = 2.23^{+0.44}_{-0.31} \times 10^{25}$ y. The result is based on a re-analysis of the HDM data. Such a claim has raised some criticism but cannot be dismissed out of hand. On the other hand, none of the existing experiments can rule out it, and the only certain way to confirm or refute it is with additional sensitive experiments. In particular, next generation experiments should easily achieve this goal.

**Table 1.** Best reported results on $\beta\beta$ processes. $\beta\beta(0\nu)$ limits are at 90% CL while $\beta\beta(2\nu)$ results are averaged values, taken from ref.[18] Except when explicitely noted $\beta\beta(2\nu)$ results are taken from ref.[18]. Future developments of the technique are shown in the alst columns togethere with their expected location and isotope mass.

| Isotope | $T^{2\nu}_{1/2}$ ($10^{19}$y) | $T^{0\nu}_{1/2}$ ($10^{24}$y) | Future Experiment | Mass (kg) | Lab |
|---|---|---|---|---|---|
| $^{48}$Ca | $(4.2^{+2.1}_{-1.0})$ | $> 0.0014$[19] | CANDLES | | OTO |
| $^{76}$Ge | $(150 \pm 10)$ | $> 19$[10] | GERDA | 18-40 | LNGS |
| | | $22.3^{+4.4}_{-3.1}$[17] | | | |
| | | $> 15.7$[11] | MAJORANA | 60 | SUSEL |
| $^{82}$Se | $(9.6 \pm 1.0)$[13] | $> 0.36$ [13] | SuperNEMO | 100 | LSM |
| $^{96}$Zr | $(2.35 \pm 0.21)$[13] | $> 0.0092$[13] | | | |
| $^{100}$Mo | $(0.711 \pm 0.054)$ | $> 1.1$[13] | MOON | | OTO |
| $^{116}$Cd | $(2.8 \pm 0.3)$ | $> 0.17$[20] | | | |
| $^{130}$Te | $(69 \pm 13.4)$[13] | $> 2.94$ | CUORE | 204 | LNGS |
| $^{136}$Xe | $> 81$[21] | $> 0.12$[22] | EXO | 160 | WIPP |
| | | | KAMLAND | 200 | KAMIOKA |
| $^{150}$Nd | $(0.911 \pm 0.067)$[13] | $> 0.0036$[23] | SNO+ | 56 | SNOLAB |

### 3. Forthcoming experiments

The performance of the different $\beta\beta(0\nu)$ experiments is usually expressed in terms of an experimental *sensitivity* or detector *factor of merit*, defined as the process half-life corresponding to the maximum signal $n_B$ that could be hidden by the background fluctuations at a given statistical C.L. At $1\sigma$ level ($n_B=\sqrt{BTM\Delta}$), one obtains:

$$F_{0\nu} = \tau_{1/2}^{Back.Fluct.} = \ln 2\ N_{\beta\beta}\epsilon \frac{T}{n_B} = \ln 2 \times \frac{x\ \eta\ \epsilon\ N_A}{A}\sqrt{\frac{M\ T}{B\ \Delta}}\ (68\%CL)$$

where B is the background level per unit mass and energy, M is the detector mass, T is the measure time, $\Delta$ is the FWHM energy resolution, $N_{\beta\beta}$ is the number of $\beta\beta$ decaying nuclei under observation, $\eta$ their isotopic abundance, $N_A$ the Avogadro number, A the compound molecular mass, $x$ the number of $\beta\beta$ atoms per molecule, and $\epsilon$ the detection efficiency.

Despite its simplicity, equation (1) has the unique advantage of emphasizing the role of the essential experimental parameters: mass, measuring time, isotopic abundance, background level and detection efficiency. Most of the criteria to be considered when optimizing the design of a new $\beta\beta(0\nu)$ experiment follow directly from it: i) a well performing detector (e.g. good energy resolution and time stability) giving the maximum number of informations (e.g. electron energies and event topology); ii) a reliable and easy to operate detector technology requiring a minimum level of maintenance (long underground running times); iii) a very large (possibly isotopically enriched) mass, of the order of one ton or larger; iv) an effective background suppression strategy. Unfortunately, these simple criteria are often conflicting and simultaneous optimisation is rarely possible.

Calorimetric detectors are usually preferred since they have produced so far the best results. The calorimetric approach suffers however from a strong limitation: it can be applied only to a small number of $\beta\beta(0\nu)$ isotopes (e.g. $^{76}$Ge, $^{136}$Xe, $^{48}$Ca), thus limiting the number of experimentally accessible isotopes. As suggested by the NEMO3 and CUORICINO experience however, a possible way out exists.

A series of new proposals has been boosted in recent years by the renewed interest in $\beta\beta(0\nu)$ following neutrino oscillation results. The ultimate goal is to reach sensitivities such to allow an investigation of the inverse hierarchy (IH) of neutrino masses ($\langle m_\nu \rangle$ ~10-50 meV). From an experimental point of view this corresponds however to active masses of the order of 1 ton with background levels of the order of 1 c/keV/ton/y. A challenge that can hardly be faced by the current technology. Phased programs have been therefore proposed in USA and Europe[24, 25].

Second generation experiments are all characterized by hundred kg detectors and 1-10 c/kev/ton background rates. Their goal is to select the best technology and approach the IH region.

A list of some of the forthcoming $\beta\beta(0\nu)$ projects is given in table 1. They can be classified in three broad classes: i) dedicated experiments using a conventional detector technology with improved background suppression methods (e.g. GERDA, MAJORANA); ii) experiments using unconventional detector (e.g. CUORE) or background suppression (e.g. EXO, SuperNEMO) technologies; iii) experiments based on suitable modifications of an existing setup aiming at a different search (e.g. SNO+, KAMLAND). In some cases technical feasibility tests are required, but the crucial issue is still the capability of each project to pursue the expected background suppression. Although all proposed projects show interesting features for a second generation experiment, only few of them are characterized by a reasonable technical feasibility within the next few years.

MAJORANA and GERDA are both phased programs representing large scale extensions of past successful experiments on $^{76}$Ge $\beta\beta(0\nu)$ .

Evolved from the HM experiment, GERDA[26] aims at implementig the concept of Ge diodes immersed in a LAr bath[27] for a radical background suppression. The GERDA setup construction is presently being completed in Gran Sasso. 18 and 40 kg of Germanium detectors enriched in $^{76}$Ge are foreseen for the first and second phase respectively. GERDA-I will scrutinize the KHDK claim starting in 2010 and reaching a sensitivity $T_{1/2} > 2 \times 10^{25}$ y (90 % CL) after two years of data taking. 40 kg of germanium isotopically enriched in $^{76}$Ge are already available for GERDA-II. A large part of the efforts are presently directed to develop the segmented detectors crucial for the targeted $10^{-3}$ c/keV/kg background level. The expected 5y sensitivity is $\sim 2.5 \times 10^{26}$ y. Depending on the physics results of the previous phases, a third phase using 500 to 1000 kg of enriched germanium detectors is planned, merging GERDA with the US lead Majorana collaboration.

MAJORANA, a mainly USA proposal with important Canadian, Japanese, and Russian contributions, is an evolution of the IGEX experiment. The proposed initial configuration[28] would consist of 171 segmented n-type germanium crystals (180 kg), distributed in 3 independent ultra-clean electro-formed conventional cryostats of 57 crystals each. The whole assembly would be enclosed in a low-background passive shield and active veto and be located deep underground. A 60 kg demonstrator (single cryostat) is presently being developed to demonstrate the viability of the technique. The completion of this phase is expected in 2014.

CUORE[29] (*Cryogenic Underground Detector for Rare Events*) is a very large extension of the $TeO_2$ bolometric array concept pioneered by the Milano group at the Gran Sasso Laboratory since the eighties. CUORE will consist of a rather compact cylindrical structure of 988 cubic natural $TeO_2$ crystals of 5 cm side (750 g), arranged into 19 separated *towers* (13 *planes* of 4 crystals each) and operated at a temperature of 10 mK. The expected energy resolution is $\sim$5 keV FWHM at the $\beta\beta(0\nu)$ transition energy ($\sim$2.53 MeV). A background level of of the order of $\sim$0.01 c/keV/kg/y is expected by extrapolating the CUORICINO background results and the dedicated CUORE R&D measurements. The expected 5y sensitivity is $2.1 \times 10^{26}$ y. CUORE will therefore allow a close look at the IH region of neutrino masses. CUORE is fully funded and presently under construction at LNGS at a relatively low cost thanks to the high natural abundance of $^{130}$Te . Setup completion is expected in 2012. Thanks to the bolometer's versatility, alternative options with respect to $TeO_2$ are also possible. In particular, promising results have been recently obtained with scintillating bolometers[30] which could allow to study in the future new $\beta\beta(0\nu)$ active isotopes with improved sensitivity.

EXO[31] (*Enriched Xenon Observatory*) is a challenging project based on a large mass ($\sim$ 1–10 tons) of isotopically enriched (85% in $^{136}$Xe) Xenon. An ingenuous tagging of the doubly charged Ba isotope produced in the decay ($^{136}Xe \rightarrow ^{136}Ba^{++} + 2e^-$) would allow an excellent background suppression. The technical feasibility of such an ambitious project aiming at a complete suppression of all the backgrounds requires a hard, still ongoing R&D phase. The unavoidable $\beta\beta(2\nu)$ contribution is a serious concern due to the poor energy resolution of Xe detectors. A smaller prototype experiment with a Xe mass of 200 kg (80% $^{136}$Xe), is presently being installed at WIPP. The prototype has no barium tagging. The primary goal is to measure $^{136}$Xe $\beta\beta(2\nu)$ and to study $\beta\beta(0\nu)$ with a sensitivity of $\sim 10^{25}$ y in two years of data taking.

The proposed Super-NEMO experiment is the only based on an inhomogeneous approach. It is an extension of the NEMO3 concept, properly scaled in order to accommodate $\sim$100 kg of $^{82}$Se foils spread among 20 detector modules. The proposed geometry is planar. The energy resolution will be improved from 12% FWHM to 7% FWHM to improve the signal detection efficiency from 8% to 40% and reduce the $\beta\beta(2\nu)$ contribution. The detector modules will have an active water shield to further reduce cosmic ray backgrounds. The proposed detector dimensions will require a larger hall than is currently available at Frejus and an expansion of the facility is therefore required and actively pursued. A demonstrator (single module) is presently fully funded to be completed in 2011 with a test run in the current NEMO3 site. If funded, Super-NEMO construction should immediately start.

New developments have been recently proposed concerning the possibility to disperse $\beta\beta(0\nu)$ active isotopes in large masses of low-activity scintillators. SNO+ is pursuing the goal of studying $^{150}$Nd with 50 to 500 kg of isotopically enriched Neodimium depending on the results of the currently ongoing R&D program.

A similar approach is proposed by KAMLAND, but for $^{136}$Xe. Their program should start in 2011 (first phase) with 200-400 kg of isotope and continue in 2013 with 1 ton of Xenon enriched to 90% in $^{136}$Xe. A preliminary estimate of the 5y sensitivity for phase I amounts to $\sim 10^{26}$ y.

## 4. Conclusions

A renewed interest in $\beta\beta(0\nu)$ searches has been stimulated by neutrino oscillation results. Neutrinoless $\beta\beta$ is finally recognized as a unique tool to measure neutrino properties (nature, mass scale, intrinsic phases) unavailable to the other neutrino experiments. Present $\langle m_\nu \rangle$ sensitivities are still outside the range required to test the inverted neutrino mass hierarchy. An international effort is however supporting a phased $\beta\beta(0\nu)$ program based on a number of newly proposed experiments to pursue such a goal. The success of such a program strongly depends on the true capability of the proposed projects to reach the required background levels in the $\beta\beta(0\nu)$ region. The claimed evidence for a $\beta\beta(0\nu)$ signal in the HM data could be soon verified by the presently running experiments and in any case, by the forthcoming next generation experiments.

## References

[1] M. Goeppert-Mayer, Phys. Rev. **48** (1935) 512.
[2] S.R. Elliott and P. Vogel, *Annu. Rev. Nucl. Part. Sci.* **52**, 115 (2002); hep-ph/0202264; J.D. Vergados, *Phys. Rep.* **361**, 1 (2002); J. Suhonen and O. Civitarese, *Phys. Rep.* **300**, 123 (1998); V.I. Tretyak and Y. Zdesenko, *Atomic Data and Nuclear Data Tables* **80**, 83 (2002) and **61**, 43 (1995); F.T. Avignone III,S.R. Elliott and J. Engel, *RRev. Mod. Phys* 80 (2008) 2013.
[3] W.H. Furry, *Phys. Rev.* **56**, 1184 (1939).
[4] A. Strumia and F. Vissani, hep-ph/0503246.
[5] S. Pascoli, S.T. Petcov and T. Schwetz, hep-ph/0505226.
[6] F.Simkovic, A.Faessler, V.Rodin, P.Vogel, and J.Engel, *Phys. Rev.* C **77** (2008) 045503; J. Suhonen and M. Kortelainen, *Int. J. Mod. Phys.* E **17**, 1 (2008).
[7] E.Caurier, J.Menendez, F.Nowacki, and A.Poves, *Phys. Rev. Lett.* **100** (2008) 052503.
[8] J.Barea and F.Iachello, *Phys. Rev.* C **79**, 044301 (2009).
[9] A.Faessler et al., hep-ph/0810.5733v2
[10] H.V. Klapdor-Kleingrothaus et al. (HM Coll.), *Eur. Phys. J.* A **12**, 147 (2001) and hep-ph/0103062
[11] C.E. Aalseth et al., *Phys. Rev.* D **65** (2002) 092007
[12] A.S. Barabash, arXiv:hep-ex/0610025.
[13] K. Lang (NEMO3 coll.), Proc. fo the WIN09 conference, Perugia September 2009, Italy.
[14] C. Arnaboldi et al., *Phys. Rev.* C **78** (2008) 035502
[15] E. Fiorini and T. Niinikoski, *Nucl. Instr. and Meth.* A **224**, 83 (1984).
[16] A. Alessandrello et al., *Nucl. Instr. and Meth.* A **440**, 397 (1998) and *Phys. Lett.* B **420**, 109 (1998)and *Phys. Lett.* B **433**, 156 (1998).
[17] H.V. Klapdor-Kleingrothaus, *Phys. Scr.* T **127** (2006) 40; H.V. Klapdor-Kleingrothaus and I.V. Krivosheina, *Modern Physics Letters* A **21** (2006) 1547.
[18] A.S. Barabash, *Czech. J. Phys.* **56** (2006) 437.
[19] I. Ogawa et al., *Nucl. Phys.* A **730** (2004) 215
[20] F.A. Danevich et al., *Phys. Rev.* C **68** (2003) 035501
[21] J.M. Gavriljuk et al., *Phys. Rev.* C **61**, 035501 (2000).
[22] R. Bernabei, *Phys. Lett.* B **546**, 23 (2002).
[23] A.S. Barabash, *Phys. At. Nucl.* **68** (2005) 414.
[24] Joint Study on the Future of Neutrino Physics: The Neutrino Matrix, http://www.aps.org/policy/reports/multidivisional/neutrino/
[25] European strategy for astroparticle physics, http://www.aspera-eu.org/
[26] I. Abt et al., hep-ex/0404039.
[27] H.V. Klapdor-Kleingrothaus HV, hep-ph/0103074.
[28] C.E. Aalseth et al., hep-ex/0201021.
[29] C. Arnaboldi et al., hep-ex/0501010.
[30] S.Pirro et al., *Phys. Atom. Nucl.* **69** (2006) 2109, nucl-ex/0510074
[31] M. Danilov et al., *Phys. Lett.* B **480** 12 (2000).

Journal of Physics: Conference Series **202** (2010) 012038 doi:10.1088/1742-6596/202/1/012038

# Search for double beta decay of zinc and tungsten with low background $ZnWO_4$ crystal scintillators

**R. Bernabei[1,2], P. Belli[2], F. Cappella[3,4], R. Cerulli[5], F.A. Danevich[6], B.V. Grinyov[7], A. Incicchitti[4], V.V. Kobychev[6], V.M. Mokina[6], S.S. Nagorny[6], L.L. Nagornaya[7], S. Nisi[5], F. Nozzoli[1,2], D.V. Poda[6], D. Prosperi[3,4], V.I. Tretyak[6], S.S. Yurchenko[6]**

[1] *Dip. di Fisica, Università di Roma "Tor Vergata", I-00133 Rome, Italy*
[2] *INFN, sez. Roma "Tor Vergata", I-00133 Rome, Italy*
[3] *Dip. di Fisica, Università di Roma "La Sapienza", I-00185 Rome, Italy*
[4] *INFN, sez. Roma, I-00185 Rome, Italy*
[5] *INFN, Laboratori Nazionali del Gran Sasso, 67010 Assergi (AQ), Italy*
[6] *Institute for Nuclear Research, MSP 03680 Kyiv, Ukraine*
[7] *Institute for Scintillation Materials, 61001 Kharkiv, Ukraine*

**Abstract.**
Double beta processes in $^{64}$Zn, $^{70}$Zn, $^{180}$W, and $^{186}$W have been searched for with the help of low background $ZnWO_4$ crystal scintillators at the Gran Sasso National Laboratories of the INFN (Italy). The total measurement time exceeds ten thousand hours. New improved half-life ($T_{1/2}$) limits on double electron capture and electron capture with positron emission in $^{64}$Zn have been set. New $T_{1/2}$ bounds were also set on different modes of $2\beta$ processes in $^{70}$Zn, $^{180}$W, and $^{186}$W. Future perspectives are considered.

## 1. INTRODUCTION

Experimental investigations in this field are concentrated mostly on $2\beta^-$ decays, processes yielding the emission of two electrons. Results for double positron decay ($2\beta^+$), electron capture with positron emission ($\varepsilon\beta^+$), and capture of two electrons from atomic shells ($2\varepsilon$) are much more modest. The most sensitive experiments give limits on the $2\varepsilon$, $\varepsilon\beta^+$ and $2\beta^+$ processes at level of $10^{17} - 10^{21}$ yr [1]. Reasons for this situation are: (1) lower energy releases in $2\varepsilon$, $\varepsilon\beta^+$ and $2\beta^+$ processes in comparison with those in $2\beta^-$ decay, that provide lower probabilities of the processes and difficult background suppression; (2) usually lower natural abundances of $2\beta^+$ isotopes (which are typically lower than 1% with only few exceptions). Nevertheless, studies of neutrinoless $2\varepsilon$ and $\varepsilon\beta^+$ decays could help to explain the mechanism of neutrinoless $2\beta^-$ decay (is it due to non-zero neutrino mass or to the right-handed admixtures in weak interactions) [2].

In Table 1 the potentially $2\beta$ active nuclides present in $ZnWO_4$ crystals are listed.

## 2. MEASUREMENTS AND RESULTS

In Table 2 the $ZnWO_4$ crystal scintillators used in the present experiment are listed; they were produced from two crystal boules grown by the Czochralski method from $ZnWO_4$ compounds prepared from two batches of zinc oxide provided by different producers, and from the same tungsten oxide. The third scintillator ZWO-2a was cut from the ZWO-2 crystal.

To estimate the presence of naturally occurring radioactive isotopes, as well as some other elements important for growing the crystals, the $ZnWO_4$ samples were measured with the help of

**Table 1.** Potentially $2\beta$ active nuclides present in $ZnWO_4$ crystals.

| Transition | Energy release (keV) [3] | Isotopic abundance (%) [4] | Decay channels | Number of nuclei in 100 g of $ZnWO_4$ crystal |
|---|---|---|---|---|
| $^{64}Zn \to ^{64}Ni$ | 1095.7(0.7) | 48.268(0.321) | $2\varepsilon$, $\varepsilon\beta^+$ | $9.28 \times 10^{22}$ |
| $^{70}Zn \to ^{70}Ge$ | 998.5(2.2) | 0.631(0.009) | $2\beta^-$ | $1.21 \times 10^{21}$ |
| $^{180}W \to ^{180}Hf$ | 144(4) | 0.12(0.01) | $2\varepsilon$ | $2.31 \times 10^{20}$ |
| $^{186}W \to ^{186}Os$ | 489.9(1.4) | 28.43(0.19) | $2\beta^-$ | $5.47 \times 10^{22}$ |

**Table 2.** $ZnWO_4$ crystal scintillators used in the present experiments.

| Crystal scintillator | Size (mm) | Mass (g) |
|---|---|---|
| ZWO-1 | $20 \times 19 \times 40$ | 117 |
| ZWO-2 | $\oslash 44 \times 55$ | 699 |
| ZWO-2a | $\oslash 44 \times 14$ | 168 |

Inductively Coupled Plasma - Mass Spectrometry (ICP-MS, Agilent Technologies model 7500a) [5].

The $ZnWO_4$ crystals were fixed inside a cavity of $\oslash 47 \times 59$ mm in the central part of a polystyrene light-guide 66 mm in diameter and 312 mm in length. The cavity was filled up with high purity silicone oil. The light-guide was optically connected on opposite sides by optical couplant to two low radioactivity 3" photomultipliers. The light-guide was wrapped by PTFE reflection tape. The detector has been installed deep underground in the low background DAMA/R&D set-up at the LNGS of the INFN (Italy). It was surrounded by Cu bricks and sealed in a low radioactive air-tight Cu box continuously flushed with high purity nitrogen gas (stored deeply underground for a long time) to avoid presence of residual environmental Radon. The copper box was surrounded by a passive shield made of 10 cm of high purity Cu, 15 cm of low radioactive lead, 1.5 mm of cadmium and 4/10 cm polyethylene/paraffin to reduce the external background. The whole shield has been closed inside a Plexiglas box, also continuously flushed by high purity nitrogen gas. An event-by-event data acquisition system accumulates the amplitude and the arrival time of the events. The sum of the signals from the PMTs was recorded with the sampling frequency of 20 MS/s over a time window of 100 $\mu$s by a 8 bit transient digitizer (DC270 Acqiris).

The measurements were carried out in four runs (see Table 3 for details). First, the energy interval of data taking was chosen as $0.01 - 1$ MeV with the aim to search for the neutrino accompanied double electron capture in $^{64}Zn$. Then the energy interval was changed to $0.05 - 4$ MeV in order to search for other possible double beta processes in $^{64}Zn$ (Run 2). Then the large ZWO-2 crystal was installed and the experiment was carried out in the same energy range (Run 3). Finally the ZWO-2 crystal was cleaved in two parts; one (ZWO-2a) was used in the Run 4 with the energy interval of data taking $0.01 - 1$ MeV.

The energy scale and resolution of the $ZnWO_4$ detectors have been measured with $^{22}Na$, $^{133}Ba$, $^{137}Cs$, $^{228}Th$ and $^{241}Am$ $\gamma$ sources. In addition, the energy scale of the detectors was checked by using background $\gamma$ lines of 609 keV of $^{214}Bi$ (in all Runs), 1461 keV ($^{40}K$), 1764 keV ($^{214}Bi$), 2615 keV ($^{208}Tl$) in Runs 2 and 3.

The knowledge of the radioactive contamination of the $ZnWO_4$ crystals is necessary to describe the measured background spectra. The time-amplitude analysis, the pulse-shape discrimination, and the Monte Carlo simulation were applied to reconstruct the background spectra and to estimate the radioactive contamination of the $ZnWO_4$ detectors. For details see ref. [5].

**Table 3.** Description of low background measurements with $ZnWO_4$ crystal scintillators. Times of measurements ($t$), energy intervals of data taking ($\Delta E$), energy resolutions at the 662 keV $\gamma$ line of $^{137}$Cs (FWHM), and background counting rates (BG) in different energy intervals are specified.

| Run | Crystal scintillator | $t$ (h) | $\Delta E$ (MeV) | FWHM (%) | BG (counts/(day×keV×kg) in energy interval (MeV) | | | |
|---|---|---|---|---|---|---|---|---|
| | | | | | 0.2–0.4 | 0.8–1.0 | 2.0–2.9 | 3.0–4.0 |
| 1 | ZWO-1 | 1902 | 0.01–1 | 11.5 | 1.93(3) | 0.27(1) | | |
| 2 | ZWO-1 | 2906 | 0.05–4 | 12.6 | 1.71(2) | 0.25(1) | 0.0072(7) | 0.0003(1) |
| 3 | ZWO-2 | 2130 | 0.05–4 | 14.6 | 1.07(1) | 0.149(3) | 0.0072(4) | 0.00031(7) |
| 4 | ZWO-2a | 3292 | 0.01–1 | 11.0 | 1.52(2) | 0.211(7) | | |

The measured background spectra were fitted by the model built from the simulated distributions. Activities of U/Th daughters in the crystals were restricted taking into account the results of the time-amplitude and pulse-shape analyzes [5]. Activities of $\beta$ active $^{87}$Rb, $^{113}$Cd, and $^{115}$In were bounded taking into account the results of the ICP-MS analysis. The values of the $^{40}$K, $^{232}$Th and $^{238}$U activities inside the PMTs were taken from [6] where the radioactive contamination of PMTs of the same model was measured. The peak in the spectrum of Run 3 at the energy $1133 \pm 8$ keV cannot be explained by a contribution from external $\gamma$ rays (the 1120 keV $\gamma$ line of $^{214}$Bi is not intense enough to provide the whole peak area). We assume the presence of $^{65}$Zn ($T_{1/2} = 244.26$ d, $Q_\beta = 1351.9$ keV [7]) in the crystal to explain the peak. $^{65}$Zn can be produced from $^{64}$Zn by thermal neutrons (the cross section of $^{64}$Zn to thermal neutrons is 0.76 barn [7]) or/and by cosmogenic activation. The result of the fit of the spectra of Run 2 and Run 3 in the energy region $0.1 - 2.9$ MeV and the main components of the background as well as the summary of the radioactive contamination of the $ZnWO_4$ crystal scintillators (or limits on their activities) are given in [5].

There are no clear peculiarities in the measured energy spectra of the $ZnWO_4$ detectors, which can be interpreted as double beta decay of Zinc or Tungsten isotopes. Therefore only lower half-life limits can be set according to the formula: $\lim T_{1/2} = N \cdot \eta \cdot t \cdot \ln 2 / \lim S$, where $N$ is the number of potentially $2\beta$ unstable nuclei, $\eta$ is the detection efficiency, $t$ is the measuring time, and $\lim S$ is the number of events of the effect searched for which can be excluded at a given confidence level (C.L.).

We have used different combinations of the accumulated data to reach maximal sensitivity to the sought double beta processes. The response functions of the $ZnWO_4$ detectors for the $2\beta$ processes were simulated with the help of the GEANT4 code [8]. The initial kinematics of the particles emitted in the decays was generated with the DECAY0 event generator [9].

In ref. [5] the procedures followed to extract the limits on the various decay modes are described in details. All the half-life limits on $2\beta$ decay processes in Zinc and Tungsten obtained in the present experiment are summarized in Table 4 where results of the most sensitive previous experiments are given for comparison.

The obtained bounds are well below the existing theoretical predictions [16, 17]; nevertheless most of the limits are near one order of magnitude higher than those established in previous experiments. It should be stressed that in contrast to a level of sensitivity obtained for double $\beta^-$ decay ($10^{23} - 10^{25}$ years in the best experiments), only two nuclei ($^{40}$Ca and $^{78}$Kr) among potentially $2\varepsilon$, $\varepsilon\beta^+$, $2\beta^+$ active isotopes were investigated at the level of $\sim 10^{21}$ yr. Moreover, the positive indication on the $(2\nu + 0\nu)\varepsilon\beta^+$ decay of $^{64}$Zn with $T_{1/2} = (1.1 \pm 0.9) \times 10^{19}$ yr suggested in [14] is completely ruled out.

Further improvements in sensitivity can be reached by increasing the mass of the $ZnWO_4$ detector, suppression of external background and development of $ZnWO_4$ scintillators with lower level of radioactive contamination. High abundance of $^{64}$Zn (48.3%) allows to build a large scale

**Table 4.** Half-life limits on $2\beta$ processes in Zn and W isotopes.

| Transition | Decay channel | Level of daughter nucleus | $T_{1/2}$ limit (yr) 90%(68%) C.L. Present work | Previous results |
|---|---|---|---|---|
| $^{64}$Zn→$^{64}$Ni | $0\nu 2\varepsilon$ | g.s. | $\geq 1.1(2.8) \times 10^{20}$ | $\geq 0.7(1.0) \times 10^{18}$ [10] |
| | | | | $\geq 3.4(5.5) \times 10^{18}$ [11] |
| | $0\nu\varepsilon\beta^+$ | g.s. | $\geq 4.3(5.7) \times 10^{20}$ | $\geq 2.8 \times 10^{16}$ [12] |
| | | | | $\geq 2.4(3.6) \times 10^{18}$ [10] |
| | | | | $\geq 1.3 \times 10^{20}$ [13] |
| | | | | $\geq 2.2(6.1) \times 10^{20}$ [11] |
| | $2\nu\varepsilon\beta^+$ | g.s. | $\geq 0.70(2.1) \times 10^{21}$ | $= (1.1 \pm 0.9) \times 10^{19}$ [14] |
| | | | | $\geq 4.3(8.9) \times 10^{18}$ [10] |
| | | | | $\geq 1.3 \times 10^{20}$ [13] |
| | | | | $\geq 2.1(7.4) \times 10^{20}$ [11] |
| $^{70}$Zn→$^{70}$Ge | $0\nu 2\beta^-$ | g.s. | $\geq 1.8(3.0) \times 10^{19}$ | $\geq 0.7(1.4) \times 10^{18}$ [10] |
| | $2\nu 2\beta^-$ | g.s. | $\geq 2.3(4.0) \times 10^{17}$ | $\geq 1.3(2.1) \times 10^{16}$ [10] |
| | $0\nu 2\beta^-$M1 | g.s. | $\geq 1.0(1.4) \times 10^{18}$ | |
| $^{180}$W→$^{180}$Hf | $0\nu 2\varepsilon$ | g.s. | $\geq 0.86(1.2) \times 10^{18}$ | $\geq 0.9(1.3) \times 10^{17}$ [15] |
| | $2\nu 2K$ | g.s. | $\geq 6.6(9.4) \times 10^{17}$ | $\geq 0.7(0.8) \times 10^{17}$ [15] |
| $^{186}$W→$^{186}$Os | $0\nu 2\beta^-$ | g.s. | $\geq 2.1(4.2) \times 10^{20}$ | $\geq 1.1(2.1) \times 10^{21}$ [15] |
| | $0\nu 2\beta^-$ | $2^+$ (137.2 keV) | $\geq 2.1(4.2) \times 10^{20}$ | $\geq 1.1(2.0) \times 10^{21}$ [15] |
| | $0\nu 2\beta^-$M1 | g.s. | $\geq 5.8(8.6) \times 10^{19}$ | $\geq 1.2(1.4) \times 10^{20}$ [15] |
| | $2\nu 2\beta^-$ | g.s. | $\geq 2.3(2.8) \times 10^{19}$ | $\geq 3.7(5.3) \times 10^{18}$ [15] |
| | $2\nu 2\beta^-$ | $2^+$ (137.2 keV) | $\geq 1.8(3.6) \times 10^{20}$ | $\geq 1.0(1.3) \times 10^{19}$ [15] |

experiment without expensive isotopical enrichment. An experiment involving ≈10 tons of non-enriched crystals ($9 \times 10^{27}$ nuclei of $^{64}$Zn) could reach the half-life sensitivity $\sim 3 \times 10^{28}$ yr (supposing zero background during ten years of measurements). Such a sensitivity could contribute to our understanding of the neutrino mass mechanism and right-handed currents in neutrinoless processes [2]. The two neutrino double electron capture should be surely observed: in accordance with theoretical expectations [16, 17], $T_{1/2}$ for the $2\nu 2\varepsilon$ process is predicted on the level of $10^{25} - 10^{26}$ yr.

## References

[1] V.I. Tretyak, Yu.G. Zdesenko, At. Data Nucl. Data Tables **61**, 43 (1995); **80**, 83 (2002).
[2] M. Hirsch et al., Z. Phys. A **347**, 151 (1994).
[3] G. Audi, A.H. Wapstra, C. Thibault, Nucl. Phys. A **729**, 337 (2003).
[4] J.K. Bohlke et al., J. Phys. Chem. Ref. Data **34**, 57 (2005).
[5] P. Belli et al., Nucl. Phys. A **826** 256 (2009).
[6] R. Bernabei et al., Il Nuovo Cim. A **112**, 545 (1999).
[7] R.B. Firestone et al., *Table of isotopes*, 8-th ed., John Wiley, New York, 1996 and CD update, 1998.
[8] S. Agostinelli et al., Nucl. Instr. Meth. A **506**, 250 (2003);
J. Allison et al., IEEE Trans. Nucl. Sci. **53**, 270 (2006).
[9] O.A. Ponkratenko et al., Phys. At. Nucl. **63**, 1282 (2000);
V.I. Tretyak, to be published.
[10] F.A. Danevich et al., Nucl. Instr. Meth. A **544**, 553 (2005).
[11] P. Belli et al., Phys. Lett. B **658**, 193 (2008).
[12] H. Kiel et al., Nucl. Phys. A **723**, 499 (2003).
[13] H.J. Kim et al., Nucl. Phys. A **793**, 171 (2007).
[14] I. Bikit et al., Appl. Radiat. Isot. **46**, 455 (1995).
[15] F.A. Danevich et al., Phys. Rev. C **68**, 035501 (2003).
[16] P. Domin, S. Kovalenko, F. Simkovic, S.V. Semenov, Nucl. Phys. A **753**, 337 (2005).
[17] E.-W. Grewe et al., Phys. Rev. C **77**, 064303 (2008).

Nuclear Physics in Astrophysics IV IOP Publishing
Journal of Physics: Conference Series **202** (2010) 012039 doi:10.1088/1742-6596/202/1/012039

# Searches for processes violating the Pauli Exclusion Principle in Sodium and Iodine

**R. Bernabei[1,2], P. Belli[2], F. Cappella[3,4], R. Cerulli[5], C.J. Dai[6], A.d'Angelo[3,4], H.L. He[6], A. Incicchitti[4], H.H. Kuang[6], X.H. Ma[6], F. Montecchia[7,2], F. Nozzoli[1,2], D. Prosperi[3,4], X.D. Sheng[6], Z.P. Ye[6,8]**

[1] *Dip. di Fisica, Università di Roma "Tor Vergata", I-00133 Rome, Italy*
[2] *INFN, sez. Roma "Tor Vergata", I-00133 Rome, Italy*
[3] *Dip. di Fisica, Università di Roma "La Sapienza", I-00185 Rome, Italy*
[4] *INFN, sez. Roma, I-00185 Rome, Italy*
[5] *Laboratori Nazionali del Gran Sasso, I.N.F.N., Assergi, Italy*
[6] *IHEP, Chinese Academy, P.O. Box 918/3, Beijing 100039, China*
[7] *Lab. Sperimentale Policentrico di Ingegneria Medica, Università di Roma "Tor Vergata"*
[8] *University of Jing Gangshan, Jiangxi, China*

**Abstract.** Searches for non-paulian nuclear processes, i.e. processes normally forbidden by the Pauli–Exclusion–Principle (PEP) with highly radiopure NaI(Tl) scintillators allow the test of this fundamental principle with high sensitivity. Status and perspectives are briefly addressed.

## 1. Introduction

The exclusion principle was postulated by W. Pauli in 1925 to explain atomic spectra and regularities of the Periodic Table of the elements. In modern Quantum Field Theory the Pauli–Exclusion–Principle (PEP) is related to the spin statistics [1] and automatically arises from the anti-commutation property of the fermion creation and destruction operators. Despite the foundation of PEP lies deep in the structure of the Quantum Field Theory, a simple explanation is still missing. Thus, the exact validity of the PEP is still an open question in spite of all its known successes. In fact, the general principles of the quantum theory allow to go beyond the Bose and Fermi statistics and to also consider generalized statistics [2, 3, 4]. Many experimental tests of the PEP validity following the first pioneering experiments [5, 6] have been carried out so far by various approaches: i) searches for PEP-forbidden electronic states [7]; ii) searches for PEP-forbidden nuclear states [8]; iii) searches for PEP-forbidden electronic transitions [9, 10, 11, 12]; iv) searches for PEP-forbidden nuclear transitions [13, 14, 15, 16, 12]. It is worth noting that in 1980 Amado and Primakoff [17] criticized – on the basis of the assumption that the total Hamiltonian, describing the atoms, is completely symmetric in the electrons – the possibility of testing the Pauli principle by searching for PEP-forbidden transitions. However, their arguments can be evaded as demonstrated in refs. [18, 19]. In fact, in ref. [18] a theory has been developed where apparently non-paulian transitions could occur owing to the possible substructure of the electrons in composite models of quarks and leptons. In ref. [19] instead a theory, where extra dimensions could lead to apparent PEP violations, is discussed.

In the following we just introduce the present status of the search for non-paulian nuclear processes in highly radiopure NaI(Tl) and the related perspectives. In particular, since the nucleus is a fermion system, the nuclear structure is stable in case of an exact PEP, unless channels for strong, weak or electromagnetic decays are open. Thus, the usually considered stable nucleus would not a priori be absolutely stable in case small violations of the PEP would exist and some exotic transitions (normally forbidden by PEP) in a stable nucleus may occur. As an example, adopting a nuclear shell description, one of the nucleons in a higher energy shell may fall into a lower energy state, normally occupied, and as a consequence another nucleon can acquire enough energy to reach the unbound region and to escape from the nucleus. The value of the average mixing probability of non-fermion statistics $\delta^2$ – which quantifies the possible PEP violations – can be derived considering the relation: $\tilde{\Gamma} = \tilde{\Gamma}(^{23}Na) + \tilde{\Gamma}(^{127}I) = \Gamma/\delta^2$, where: i) $\Gamma$ is the total width of the PEP violating nucleon transition; ii) $\tilde{\Gamma}$ is the total width of the corresponding PEP allowed nucleon transition in case final states are empty, calculated for $^{23}$Na and $^{127}$I according to possible models for the momentum distribution functions of the nucleons in the bound state.

## 2. Present status in NaI(Tl) and the perspectives

Following previous studies [9, 14, 10], the present best limit on the investigation of non-paulian nuclear processes in NaI(Tl) scintillators has recently been set by means of the DAMA/LIBRA set-up at the Gran Sasso National Laboratory of the I.N.F.N. [12]; in particular, the non-paulian emissions of protons with $E_p \geq 10$ MeV in $^{23}$Na and in $^{127}$I have been studied.

The performances of the highly radiopure DAMA/LIBRA set-up have been discussed in details in Ref. [20]; here we just remind that the sensitive part of the set-up is made of a $5 \times 5$ matrix of highly radiopure NaI(Tl) crystal scintillators, 9.70 kg each one. In each detector two 10 cm long special quartz light guides act also as optical windows on the two end faces of the crystal and are coupled to two low background photomultipliers working in coincidence at single photoelectron level. The detectors are housed in a sealed low-radioactive copper box installed in the center of a low-radioactive Cu/Pb/Cd-foils/polyethylene/paraffin shield; moreover, about 1 m concrete (made from the Gran Sasso rock material) almost fully surrounds (mostly outside the barrack) this passive shield, acting as a further neutron moderator. The copper box is maintained in HP Nitrogen atmosphere in slightly overpressure with respect to the external environment; it is part of the threefold-levels sealing system which excludes the detectors from environmental air. A hardware/software system to monitor the running conditions is operative and self-controlled computer processes automatically control several parameters and manage alarms. Moreover, additional dedicated data taking optimized for the very high energy region can be performed as the running period of 23.7 days considered in ref. [12] to search for non-paulian proton spontaneous emission with energy above 10 MeV. It is worth noting that in that period one of the more external detectors was out of operation [20].

The process is investigated by analysing – in the MeV energy region – the events where just one detector fires (i.e. each detector has all the others as veto). The presence of identified alphas from residual U/Th contamination has also offered references for the energy scale [20]. The obtained results are discussed in details in ref. [12]. Here Table 1 just summarizes the experimental results, obtained with the DAMA/LIBRA detectors (24 in operation during the considered running period) by combining them in independent groups. The expected background events are also quoted; in this energy region the background is essentially due to very high energy muons possibly surviving the mountain. Generally the very high energy muons give rise in DAMA/LIBRA to events, in which several detectors fire, and thus they can be easily identified and are not competing background in the search for the PEP violating processes. Thus, only muons, impinging the sensitive volume of the set-up with a direction that forbids them to hit more than one detector, can play the role of background for the processes searched for and have to be considered. Therefore, a suitable MonteCarlo simulation [12] has been realized on the basis of the features of the DAMA/LIBRA set-up [20], of the vertical muon intensity distribution and of the Gran Sasso rock overburden map of ref. [21]. From the final combined result given in Table 1: $\lambda \leq 1.63 \times 10^{-33}$ s$^{-1}$, the limit on the non-paulian nuclear transition width has been

**Table 1.** Comparison between the number of background events expected above 10 MeV from high energy muons surviving the mountain and the number of measured events, considering various groups ($J$) of detectors in the experimental set-up (5 rows by 5 columns detectors matrix, but an external detector out of operation); see text. Upper limits on the rates of non-paulian processes leading to the emission of protons with $E_p \geq 10$ MeV are also reported for the various configurations; the final combined result is given. See ref. [12]

| Group ($J$) of considered detectors | Corresponding exposure ($N_J t$) (nuclei × s) | Expected background events ($b_J$) | Measured events ($n_J$) | Upper Limit on $\lambda$ (90% C.L.) ($s^{-1}$) |
|---|---|---|---|---|
| Just the 4 detectors at corners (I) | $3.2 \times 10^{32}$ | 12.1 | 11 | $1.99 \times 10^{-32}$ |
| Just the remaining 6 detectors in the upper and lower rows (II) | $4.8 \times 10^{32}$ | 8.7 | 6 | $9.33 \times 10^{-33}$ |
| Just the 14 central detectors (III) | $1.1 \times 10^{33}$ | 2.2 | 0 | $2.06 \times 10^{-33}$ |
| Just the 9 core detectors (IV) | $7.2 \times 10^{32}$ | 0.057 | 0 | $3.19 \times 10^{-33}$ |
| Combined analysis (I+II+III): | | | | $1.63 \times 10^{-33}$ |

derived to be: $\Gamma = \Gamma(^{23}Na)+\Gamma(^{127}I) = \hbar\lambda \leq 1.1\times10^{-54}$ MeV (90% C.L.). It is an improvement of about a factor 3 with respect to the limits previously available [13, 14]. From this value – considering the same nuclear Physics frameworks as in ref. [14] – a cautious estimate of the average probability, $\delta^2$, which quantifies the possible PEP violations, has been derived to be: $\delta^2 \leq 3 - 4 \times 10^{-55}$, and a lower limit on the mean life for non-paulian proton emission has also been set: $\tau_{Na} \geq 2 \times 10^{25}$ yr and $\tau_I \geq 2.5 \times 10^{25}$ yr. These latter limits improve those previously available [13, 14].

Let us now discuss the perspectives to further investigate non-paulian nuclear transitions in DAMA/LIBRA with dedicated long data taking optimized for very high energy. In fact, it is worth noting that a much larger sensitivity can be obtained by suitably increasing the dedicated collected exposure, considering the high self-veto efficiency of the 9 inner core detectors. Fig. 1

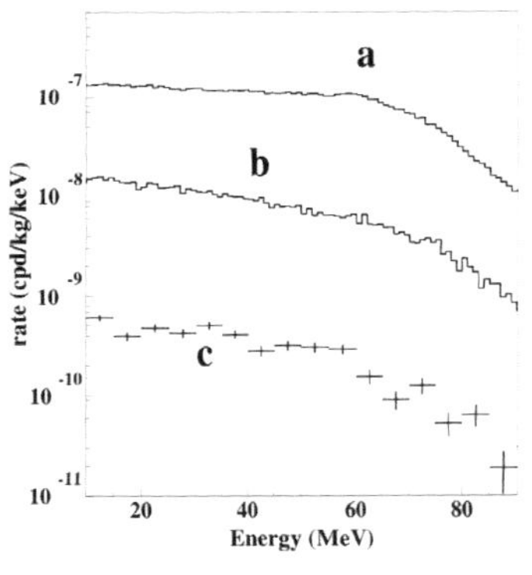


**Figure 1.** Background energy distributions expected for different sets of DAMA/LIBRA crystals considering the events where just one detector fires. Case a): average contribution of the 10 upper and lower crystals in the 5 × 5 matrix. Case b): average contribution of the remaining 15 crystals. Case c) average contribution of the 9 inner crystals.

shows the background energy distributions expected for different sets of DAMA/LIBRA crystals considering the events where just one detector fires. In particular, it is shown the average background contribution of: i) the 10 upper and lower crystals in the $5 \times 5$ matrix (case a); ii) the 15 central crystals (case b); iii) the 9 inner detectors (case c).

It is worth noting that the maximum expected energy release in the PEP violating nuclear transition can be written as: $E_{max} = V - B_N$, where $V$ is the nuclear potential and $B_N$ is the nucleon binding energy. Considering $V \simeq 40$ MeV and $B_N \simeq 6 - 12$ MeV for $^{23}$Na and $^{127}$I, the maximum energy release is expected to be below $\sim 34$ MeV [13]. As it can be inferred from the case c) of Fig. 1, less than 1 background event is expected in the 9 NaI(Tl) detectors in the inner core of DAMA/LIBRA during $\sim 1000$ days exposure in the 10 – 35 MeV energy interval. This can allow to explore the probability of admixed symmetric component at level of $\delta^2 \leq 1 - 2 \times 10^{-56}$ (at 90% C.L.). Therefore in $\sim 3$ years exposure the sensitivity of DAMA/LIBRA to non-paulian nuclear transitions in $^{23}$Na and $^{127}$I can be improved of about one order of magnitude without any simulated muon background subtraction.

## 3. Conclusions

The highly radiopure DAMA/LIBRA set-up [20, 22, 12] is in operation at the Gran Sasso National Laboratory of the INFN; it has a sensitive mass of about 250 kg of highly radiopure NaI(Tl). By means of this set-up new searches for non-paulian nuclear processes, i.e. processes normally forbidden by the Pauli exclusion principle, have been carried out achieving the presently most stringent results in $^{23}$Na and $^{127}$I. The possibility to achieve significantly increased sensitivity with new long-term optimized data taking has been introduced.

For completeness, we remind that DAMA/LIBRA has also investigated with a very large exposure PEP violating electron transitions in Iodine atoms in ref. [12].

## References

[1] W. Pauli, Phys. Rev. 58 (1940) 716.
[2] A. M. Messiah and O. W. Greenberg, Phys. Rev. 136 (1964) B248.
[3] O. W. Greenberg, Phys. Rev. Lett. 64 (1990) 705; R. N. Mohapatra, Phys. Lett. B 242 (1990) 407; O. W. Greenberg and R. C. Hilborn, Fund. Phys. 29 (1999) 397.
[4] G. Gentile, Nuovo Cimento 17 (1940) 493; H. S. Green, Phys. Rev. 90 (1953) 270; A.Yu. Ignatiev and V.A. Kuzmin, Sov. J. Nucl. Phys. 461 (1987) 786; V. N. Gavrin, A. Yu. Ignatiev and V. A. Kuzmin, Phys. Lett. B 206 (1988) 343; O. W. Greenberg and R. N. Mohapatra, Phys. Rev. Lett. 59 (1987) 2507.
[5] M. Goldhaber and G. Goldhaber, Phys. Rev. 73 (1948) 1472; E. Fischbach, T. Kirsten, and O. A. Schaeffer, Phys. Rev. Lett. 20 (1968) 1012; B. A. Logan and A. Ljubicic, Phys. Rev. C 20 (1979) 1957.
[6] F. Reines and H. W. Sobel, Phys. Rev. Lett. 23 (1974) 954.
[7] V. M. Novikov et al., Phys. Lett. B 240 (1990) 227; K. Deilamian et al., Phys. Rev. Lett. 74 (1995) 4787; A. S. Barabash, et al., JETP Lett. 68 (1998) 112; D. Javorsek II et al., Phys. Rev. Lett. 85 (2000) 2701.
[8] E. Nolte et al., J. Phys. G 17 (1991) S355.
[9] H. Ejiri et al., Nucl. Phys. B (Proc. Suppl.) 28A (1992) 219.
[10] P. Belli et al., Phys. Lett. B 460 (1999) 236.
[11] S. Bartalucci et al., Phys. Lett. B 641 (2006) 18; E. Milotti et al., Int. J. Mod. Phys. A 22 (2007) 242.
[12] R. Bernabei et al., Eur. Phys. J. C 62 (2009) 327.
[13] H. Ejiri and H. Toki, Phys. Lett. B 306 (1993) 218.
[14] R. Bernabei et al., Phys. Lett. B 408 (1997) 439.
[15] R. Arnold et al., Eur. Phys. J. A 6 (1999) 361; Nucl. Phys. B (Proc. Suppl.) 87 (2000) 510.
[16] H. O. Back et al., Eur. Phys. J. C 37 (2004) 421.
[17] R. D. Amado and H. Primakoff, Phys. Rev. C 22 (1980) 1338.
[18] K. Akama et al., Phys. Rev. Lett. 68 (1992) 1826.
[19] see e.g. O. W. Greenberg and R. N. Mohapatra, Phys. Rev. D 39 (1989) 2032.
[20] R. Bernabei et al., Nucl. Instr. & Meth. A 592 (2008) 297.
[21] M. Ambrosio et al, Phys. Rev. D 52 (1995) 3793.
[22] R. Bernabei et al., Eur. Phys. J. C 56 (2008) 333.

# TACTIC - Ion tracking in nuclear physics

**L Martin[1], G Ruprecht[1], U Hager[1], L Buchmann[1], P A Amaudruz[1], S Fox[2], A M Laird[2], K Chipps[2], P Machule[1], R Openshaw[1], P Walden[1], M Walter[1], P Bruskiewich[1] and A C Shotter[1,3]**

[1] TRIUMF national laboratory, Canada
[2] University of York, UK
[3] The University of Edinburgh, UK

E-mail: `lars.martin@triumf.ca`

**Abstract.** Ion tracking is still an unusual detection method in nuclear physics. TACTIC (TRIUMF annular chamber for tracking and identification of charged particles) is suitable for tracking low-energy ejectiles. The cylindrical design allows for an electrical separation of target and drift region without hindering the ejectiles to enter the drift region. The first experiment with radioactive beam – the reaction $^{8}$Li($\alpha$,n)$^{11}$B which is important for the $\alpha$-process in neutrino-driven supernova winds – took place in June '09. Since no off-the-shelf analysis tools and methods are available for this kind of tracking, TRIUMF's TACTIC group developed graphical tool for analysing data that can also be used for other modern nuclear physics detectors.

## 1. Introduction

Time projection chambers (TPC) are well known and widely used tools in particle physics [1, 2], but not very common in nuclear physics [3, 4]. TACTIC is an active target TPC [5, 6], designed for charged particle reactions preferably on helium like $^{8}$Li($\alpha$,n)$^{11}$B, an astrophysically interesting reaction[8].

TACTIC is a cylindrical chamber with a blind centre region (details of the geometry in [5]), so only particles from reactions or scattered beam particles are detected, increasing the signal to background ratio, but sacrificing part of the ion tracks. Because all planned experiments are symmetric around the beam axis, TACTIC was built with a poor resolution in $\phi$ (8 anode sectors) in favour of better resolution in $z$ (60 anode pads per sector). The $r$-coordinate can be derived from electron drift times.

## 2. Data acquisition

The signals from the 480 anode pads are preamplified and then digitized in 1 flash-ADC-modules (type VF48 [7]). These sample the signals with a frequency of 40 MHz and provide triggers, maximum position, pulse height and –optionally– full signals for all channels, eliminating the need for a separate timing- and energy-branch. The modules are read out by a VME computer which transmits the data via private ethernet to a PC running a MIDAS-server [9] (see figure 1). Analyzers can run on this PC or any other computer on the network, usually most analyzing is done remotely on a computer running MacOS X ("iTactic" in the figure). Multiple analyzers can connect at the same time.

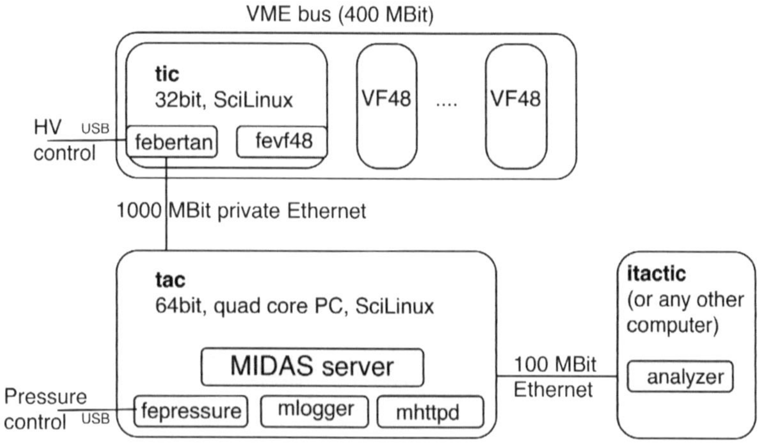


**Figure 1.** DAQ-setup: Everything before the analyzer is modular.

## 3. The FlowAnalyzer

A new analysis software is being developed to process the TACTIC data [10]. It is comprised of a manager and a multitude of plugins or modules. This clean object-oriented concept allows to connect and disconnect plugins in runtime, thereby making recompilation unnecessary for changes in the analysis. This makes a simple graphical user interface possible, organizing data flow visually and intuitively (see figure 2). It also facilitates the collaboration of several developers. Additionally exchanging the input plugin allows to send real (online and offline) and simulated data through exactly the same analyzer, facilitating checks and debugging.

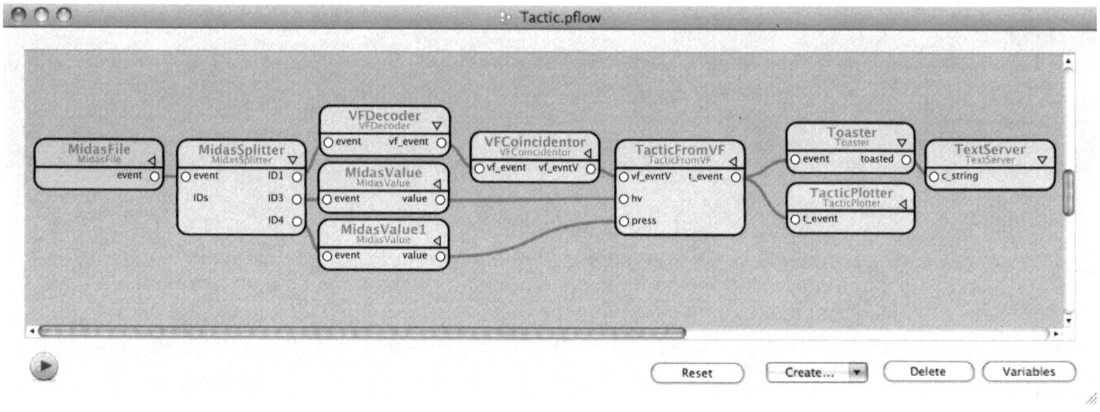


**Figure 2.** FlowComposer: In this program, an analyzer can be assembled from existing modules at runtime even without programming skills.

All modules can be loaded dynamically into the running application. Each module is free to use a custom system for visualization like histogramming or tracking. However, 2 quasi-standardized interfaces are supported, **AIDA** [11] for histogramming, and **OpenInventor** for 3D visualization (see figure 3). The concrete implementations are not fixed and can be dynamically changed. **OpenScientist** [12] provides both interfaces and has therefore been chosen as the default option. Another AIDA implementation for CERN's "ROOT" is in

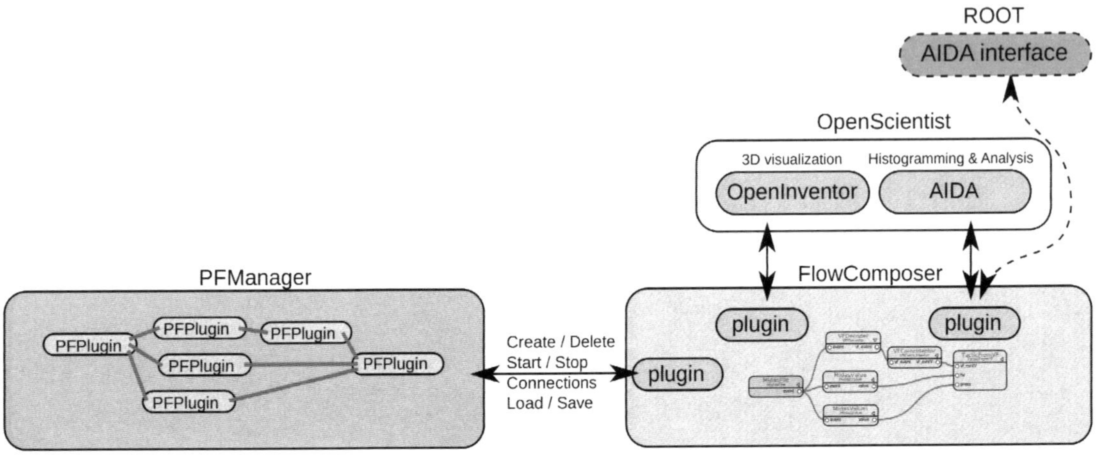


**Figure 3.** A FlowAnalyzer consists of one manager controlling several plugins. These can interface to the outside world by text output or interfaces like AIDA and OpenInventor.

preparation.

Therefore, the central program, here "FlowComposer" for Mac OS X, is just a hub for loading all the plugins for the analyzer and for the histogramming/3D visualization and can easily be implemented on other systems. We also have a version for Linux using the Qt GUI system.

## 4. Radioactive beam in TACTIC

The first radioactive beam run with TACTIC took place in June 2009 at TRIUMF. It was also the first run using a BGO array from DRAGON (10 of DRAGON's 30 BGOs)[13] in connection with TACTIC.

There were two main concerns about the $^{8}$Li beam prior to this run: the $2\alpha$-decay inside TACTIC and the background bremsstrahlung from decay betas in the BGO array.

*$2\alpha$-decay:* We do see tracks from the double-$\alpha$ decay of $^{8}$Be in TACTIC, but we can easily distinguish them from other tracks (see figure 4) by which sectors are involved and by the difference in specific energy loss and they are not so numerous as to drown our DAQ. Further analysis might even allow us to reconstruct a complete decay event (i. e. the track of a scattered $^{8}$Li up to the stopping point (or point of in-flight decay), decay gammas and the track of both alphas.) Although not being the point of the experiment this would be a good test of the system.

*bremsstrahlung:* The high energy betas from the decay of $^{8}$Li to $^{8}$Be can produce bremsstrahlung up into the energy range of the reaction gammas. Of course the actual spectrum depends on the stopping material, which makes this problem very difficult to estimate. As a precaution the beam dump was shielded with lead bricks. A quick preanalysis shows a clear coincidence between tracks in TACTIC and BGO events, indicating a ratio of real to accidental coincidences of the order of $10^{3}$. Since TACTIC is self-triggering, the time between a trigger and the actual physical event depends on the electron drift time, i. e. the $r$-coordinate of the ionization event. It is therefore very difficult to set up a hardware coincidence, so coincidences have to be determined in analysis.

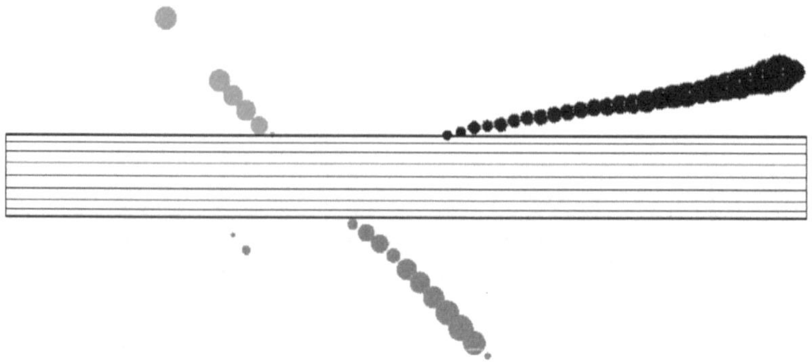

**Figure 4.** Tracks inside TACTIC: *color:* 2-$\alpha$-decay of $^8$Be (some channels in the orange sector didn't work in this run), *grayscale:* scattered $^8$Li

A different problem turned up, however. The energy resolution per point seems to be non-linear and sector dependent. We are currently examining if this is a problem with the GEMs. Unfortunately there is very little literature on the characteristics of GEMs in conditions as ours (He/$CO_2$ gas and highly ionizing radiation).

## 5. Conclusions

There are no major problems with running TACTIC with a $^8$Li beam and the coincidence setup with the BGO array works. The next beamtime can focus on the more interesting lower beam energies. Improvement is needed to check and handle the non-linearity of the GEM gain which probably requires a laser calibration system and a detailed simulation of the electron cascade through the GEM, including charge recombination.

## References

[1] Marx J N, Nygren D R 1978 *Physics Today* **31**(10) 46–53
[2] Musa L and ALICE Collaboration 2003 *Nucl. Phys.* A **715** 843–48
[3] Christie W B et al. 1987 *Nucl. Instrum. Methods Phys. Res.* A **255**(3) 466–76
[4] Demonchy C E et al. 2007 *Nucl. Instrum. Methods Phys. Res.* A **573**(1-2) 145–48
[5] Ruprecht G et al. 2006 *Eur. Phys. J.* A **27** 315
[6] Laird A M et al. 2007 *Nucl. Instrum. Methods Phys. Res.* A **573** 306–9
[7] Martin J P and Amaudruz P A 2006 *IEEE Transactions on Nuclear Science* **53**(3)
[8] Terasawa M, Sumiyoshi K, Kajino T, Mathews G J and Tanihata I 2001 *Astrophys. J.* **562** 470–9
[9] MIDAS data acquisition system, `http://midas.psi.ch` (PSI)
[10] FlowAnalyzer analysis system (beta version), `http://tactic.triumf.ca/software/FlowAnalyzer`
[11] AIDA Abstract Interface for Data Analysis, `http://aida.freehep.org` (CERN)
[12] OpenScientist, `http://midas.psi.ch` (PSI)
[13] Gigliotti D G 2003 *Nucl. Instrum. Methods Phys. Res.* B **204** 671–7

# Influence of the target surface contamination on UHV screening energies

**N. Targosz-Ślęczka[1], K. Czerski[1], A. Huke[2], L. Martin[2], P. Heide[2], A.i. Kılıç[1], D. Blauth[3] and H. Winter[3]**

[1] Institute of Physics, University of Szczecin, Szczecin, Poland
[2] Institut für Atomare Physik und Optik, Technische Universität Berlin, Berlin, Germany
[3] Institut für Physik, Humboldt-Universität zu Berlin, Berlin, Germany

E-mail: `natalia.targosz@wmf.univ.szczecin.pl`

**Abstract.** The $d + d$ fusion reactions have been investigated in the Zirconium environment under ultra high vacuum (UHV) conditions for projectile energies below 30 keV. The experimentally determined screening energy value of $497 \pm 7$ eV is larger than the previous results by a factor of almost two. Despite the UHV conditions a small deviation between experimental data and the theoretical curve arising from the target surface contamination could be still observed at the lowest projectile energies. Calculations made under the assumption of formation of a Zirconium oxide contamination, show that every atomic monolayer reduces the estimated screening energy significantly.

## 1. Introduction

The enhanced electron screening effect recently observed in $d + d$ fusion and some other nuclear reactions in metallic environments is extremely important for general understanding of stellar processes since it can strongly increase the nuclear reaction rates in dense astrophysical plasmas [1]. For this reason, the electron screening effect in metallic environments has been experimentally investigated by many research groups [2, 3, 4, 5]. Despite the large amount of data, the results obtained by various groups are inconsistent because of some systematic errors. The electron screening measurement have been performed up to now under high vacuum conditions. In those experimental studies one came across some difficulties concerning the target surface cleanness. Metal oxide or carbon deposits reduce the experimental screening energy values. In extreme case no screening is observed [6].

The screened reaction cross section reads as follows [7]:

$$\sigma_{scr}(E_{cm}) = \frac{1}{\sqrt{E_{cm}(E_{cm}+U_e)}} S(E_{cm}) \exp\left(-\sqrt{\frac{E_G}{E_{cm}+U_e}}\right) \tag{1}$$

where $E_G = 986$ keV is the Gamow energy for the d+d system, $S(E_{cm})$ is the astrophysical S-factor and $U_e$ denotes the screening energy. The strength of the screening effect can be described by means of the thick-target enhancement factor $F(E)$ defined as the ratio between the thick-target yields for screened and bare nuclei [8]:

$$F(E) = \frac{Y_{scr}(E)}{Y_{bare}(E)} = \frac{\int_0^E \frac{\sigma_{scr}(E)}{\sqrt{E}} \, dE}{\int_0^E \frac{\sigma_{bare}(E)}{\sqrt{E}} \, dE} \quad (2)$$

The bare nuclei cross section is known from the precise measurements performed with the gas target [9]. To address the contamination layer problem and improve the accuracy of the measurements we performed the first ultra high vacuum (UHV) experiment on d+d fusion reactions in metals.

## 2. Experiment

We measured the $^2H(d,p)^3H$ and $^2H(d,n)^3He$ total cross sections and angular distributions in Zirconium foils (1 mm thick) at deuteron bombarding energies ranging from 9 to 30 keV. Details of the experimental set-up, data-taking procedure and data-analysis methods are given in our previous papers [10]. In brief, the experiment has been performed with an electrostatic UHV accelerator using an electron cyclotron resonance (ECR) ion source and a highly stabilized power supply. The ion-beam was magnetically analyzed and focused on the Zr target to a spot of 5 mm in diameter. The charged products of $d + d$ reactions were detected at three fixed backward angles 90°, 125° and 150° with silicon detectors. Since the dominating error source of that kind of measurement comes from contamination layer formation, we cleaned the target surface by means of Argon ion sputtering. Additionally, we applied the Auger electron spectroscopy to check the target surface cleanness. Contamination of carbon and oxygen could be clearly observed during longer measurements at the lowest projectile energies. A differential pumping system and use of $LN_2$ cooling system allowed to reduce the gas pressure to the value of $5 \times 10^{-10}$ mbar in the target chamber which is one order of magnitude better than in our previous UHV experiment [10].

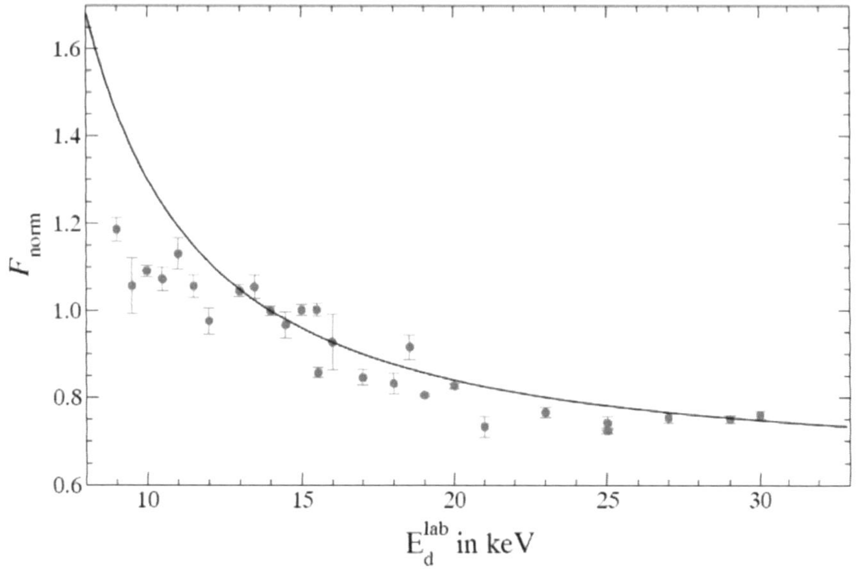


**Figure 1.** Experimental normalized enhancement factor obtained for Zr target under UHV conditions due to liquid $LN_2$ cooling. The value of the electron screening energy $U_e$ amounts to $497 \pm 7$ eV.

## 3. Results

### *3.1. Experimental screening energy*

The enhancement factor $F(E)$ in equation (2) is independent of both the deuteron density in the target and the stopping power. The screening energy value has been determined by fitting an increase of $F(E)$ towards lower projectile energies to experimental results. The normalized thick-target enhancement factor is defined as the ratio between enhancement factors for a certain measurement and for the monitor measurement (here at the projectile energy $E_0 = 14$ keV):

$$F_{norm}(E) = \frac{F(E)}{F(E_0)} \tag{3}$$

The experimental thick target enhancement factors are shown in Figure 1. The error bars of presented data are dominated by statistical uncertainties and include also systematic uncertainties resulting from the alternating deuteron density in the target. The experimental data analysis gives the screening energy value of $U_e = 497 \pm 7$ eV. The fitted curve according to equations (2) and (3) takes on unity for the projectile energy $E_0 = 14$ keV. Low energy experimental points (below 12 keV) deviate from the fitted curve, most probably because of the contamination layer formation. Thus, we decided to study quatntitatively influence of the target surface contamination on the experimentally estimated UHV screening energies.

### *3.2. Influence of contamination layer*

We considered the proton branch of deuteron fusion reactions in the Zirconium target. Calculations assumed formation of the Zirconium dioxide layer on the target surface where the deuteron density can be neglected compared to that in the Zr target. Therefore, the $ZrO_2$ layer

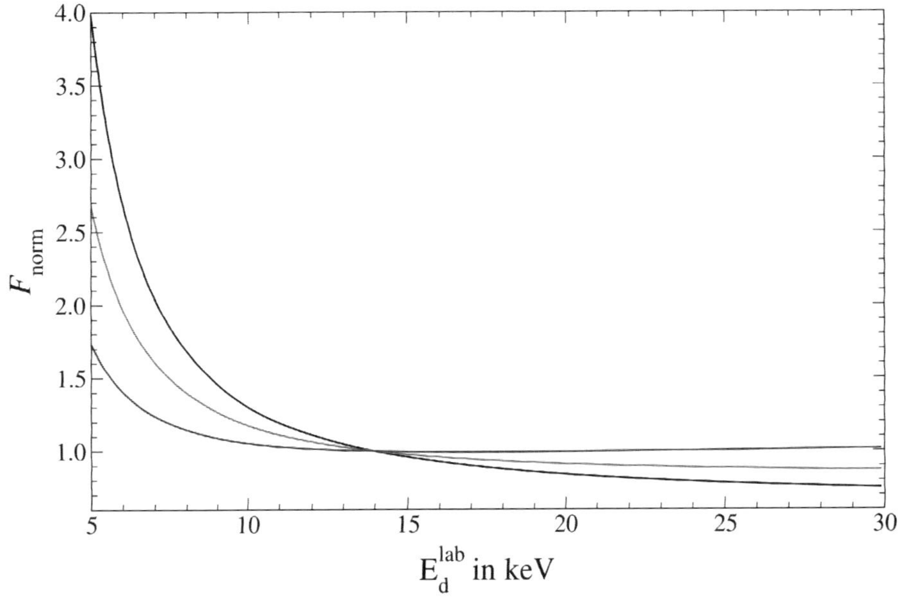


**Figure 2.** Theoretical normalized enhancement factor for Zr target with $U_e = 497$ eV. The black line stands for the clean target, the red line is the case when the target is contaminated by 10 monolayers of Zirconium dioxide and the blue line is for 20 contamination layers.

reduces only the projectile energy but does not contribute to the experimental reaction yield. The theoretical normalized thick-target enhancement factors were calculated under assumption that, the stopping power of deuterium in Zirconium and in the oxidation layer are proportional to the square root of the energy [11]. The astrophysical S-factor for considered reaction has been taken from [9]. The calculations were carried out for the experimentally found screening energy value of $U_e = 497$ eV.

Figure 2 shows the theoretical normalized enhancement factors for a different number of the contamination monolayers. The curve in black corresponds to the virgin target case. The red and the blue curves were plotted for 10 and 20 $ZrO_2$ monolayers formed on the target surface, respectively. It can be noticed that the enhancement factor is significantly reduced toward lower projectile energies.

The calculations showed that even a single contamination monolayer can change the enhancement factor value at the deuteron energies below 12 keV. Our experimental results suggest formation of about 10 contamination monolayers during the measurement. However, the number of contamination monolayers can strongly vary for different deuteron energies. The largest number can be expected for the lowest projectile energies with the longest measurement time. Thus, the final quantitative analysis of systematic errors of the determined screening energy due to the target surface contamination remains ambiguous [12].

## 4. Conclusions

We have measured the deuteron fusion reaction cross sections and angular distributions in the Zirconium host target under UHV conditions with the additional $LN_2$ cooling system at deuteron bombarding energies below 30 keV. The observed enhancement of the thick target yield leads to the screening energy value of $497 \pm 7$ eV which is much larger than our previous values measured for Zr under high vacuum conditions ($5 \times 10^{-7}$ mbar) [7] and ultra high vacuum conditions (without $LN_2$ cooling, $5 \times 10^{-9}$ mbar) [10], both being of about 300 eV. As we could experimentally and theoretically show, even at the vacuum pressure in the target chamber of the present experiment of $5 \times 10^{-10}$ mbar, a small oxide contamination of the target surface observed for the deuteron energies below 12 keV and longer measurement times still decreases experimental thick target yields, significantly. Thus, our new value of the screening energy for the $d+d$ fusion reactions in the Zr environment, although the systematic uncertainties could be strongly reduced, probably represents its lower limit. Further improvement of the experiental set-up is necessary.

## References

[1] Ichimaru S and Kitamura H 1999 *Phys. Plasmas* **6** 2649
[2] Czerski K, Huke A, Biller A, Heide P, Hoeft M and Ruprecht G 2001 *Europhys. Lett.* **54** 449-55
[3] Kasagi J, Yuki H, Baba T, Noda T, Ohtsuki T and Lipson A G 2002 *J. Phys. Soc. Jpn.* **71** 2281
[4] Raiola F *et al.* 2004 *Eur. Phys. J.* A **19** 283
[5] Jeppesen H B, Byskov-Nielsen J, Wright P, Correia J G, Fraile L M, Fynbo H O U, Johnston K and Riisager K 2007 *Eur. Phys. J.* A **32** 31-4
[6] Huke A, Czerski K, Heide P, Ruprecht G, Targosz N and Żebrowski W 2008 *Phys. Rev.* C **78** 015803 1-20
[7] Czerski K, Huke A, Heide P, Ruprecht G 2004 *Europhys. Lett.* **68** 363-9
[8] Czerski K, Huke A, Biller A, Heide P, Hoeft M and Ruprecht G 1998 *Proc. Int. Conf. on Nuclear Astrophysics 'Nuclei in the Cosmos' (Volos: Greece, 6-11July 1998)* ed Prantzos N and S 152 (Gif-sur-Yvette: Editions Frontières)
[9] Brown R E and Jarmie N 1990 *Phys. Rev.* C **41** 1391-400
[10] Czerski K, Huke A, Martin L, Targosz N, Blauth D, Górska A, Heide P and Winter H 2008 *J. Phys.* G **35** 014012
[11] Andersen H H and Ziegler J F 1977 *Hydrogen: Stopping Powers and Ranges in All Elements* (The Stopping and Ranges of Ions in Matter vol 3) (New York: Pergamon Press)
[12] Huke A, Czerski K and Heide P 2007 *Nucl. Instrum. and Meth. in Phys. Res.* B **256** 599-618

# Study of astrophysically important resonant states in $^{30}$S using the $^{32}$S(p,t)$^{30}$S reaction

**K. Setoodehnia[1], A. A. Chen[1], J. Chen[1], J. A. Clark[2], C. Deibel[2], D. Kahl[1,3], W. N. Lennard[4], P. D. Parker[5] and C. Wrede[6]**

[1] Department of Physics & Astronomy, McMaster University, Hamilton, ON L8S 4M1, Canada
[2] Physics Division, Argonne National Laboratory, Argonne, IL 60439, USA
[4] Department of Physics & Astronomy, University of Western Ontario, London, ON N6A 3K7, Canada
[5] Wright Nuclear Structure Laboratory, Yale University, New Haven, CT 06520, USA
[6] Department of Physics, University of Washington, Seattle, WA 98195-1560, USA

E-mail: setoodk@mcmaster.ca

**Abstract.** A small fraction ($< 1\%$) of presolar SiC grains is suggested to have been formed in the ejecta of classical novae. The $^{29}$P(p,$\gamma$)$^{30}$S reaction plays an important role in understanding the Si isotopic abundances in such grains, which in turn provide us with information on the nature of the probable white dwarf progenitor's core, as well as the peak temperatures achieved during nova outbursts, and thus the nova nucleosynthetic path. This rate at nova temperatures is determined by two low-lying $3^+$ and $2^+$ resonances above the proton threshold at 4399 keV in $^{30}$S. Despite several experimental studies in the past, however, these two states have only been observed very recently. We have studied the $^{30}$S nuclear structure via the $^{32}$S(p,t)$^{30}$S reaction at 5 laboratory angles between 9$^\circ$ to 62$^\circ$. We have observed 14 states, eleven of which are above the proton threshold, including two levels at 4692.7 $\pm$ 4.5 keV and 4813.8 $\pm$ 3.4 keV that are candidates for the $3^+$ and the previously "missing" $2^+$ state, respectively.

## 1. Introduction

A small fraction ($< 1\%$) of presolar SiC grains is suggested to have been formed in the ejecta of classical novae [1]. Classical novae are stellar explosions powered by thermonuclear runaway, which occur in close binary systems consisting of a compact white dwarf and a low-mass Main Sequence companion. The temperature that can be reached during these outbursts is on order of 0.1 to 0.4 GK. Analyzing the Si isotopic abundance ratios ($^{29}$Si/$^{28}$Si and $^{30}$Si/$^{28}$Si) in presolar grains of nova origins help us understand the chemical composition of the white dwarf and the thermal history of the white dwarf's convective envelope [1], which will in turn help us to determine the dominant nova nucleosynthetic paths followed by the thermonuclear runaway.

To calculate the Si isotopic abundances in presolar grains, it is critical to know the rates of the thermonuclear reactions which affect the $^{29,30}$Si production and destruction in novae. One such reaction is the $^{29}$P(p,$\gamma$)$^{30}$S reaction [2]. The $^{29}$P(p,$\gamma$) reaction rate depends significantly on the $^{30}$S resonances above the proton threshold ($S_p = 4399 \pm 3$ keV [3]), whose properties are not well understood. Iliadis *et al.* [2] concluded that at nova temperatures the $^{29}$P(p,$\gamma$) reaction

[3] Present address: Center for Nuclear Study (CNS), Graduate School of Science, University of Tokyo, Wako Branch at RIKEN, 2-1 Hirosawa, Wako, Saitama, 351-0198, Japan

rate is dominated by low energy $3^+$ and $2^+$ resonances, whose energies were predicted to be 4733 ± 40 keV and 4888 ± 40 keV, respectively [2]. Prior to 2007, despite several experiments that attempted to observe these two levels, no evidence of their existence was published. Thus, the $^{29}$P(p,$\gamma$) reaction rate, which depends exponentially on the resonance energy[4], remained quite uncertain.

In 2007, to search for these unobserved resonances, two experiment were performed separately by Bardayan *et al.* [3] and by Galaviz *et al.* [4]. In the same year, we also performed a $^{32}$S(p,t) experiment to measure the level energies of $^{30}$S at the Wright Nuclear Structure Laboratory at Yale University. We were interested in the region just above the proton threshold at 4399 keV in $^{30}$S. Our energy resolution was $\sim$ 30 keV, which is a factor of $\sim$ 3 better than that of Bardayan *et al.* (80 – 120 keV) [3]. Thus, we were able to resolve levels better.

This contribution aims to describe the experimental setup and the detection system used to study the level properties of $^{30}$S, as well as to present our results.

## 2. Experimental Procedure and Data Analysis

The $^{32}$S(p,t)$^{30}$S reaction was studied using proton beams, which were accelerated by the Yale ESTU tandem Van de Graaff accelerator to fixed energies of 33.5 and 34.5 MeV. Our main target was a 249 $\mu$g/cm$^2$ ± 10% CdS foil supported by a 20 $\mu$g/cm$^2$ natural carbon substrate. In addition to this target, a free standing 311 ± 31 $\mu$g/cm$^2$ natural Si foil as well as an 75 $\mu$g/cm$^2$ isotopically pure $^{13}$C target foil were used for calibration and contamination subtraction purposes, respectively. The Yale high resolution Enge split pole magnetic spectrograph accepted light reaction products through a rectangular aperture and momentum analyzed them. Tritons were focused at the focal plane of the spectrograph, where a position sensitive ionization drift chamber measured the position and energy loss, ΔE, of the particles which deposited their residual energies, E, into a plastic scintillator.

The $^{32}$S(p,t)$^{30}$S, $^{28}$Si(p,t)$^{26}$Si and $^{13}$C(p,t)$^{11}$C reactions were measured over seven days at spectrograph angles $\theta_{\text{lab}} = 9^\circ, 10^\circ, 20^\circ, 22^\circ$ and $62^\circ$. The spectrograph angles were selected so that the location of the background peaks would allow a clear observation of important resonances of $^{30}$S.

Tritons were identified by plotting focal plane position (equivalent to momentum), ΔE and E in 2D histograms, and were selected by applying software gates in such histograms. Thus, the $^{30}$S spectra at each spectrograph angle were plotted (see Fig. 1). Background peaks from the $^{12}$C(p,t)$^{10}$C$^{\text{g.s.}}$ (higher states of $^{10}$C from the $^{12}$C(p,t)$^{10}$C reaction were not on the focal plane) and $^{13}$C(p,t)$^{11}$C reactions were produced by the natural carbon backing in the CdS target and were identified kinematically. In addition, a roughly flat background existed which was attributed to the Cd component of the target as well as other stable sulfur isotopes ($^{33,34,36}$S) present in the CdS target. The spectra of $^{30}$S were analyzed using a least-square fit of multiple $\sim$ 30 keV-FWHM Gaussian function (our energy resolution $\sim$ 30 keV), from which the peak centroids were determined. Then the focal plane spectra at each angle was calibrated by isolated easily identifiable peaks corresponding to the first, second and the 3rd excited states of $^{26}$Si produced from the $^{28}$Si(p,t)$^{26}$Si reaction and thus the excitation energies of $^{30}$S at all angles were determined. For the excitation energies of $^{26}$Si, we used a weighted average between the excitation energies listed in the Nuclear Structure and Decay Databases [5], and those

[4] In the case of narrow resonances for which the total width Γ of the resonance is much smaller than the resonance energy (typically, if $\frac{\Gamma(E)}{E_R} < 0.1$), the resonance reaction rate can be calculated using the following formula:

$$< \sigma v >_R = \left(\frac{2\pi}{\mu kT}\right)^{\frac{3}{2}} \hbar^2 \, (\omega\gamma)_R \exp\left(-\frac{E_R}{kT}\right) \qquad (1)$$

Where $< \sigma v >_R$ is the reaction rate, $\mu$ is the reduced mass, k is the Boltzmann factor, T is the temperature in GK, $(\omega\gamma)_R$ is the resonance strength and $E_R$ is the resonance energy in keV.

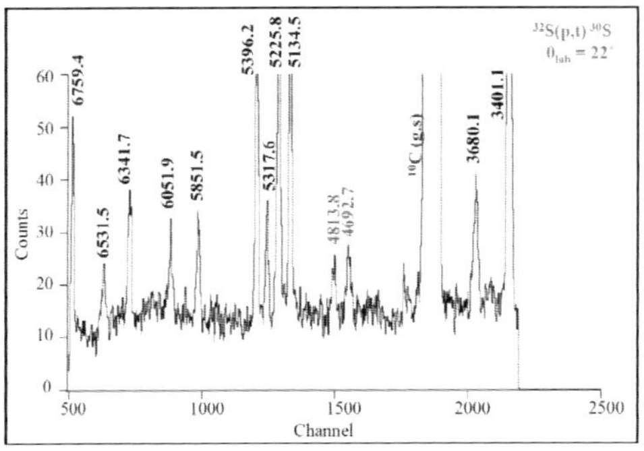


**Figure 1.** Focal plane spectra of the $^{32}S(p,t)$ reaction corresponding to the states of $^{30}S$, after gating on tritons, shown at selected angle. The background peak of $^{10}C^{g.s.}$ is labeled. The states of $^{30}S$ are labeled by their energies averaged over all angles obtained in this work. All energies are in keV. The states of interest are labeled in red.

measured by Seweryniak *et al.* [6]. After determining the excitation energies of $^{30}S$ states at each angle, a statistically weighted average excitation energy was calculated for each level. In addition to uncertainties in location of the peaks on the focal plane spectra, which translated into uncertainties in excitation energies, a universal uncertainty of 3.0 keV was also present due to the uncertainties in masses of $^{26}Si$ ($\pm$ 0.17 keV) [7] and $^{30}S$ ($\pm$ 3 keV) [8]. These uncertainties were mutually independent and were added together in quadrature.

## 3. Results and Future Outlook

With the 30 keV energy resolution obtained in our experiment, at all angles we were able to observe two peaks, most visible at $22^{\circ}$ (see Fig. 1), just above the proton threshold. In spite of low statistics, since the excitation energies of these two peaks did not change with angle, we concluded that they belong to $^{30}S$. For these two levels, we extracted excitation energies averaged over all angles of 4692.7 $\pm$ 4.5 keV and 4813.8 $\pm$ 3.4 keV, respectively. In addition to these two states, we have also observed nine more levels of $^{30}S$ up to 6.7 MeV (see Fig. 2). A more comprehensive description of these levels, including data from the methods outlined below, is underway, and those results will be published in the future. Our result for the energy of the $3^+$ state agrees very well with the energy of this state obtained in Galaviz's experiment [9] and it differs from that of Bardayan *et al.* by 1.8 keV. Our result for the energy of the $2^+$ state is close to that of Galaviz *et al.* result [9].

Our CdS target produced a high roughly flat background. To obtain cleaner spectra, we implanted $^{32}S$ into 40 $\mu g/cm^2$ isotopically pure $^{12}C$ foils. Rutherford backscattering spectroscopy revealed that there is 10.4 $\pm$ 0.3 $\mu g/cm^2$ of $^{32}S$ in this target. We then repeated the $^{32}S(p,t)$ experiment recently with this target and were able to achieve our goal of reducing the background. The analysis of this new data is still in progress.

In late July 2009, we also performed an in-beam $\gamma$-ray spectroscopy experiment using the $^{28}Si(^3He,n\gamma)^{30}S$ reaction at the University of Tsukuba Tandem Accelerator Complex (UTTAC) in Japan. In the latter experiment, the energies of the excited states of $^{30}S$ were studied via the n-$\gamma$ and $\gamma\gamma$ coincidence measurements using a liquid scintillator to detect neutrons as well as two high energy resolution Ge-detectors. The analysis of this data is currently in progress.

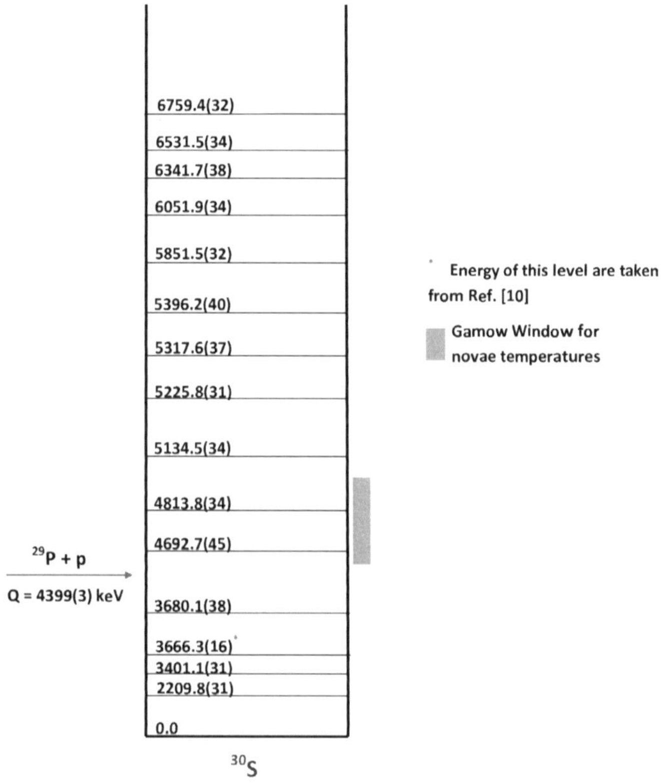


**Figure 2.** Level Scheme of $^{30}$S with energies in keV from this work unless indicated by an asterisk. The uncertainties are in energies are in two last digits, e.g. a level with $E_x = 2209.8$ keV has an uncertainty of 3.1 keV. The spins and parities of the $^{30}$S levels extracted from this work are still preliminary so they are not shown here. For information on the properties of levels with energy higher than 6.7 MeV, see Ref. [3]. Note that the level at 3666.3 keV could not be resolved in our work, and thus it is taken from Ref. [10].

## Acknowledgments

We wish to sincerely thank the staff of the Wright Nuclear Structure Laboratory at Yale University and also Mr. Jack Hendriks from the University of Western Ontario. This project was supported by Natural Sciences and Engineering Research Council of Canada and the U.S. Department of Energy under grants No. DE-FG02-91ER-40609 and No. DE-FG02-97ER-41020.

## References

[1] José J, Hernanz M, Amari S and Zinner E 2004 *Astrophys. J.* **612** 414
[2] Iliadis C, D'Auria J M, Starrfield S, Thompson W J and Wiescher M 2001 *Astrophys. J. Suppl. Ser.* **134** 151
[3] Bardayan D W *et al.* 2007 *Phys. Rev. C* **76** 045803
[4] Galaviz D *et al.* 2008 *J. Phys. G: Nucl. Part. Phys.* **35** 014030
[5] Nuclear Structure and Decay Databases *http://www.nndc.bnl.gov/databases*
[6] Seweryniak D *et al.* 2007 *Phys. Rev. C* **75** 062801
[7] Eronen T *et al.* 2009 *Phys. Rev. C* **79** 032802(R)
[8] Audi G, Wapstra A H and Thibault C 2003 *Nucl. Phys. A* **729** 337
[9] Galaviz D Private Communication
[10] Kuhlmann E, Albercht W and Hoffmann 1973 A *Nucl. Phys. A* **213** 82

# AUTHOR INDEX

# AUTHOR INDEX

# AUTHOR INDEX

# AUTHOR INDEX

**Institute of Physics**
Dirac House, Temple Back
Bristol BS1 6BE UK

proceedings
.com